Matching of Asymptotic Expansions of Solutions of Boundary Value Problems

Recent Titles in This Series

(Continued in the back of this publication)

Translations of
MATHEMATICAL
MONOGRAPHS

Volume 102

Matching of Asymptotic Expansions of Solutions of Boundary Value Problems

A. M. Il'in

American Mathematical Society
Providence, Rhode Island

А. М. ИЛЬИН

СОГЛАСОВАНИЕ АСИМПТОТИЧЕСКИХ РАЗЛОЖЕНИЙ РЕШЕНИЙ КРАЕВЫХ ЗАДАЧ

Translated from the Russian by V. Minachin
Translation edited by Simeon Ivanov

1991 *Mathematics Subject Classification.* Primary 34-02, 34E15; Secondary 41A60.

ABSTRACT. The author describes an approach to the analysis of solutions of boundary value problems for partial differential equations containing a small parameter. The asymptotic expansions of solutions are different in different regions (for example, in the boundary layer region; near the discontinuity of the limiting solution; etc.). The main problem discussed in the book is the matching problem for asymptotic solutions.

Using examples originating in various problems of fluid mechanics and continuum mechanics of solids, the author presents a rigorous construction of complete asymptotic expansions for solutions.

The book can be useful for researchers and graduate students working in various areas of analysis, partial differential equations, applied mathematics, and mechanics. It can also be used as a basis for an advanced graduate course.

Library of Congress Cataloging-in-Publication Data

Il′in, A. M.
  [Soglasovanie asimptoticheskikh razlozhenii reshenii kraevykh zadach. English]
  Matching of asymptotic expansions of solutions of boundary value problems/A. M.
Il′in; [translated from the Russian by V. Minachin].
  p. cm.—(Translations of mathematical monographs, ISSN 0065-9282; v. 102)
  Translation of: Soglasovanie asimptoticheskikh razlozhenii reshenii kraevykh zadach.
  Includes bibliographical references and index.
  ISBN 0-8218-4561-6
  1. Boundary value problems–Numerical solutions. 2. Asymptotic expansions. 3. Boundary value problems. I. Title. II. Series.
QA379.I4 1992   92-12324
515′.35—dc20   CIP

Information on Copying and Reprinting can be found at the back of this volume.

The paper used in this book is acid-free and falls within the guidelines
established to ensure permanence and durability. ♾
This publication was typeset using $\mathcal{A}_{\mathcal{M}}\mathcal{S}$-TEX,
the American Mathematical Society's TEX macro system.

10 9 8 7 6 5 4 3 2 1   96 95 94 93 92

Contents

Preface

Asymptotic methods in analysis, and, especially, in the theory of equations of mathematical physics are steadily gaining in popularity among a wide range of researchers in various areas of natural sciences. This is testified by the relative increase in the number of articles appearing in the periodic publications and the considerable growth in the number of monographs published on the subject in the last 10 to 15 years. Many of these monographs touch upon the method mentioned in the title of this book. However, the expositions available are usually of a fragmentary nature, and scarcely concern the questions of justifying the asymptotics. At the same time, in the last 5 to 10 years a common approach to a class of small parameter problems frequently arising in widely different areas has been developed. We call these problems bisingular. The reader will find the precise definition in the Introduction below.

This approach is one of the versions of the method of matching different asymptotic expansions for solutions of boundary value problems. Its description can only be found in periodic publications, and, naturally, first papers do not provide the best way of presenting the subject. The purpose of the present book is, therefore, to provide a preliminary assessment of these works and to make the method available to experts in different areas. The presentation follows an inductive scheme and is based on the analysis of a series of examples. As a rule, each next example is more complicated than the preceding one.

The idea that the asymptotic analysis includes two basic steps has gained wide acceptance. The first is the actual construction of the asymptotics. One has to choose the form in which the formal asymptotic expansion of a solution (or the formal asymptotic solution, the other names are Ansatz, FAS, f.a.s., f.a.e.) is to be sought, and specify the way of constructing this f.a.s.

The second step includes the justification of the constructed asymptotics, i.e., a proof that the f.a.s. obtained is indeed an asymptotic expansion of the solution of the problem. This is achieved by providing an estimate of the difference between the true solution and partial sums of the f.a.s.

Which of the two steps is more difficult depends on the problem. Sometimes, one of them is trivial while the other requires a lot of effort. In other cases, the difficulties are distributed more or less evenly. The first part, i.e., the construction of the asymptotics, is certainly of interest to experts in many different areas, e.g., physicists, engineers or anyone who has to deal with large or small parameters in his or her problems, while the second part is mainly of interest to a much narrower community of pure mathematicians.

With that in mind, the author has endeavored to satisfy both groups of his prospective readers. The construction of asymptotics for the problems in question is given in the main text which, as far as possible, is not overloaded with unimportant details. The material necessary for the justification of the asymptotics appears in small print. If the reader's aim is just to master the methods of constructing asymptotic expansions of solutions of bisingular problems, the main text is a sufficient reading. The full text, including the small print, contains strict mathematical justifications of the asymptotics which hitherto were to be found only in periodic publications. This attempt "to trap two rabbits in one book" seems to be worth the effort. The reader has to judge whether it is a success, or the Russian proverb "If you chase two rabbits, you won't catch one" still holds true. *Spero meliora.*

Much of the material appearing in the book has been included in the lecture course given at the Bashkir State University. Most of it is based on the three-year university mathematics course, and for the first two chapters even two years of mathematics at a university or a technical school is sufficient.

The author did not make it his goal to compile a comprehensive list of all significant publications on the subject considered in this book. The references in the text are reduced to a minimum and refer mainly to the justification of asymptotics. All mention of sources and articles relevant to the subject is relegated to the end of the book.

All the results presented in the book, with the exception of Chapter 1 which is of an auxiliary, tutorial nature, have been obtained by a group of Russian mathematicians working in the cities of Ufa and Sverdlovsk, in the Ural region. I take this opportunity to thank my colleagues and students whose research and discussions of results made an important contribution to the publication of this book. In writing the book I received direct assistance from E. F. Lelikova and Yu. Z. Shaygardanov who helped me in my work on Chapter IV, and from L. A. Kalyakin, who helped with Chapter V. T. N. Nesterova and O. B. Sokolova performed extensive work preparing the manuscript. To all of them I express my deep gratitude.

Interdependence of Chapters

Provision is made for a selective study of the book. For the reader's convenience, the following diagram indicates how the different chapters and sections depend on each other.

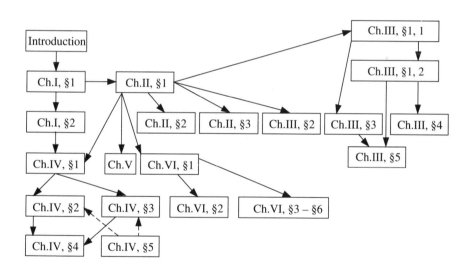

Introduction

The aim of the present book is to describe an approach to the analysis of solutions of boundary value problems for equations of mathematical physics containing a small parameter. Problems of this kind arise in widely different areas of natural sciences and technology. The mathematical treatment of these problems has a long history, but the last two or three decades witnessed a particularly active development of asymptotic approaches. In fact, the science of mathematical analysis is to a considerable extent devoted to the study of asymptotics in a wider sense of the word. Indeed, it is a rare occasion that a problem admits an exact solution expressible as a composition of elementary functions, quadratures, etc., and, therefore, a powerful device is to consider problems which are only slightly different from those having an exact solution or admitting a simpler treatment. This causes the appearance of a small parameter, and the question then is how does the solution depend on this parameter. We will not dwell on the history of the problem referring the reader to the books [121], [15], [85]. Let us now make the subject of our investigation more precise.

We will consider, as a rule, a differential equation $L(x, u, \varepsilon) = 0$, where x is a point of a domain $\Omega \subset \mathbf{R}$, $\varepsilon > 0$ is a small parameter, and $u(x, \varepsilon)$ is a desired solution of this equation. The solution should, furthermore, satisfy some boundary conditions. We will assume that for each $\varepsilon > 0$ there is a unique solution $u(x, \varepsilon)$ of the problem. Our task is to determine the behavior of the solution as $\varepsilon \to 0$.

This formulation is still rather vague. Being unable to write out an exact solution of the problem, one can, naturally, try to find a function satisfying both the equation and the boundary conditions approximately. Let $U_n(x, \varepsilon)$ be a sequence of functions satisfying the inequalities $|L(x, U_n(x, \varepsilon), \varepsilon)| < M\varepsilon^n$ and similar inequalities for the boundary conditions. If $U_n(x, \varepsilon)$ are partial sums of the series

$$U = \sum_{k=0}^{\infty} \varepsilon^k u_k(x), \qquad (0.1)$$

then the series is said to be a *formal asymptotic solution* (f.a.s.) of the differential equation, or an f.a.s. of the boundary value problem depending on whether the functions $U_n(x, \varepsilon) = \sum_{k=0}^{n} \varepsilon^k u_k(x)$ approximate just the solution of the differential equation, or the boundary conditions as well. One can expect that the functions $U_n(x, \varepsilon)$ provide a good approximation to the true solution $u(x, \varepsilon)$ for ε small. In other words, one can expect that the series (0.1) is an asymptotic series for the solution $u(x, \varepsilon)$ as $\varepsilon \to 0$.

It is natural to assume that a reader interested in this book is familiar with the notion of an asymptotic series. Nevertheless, we will, for the convenience of the reader, provide the basic definitions in the form they will be used everywhere in the book.

Suppose that two functions $f(\varepsilon)$ and $\varphi(\varepsilon)$ are defined in a neighborhood $\mathfrak{U}$ of the point 0 for $\varepsilon > 0$, and let $\varphi(\varepsilon) > 0$. Then the formulas

$$f(\varepsilon) = O(\varphi(\varepsilon)), \qquad \varepsilon \to 0 \tag{1}$$

and

$$f(\varepsilon) = o(\varphi(\varepsilon)), \qquad \varepsilon \to 0 \tag{2}$$

mean, respectively, that

$$|f(\varepsilon)| < M\varphi(\varepsilon) \quad \text{for } \varepsilon \in \mathfrak{U}, \tag{3}$$

where M is a positive constant, and

$$f(\varepsilon)/\varphi(\varepsilon) \to 0 \quad \text{as } \varepsilon \to 0, \quad \varepsilon > 0. \tag{4}$$

The same definitions apply to the case where the function f depends on some additional variables $z_1, z_2, \ldots, z_N$. The relations (1)–(4) (with $f(\varepsilon)$ replaced by $f(\varepsilon, z)$, $z = (z_1, z_2, \ldots, z_N)$) are then understood to be satisfied for each fixed z. If the additional statement is made that equalities (1) and (2) are satisfied "uniformly with respect to $z \in \mathfrak{A}$," this means that the constant M in (3) does not depend on z and that the passage to the limit in (4) is uniform with respect to $z \in \mathfrak{A}$.

The definitions of the symbols O and o are transferred, in an obvious manner, to the functions $f(x)$ and $\varphi(x)$ as $x \to a$, provided these functions are defined on a set $\mathfrak{U} \subset \mathbf{R}^m$ for which a is a limit point. Let, for example, $\mathfrak{U}$ be an unbounded domain in $\mathbf{R}^m$, and suppose that the functions $f(\xi)$ and $\varphi(\xi)$ are defined in $\mathfrak{U}$ and $\varphi(\xi) > 0$. Then the relation $f(\xi) = o(\varphi(\xi))$, $\xi \to \infty$, $\xi \in \mathfrak{U}$ means that $f(\xi)/\varphi(\xi) \to 0$ as $|\xi| \to \infty$, $\xi \in \mathfrak{U}$. When no misunderstanding is possible, the mention of the set $\mathfrak{U}$ will be omitted.

We now proceed to asymptotic series. A sequence $\varphi_k(\varepsilon)$, $k = 0, 1, 2, \ldots$, will be called *a gauge sequence*, if the functions $\varphi_k(\varepsilon)$ are defined and positive in a neighborhood $\mathfrak{U}$ of the point 0 for $\varepsilon > 0$ and if the relations

$$\varphi_{k+1}(\varepsilon)/\varphi_k(\varepsilon) \to 0 \quad \text{as } \varepsilon \to 0$$

are satisfied for all k.

(The term "asymptotic sequence" is used more often. However, as the text is already overloaded with the word "asymptotic," we prefer the rarer but convenient term "gauge sequence.")

A function $f(\varepsilon)$ defined on $\mathfrak{U}$ for $\varepsilon > 0$ is said to be *expanded in the asymptotic series*

$$\sum_{k=0}^{\infty} c_k \varphi_k(\varepsilon) \tag{5}$$

(or the series (5) is said to be an *asymptotic expansion* of the function $f(\varepsilon)$) if for any natural n one has

$$f(\varepsilon) - \sum_{k=0}^{n} c_k \varphi_k(\varepsilon) = o(\varphi_n(\varepsilon)), \qquad \varepsilon \to 0. \tag{6}$$

An asymptotic expansion is denoted by the usual formula

$$f(\varepsilon) = \sum_{k=0}^{\infty} c_k \varphi_k(\varepsilon), \qquad \varepsilon \to 0. \tag{7}$$

No convergence of the series (7) is implied by this notation in the present book (unless specifically stated). Moreover, in the most interesting cases the asymptotic series (7) diverge for all $\varepsilon > 0$. Thus, relation (7) means only that equalities (6) are satisfied.

The same definition applies to the case where the function f depends not only on ε but on the variables $z_1, z_2, \dots, z_N$ as well. The coefficients c_k in (6) and (7) then, naturally, depend on $z = (z_1, z_2, \dots, z_N)$. If relation (6) holds uniformly with respect to $z \in \mathfrak{A}$, the function $f(\varepsilon, z)$ is said to be expanded in the asymptotic series $\sum_{k=0}^{\infty} c_k(z) \varphi_k(\varepsilon)$ uniformly with respect to $z \in \mathfrak{A}$. For example, in relation (0.1) one has $\varphi_k(\varepsilon) = \varepsilon^k$, $z = x$, where the set $\mathfrak{A}$ is either $\overline{\Omega}$ or a part of $\overline{\Omega}$.

The definitions of an asymptotic series and gauge functions are transferred in an obvious manner to functions $f(x)$, $\varphi_k(x)$ defined on a set $\mathfrak{U} \subset \mathbf{R}^m$ for which a is a limit point:

$$f(x) = \sum_{k=0}^{\infty} c_k \varphi_k(x), \qquad x \to a, \qquad x \in \mathfrak{U}.$$

In particular, a may be a point at infinity.

If $f(x) \in C^{\infty}$ in a neighborhood of the point a then its Taylor expansion is evidently its asymptotic series as $x \to a$, the gauge sequence of functions being $(x - a)^k$. No convergence of the Taylor series is required.

If relations (6) are satisfied for just finitely many values of n, the asymptotic expansion of $f(\varepsilon)$ is said to be valid for these n, meaning that relations (6) are satisfied for them.

One can easily show that for any function the asymptotic expansion of the form (5) is unique (provided the gauge sequence $\varphi_k(\varepsilon)$ is fixed!). Definition (6) may be replaced by equivalent but often more convenient formulations. Namely, one can replace $o(\varphi_n(\varepsilon))$ in the right-hand side of (6) with $O(\varphi_{n+1}(\varepsilon))$, $O(\varphi_n(\varepsilon))$, or, in general, $O(\varphi_{m_n}(\varepsilon))$, where m_n is a sequence of natural numbers such that $m_n \to \infty$ as $n \to \infty$.

Asymptotic series of the form (5) can be added up, subtracted, multiplied, or divided (unless, of course, all coefficients c_k in the denominator vanish). This means that if two functions can each be expanded in an asymptotic series of the form (5), then their sum can also be expanded in an asymptotic series which is the formal sum of the asymptotic series corresponding to the summands, etc. Both addition and subtraction preserve the gauge sequence, while multiplication and division alter it in an obvious manner.

Under simple extra assumptions, asymptotic series can be integrated term-by-term although, in general, they cannot be differentiated term-by-term. In real situations, as a rule, asymptotic series admit termwise differentiation, but this has to be proved in each case separately.

For more details on asymptotic series and their properties the reader is advised to consult the books [10], [24], [84], [97], [23].

The simplest case, which is of little interest, is when the asymptotic series
(0.1) describes the solution uniformly for all $x \in \Omega$. It is natural to call such
dependence on the small parameter *regular*. To other problems, for which the
asymptotic expansion (0.1) either does not exist or does not hold everywhere
in $\overline{\Omega} \cup \partial\Omega$ the (maybe, unfortunate) name of *singular perturbation problems*
is applied.

It is often said that looking for all terms of the series (0.1) is an unnecessary extravagance, because one or, at most, two terms are quite sufficient for
applications. To a certain degree, this argument makes sense. However, there
are counter-arguments as well. First, there are cases in numerical calculations
where the knowledge of not just two, but three, or even more terms of the
series is useful (examples of this situation are to be found in Chapter II, §2,
and Chapter III, §3). Then it does not make a great difference whether one
looks for just three terms of the series or finds a general way of constructing
all $u_k(x)$. Neither can one ignore the natural desire of mathematicians to
solve the purely theoretical problem of finding the solution up to any power
of ε. Finally, there is the following technical but important consideration.
Sometimes, when only one or two terms of the series are obtained, it is very
difficult to prove that the functions constructed do indeed approximate the
true solution. On the other hand, such a proof can be carried out much
easier if an f.a.s. of the form (0.1) is available. (The reader will encounter
such a situation in Chapter III, §4.) In this book, we always construct full
asymptotic expansions, i.e., approximations of solutions up to any power of
ε.

For singular perturbation problems, the solutions can behave in different
ways as $\varepsilon \to 0$. We will consider only those problems whose solutions admit
asymptotic expansions of the form (0.1) everywhere in $\overline{\Omega}$ except a small
neighborhood of a set of lesser dimension. Denote this singular set by Γ.
The set Γ is often a part of the boundary $\partial\Omega$ so that its neighborhood is
then naturally called the *boundary layer*. In many other interesting cases the
set Γ is located inside Ω, but its neighborhood is called by the same name
(however, another name is also in use—the *interior layer*). We also note
that in some cases the small parameter is included (in an essential manner)
into the boundary conditions. The series (0.1) is called the *outer asymptotic
expansion*, or, for short, the *outer expansion*. This term owes its origin to the
problems of fluid dynamics dealing with flows past a solid boundary of a fluid
with small viscosity. Thus, in what follows, (0.1) is a uniform asymptotic
expansion of a solution $u(x, \varepsilon)$ as $\varepsilon \to 0$ everywhere outside any sufficiently
small fixed neighborhood of the set Γ. How can one find the asymptotics of
the solution in a boundary layer, i.e. in a neighborhood of Γ, and what for?

Let us start by answering the second question. First, the coefficients $u_k(x)$
of the series (0.1) are solutions of some auxiliary problems. The equations for
them are easily derived from the original equation $L(x, u, \varepsilon) = 0$ after its
expansion in a series in powers of ε. However, the corresponding boundary

conditions for $u_k(x)$ are usually not so easy to find. Without these conditions one cannot determine $u_k(x)$, and in order to do that one has to know the behavior of the solution $u(x, \varepsilon)$ in the boundary layer. Second, much of the most important and interesting information is often related to the behavior of the function on the boundary. This is, for example, the case when one studies the flow past a solid boundary, or calculates the capacitance of a thin capacitor, etc.

Consider now the first question, i.e., how the asymptotics in the boundary layer can be found. The idea, whose origins are difficult to trace, but which was apparently clearly formulated for the first time in L. Prandtl's report to the Third International Congress of Mathematicians (see [102]), is to define new, generally speaking, "stretched" coordinates $\xi = \xi(x, \varepsilon)$ in the boundary layer, and to seek the asymptotic expansion for the solution in the form

$$u(x, \varepsilon) = \sum_{i=0}^{\infty} \mu_i(\varepsilon)v_i(\xi), \qquad \varepsilon \to 0. \tag{0.2}$$

Here $\mu_i(\varepsilon)$ is a gauge sequence of functions which has to be determined, the functions $v_i(\xi)$ are called the *boundary layer functions* (or boundary functions). The coordinate functions $\xi(x, \varepsilon)$ are usually rather simple. If, for example, the set Γ consists of a single point, viz. the origin, one often sets $\xi = \varepsilon^{-\alpha}x$, where $\alpha > 0$. In other cases only some of the coordinate functions undergo the stretching procedure by setting, e.g., $\xi_1 = \varepsilon^{-\alpha}x_1$, $\xi_2 = x_2$. More complicated combinations are also possible. The series (0.2) is called the *inner expansion*, and $\xi(x, \varepsilon)$ are called *inner variables*. The differential equations for the functions $v_i(\xi) = 0$ are obtained from the original equation $L(x, u, \varepsilon) = 0$. Although the domain of definition of the functions $v_i(\xi)$ in the variables x is small (because it is a thin boundary layer), in the "stretched" variables ξ the domain becomes a large one, and the smaller ε, the larger it is. One can therefore assume that the functions $v_i(\xi)$ are defined in an unbounded domain which does not depend on ε.

We have thus delineated the class of problems to be considered below. They are sometimes called problems of the boundary layer type. Evidently, they do not include all singular perturbation problems. Another important class of phenomena is described by rapidly oscillating functions so that the series (0.1) can nowhere be considered as an approximation to the solution. We will not dwell on these problems here, although the approach we consider may turn out to be useful for problems of this kind as well. A short commentary on this can be found at the end of the book.

What are the difficulties arising in the investigation of the boundary layer type problems? First, one has to define a singular set Γ, around which the boundary layer is formed. Then the inner variables ξ and the gauge sequence $\mu_i(\varepsilon)$ are to be selected. There is no unified approach to making this selection. Nevertheless, the analysis of a number of representative examples provides some clues to the treatment of similar situations.

Once Γ, $\xi(x, \varepsilon)$, and $\mu_i(\varepsilon)$ are chosen, the equations for $u_k(x)$ and $v_i(\xi)$ are usually written out without difficulty, and the question that arises is: how should the problems for these equations be formulated? As a rule, the boundary conditions for $u_k(x)$ and $v_i(\xi)$ formally implied by the original problem fail to define the solutions $u_k(x)$ and $v_i(\xi)$ of the auxiliary problems uniquely. This brings us to the central point of the analysis: how should one choose $u_k(x)$ and $v_i(\xi)$ which are not a priori defined by their boundary value problems uniquely? The problem is that the series (0.1) and (0.2) must approximate one and the same solution $u(x, \varepsilon)$ but in different domains: the series (0.2) inside the boundary layer, and (0.1) outside the boundary layer. It turns out that there is a rather broad (as compared with the boundary layer) domain where both series approximate the solution simultaneously and thus asymptotically coincide. The series (0.1) and (0.2), therefore, are matched to each other. This matching procedure makes it possible to define the functions $u_k(x)$ and $v_i(\xi)$ uniquely, and the corresponding method is called the *matching method for asymptotic expansions*, or the method of matched asymptotic expansions. This book presents the solutions for a number of typical problems, mainly for partial differential equations, the asymptotic analysis of which is achieved by the method of matched asymptotic expansions.

The set of singular perturbation problems naturally falls into two classes. The first includes the problems for which the functions $u_k(x)$, i.e., the coefficients of the outer expansion of (0.1), are smooth in $\overline{\Omega}$. Problems of the second class involve one more singularity: the coefficients $u_k(x)$ themselves have singularities on the set Γ and the order of singularity grows with k. We shall call problems of this kind *bisingular*. These problems constitute the main subject of the present book. Problems of the first class are, as a rule, much simpler, and their brief description is given in Chapter I.

We conclude this introduction with an elementary but instructive example. It is related to the fact that the solutions to problems of the boundary layer type are functions which are approximated by different asymptotic series in different subdomains of their domains of definition. The example considered below is not formally a solution to a boundary value problem, but serves well to demonstrate how one and the same function can be expanded into different asymptotic series in different domains. The example also illustrates an important for what follows notion of a composite asymptotic expansion.

Let

$$f(x, \varepsilon) = \frac{e^x}{x + \varepsilon}, \qquad \overline{\Omega} = \{x : 0 \le x \le 1\}, \quad \varepsilon > 0, \quad \varepsilon \to 0. \qquad (0.3)$$

For each $x > 0$ this function can be expanded in an asymptotic (and even convergent) series in the powers of ε:

$$f(x, \varepsilon) = \frac{e^x}{x}\left(1 - \frac{\varepsilon}{x} + \frac{\varepsilon^2}{x^2} - \frac{\varepsilon^3}{x^3} + \cdots\right). \qquad (0.4)$$

The equality

$$g_n(x, \varepsilon) \stackrel{\text{def}}{=} \left| f(x, \varepsilon) - \frac{e^x}{x} \sum_{j=0}^{n} (-1)^j \frac{\varepsilon^j}{x^j} \right| = \frac{e^x}{x+\varepsilon} \frac{\varepsilon^{n+1}}{x^{n+1}} \qquad (0.5)$$

implies that for $x \geq \delta > 0$ the asymptotics is uniform with respect to x: $g_n(x, \varepsilon) \leq 3\delta^{-n-2}\varepsilon^{n+1}$. However, the approximation becomes poor for x small. Indeed, for example, for $x \leq \varepsilon$ the inequality $g_n(x, \varepsilon) > (2\varepsilon)^{-1}$ holds which means that the partial sums of the series (0.4) are by no means close to $f(x, \varepsilon)$ for $x \leq \varepsilon$.

For x small it is natural to introduce a new variable $\xi = \varepsilon^{-1}x$ and represent $f(x, \varepsilon)$ by another (also convergent) series

$$f(x, \varepsilon) = \frac{\exp(\varepsilon\xi)}{\varepsilon(1+\xi)} = \frac{1}{\varepsilon(1+\xi)}\left(1 + \varepsilon\xi + \varepsilon^2\frac{\xi^2}{2} + \cdots\right). \qquad (0.6)$$

In this case one can also easily obtain the error estimate for the approximation:

$$h_k(x, \varepsilon) \stackrel{\text{def}}{=} \left| f(x, \varepsilon) - \frac{1}{\varepsilon(1+\xi)} \sum_{j=0}^{k} \frac{(\varepsilon\xi)^j!}{j} \right|$$

$$\leq \frac{3}{\varepsilon(1+\xi)} \frac{(\varepsilon\xi)^{k+1}}{(k+1)!} = 3\frac{\xi^{k+1}}{(1+\xi)(k+1)!}\varepsilon^k. \qquad (0.7)$$

The estimate (0.7) is good for bounded or not too large values of ξ, but it is unsatisfactory for $\xi \sim \varepsilon^{-1}$, i.e. for finite values of x. In this example, the series (0.4) provides the outer expansion (0.1) for the function (0.3). It is uniformly asymptotic everywhere for $x \geq \delta$ $\forall \delta > 0$, but fails in a neighborhood of the point $x = 0$. The series (0.6) is the inner expansion (0.2). It is uniformly asymptotic in a thin boundary layer $\varepsilon\xi < \varepsilon^\alpha$ but fails outside a neighborhood of zero.

Let us now give a more precise description of the domains where the series (0.4) and (0.6) provide good approximations of the function $f(x, \varepsilon)$. Let α and β be any two numbers such that $0 < \alpha < \beta < 1$. One can easily see that the series (0.4) and (0.6) are uniformly asymptotic in the domains $\Omega_\beta = \{x: \varepsilon^\beta \leq x \leq 1\}$ and $\omega_\alpha = \{x: 0 \leq x \leq \varepsilon^\alpha\}$, respectively (see Figure 1, next page). Indeed, it follows from (0.5) and (0.7) that $g_n(x, \varepsilon) \leq 3\varepsilon^{(1-\beta)n-1}$ in Ω_β and $h_k(x, \varepsilon) \leq 3\varepsilon^{\alpha(k+1)-1}$ in ω_α. The domains ω_α and Ω_β overlap. *The width of their intersection $\{x: \varepsilon^\beta \leq x \leq \varepsilon^\alpha\}$ is substantially greater than the characteristic measure of the boundary layer in this problem: $0 \leq x \leq M\varepsilon$. Thus, jointly, the series (0.4) and (0.6) provide a uniform asymptotics for the function (0.3) everywhere for $0 \leq x \leq 1$.* Furthermore, in the common domain $\omega_\alpha \cap \Omega_\beta$ the partial sums of both series are close to $f(x, \varepsilon)$, and, therefore, to each other.

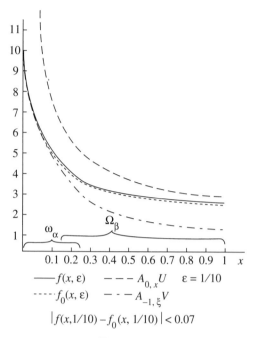

$$-f(x,\varepsilon) \quad ---A_{0,x}U \quad \varepsilon = 1/10$$
$$\cdots f_0(x,\varepsilon) \quad -\cdot-A_{-1,\xi}V$$
$$|f(x,1/10)-f_0(x,1/10)|<0.07$$

FIGURE 1

Although the closeness between (0.4) and (0.6) is not apparent at first glance, it does hold in the domain $\omega_\alpha \cap \Omega_\beta$. This enables one to write out another asymptotic approximation for the function $f(x,\varepsilon)$. The fact that the outer expansion (0.4) and the inner expansion (0.6) provide good approximations only in certain domains whose boundaries are not defined uniquely often makes their subsequent application difficult. Therefore, a sufficiently simple expression approximating the function uniformly everywhere in $\overline{\Omega}$ would be helpful. Such an expression is called a *composite asymptotic expansion*, and in the case of this example it is constructed as follows. Denote the series appearing on the right-hand sides of (0.4) and (0.6) by $U(x,\varepsilon)$ and $V(\xi,\varepsilon)$, respectively. Denote by $A_{n,x}U$ the sum of all terms of the series U containing the powers of ε not greater than n: $A_{n,x}U = \sum_{j=0}^{n}(-1)^j(e^x/x^{j+1})\varepsilon^j$. Similarly, $A_{k,\xi}V = \sum_{j=0}^{k+1}(\xi^j/(1+\xi)j!)\varepsilon^{j-1}$. We have already seen that these functions are close to each other in the domain $\omega_\alpha \cap \Omega_\beta$ (assuming that $x = \varepsilon\xi$), i.e., for x small and ξ large. It is, therefore, natural to expand each term in the sum $A_{n,x}U$ in a series as $x \to 0$, and each term in the sum $A_{n,\xi}V$ in a series as $\xi \to \infty$. Then, substituting x in the first sum with $\varepsilon\xi$, write out the sum of all terms with ε^s, where $s \le m$. Denote this sum by $A_{m,\xi}(A_{n,x}U)$. In our example

$$A_{m,\xi}(A_{n,x}U) = \sum_{\substack{0 \le j \le n \\ 0 \le i \le m+1}} (-1)^j \frac{\xi^{-j-1+i}}{i!}\varepsilon^{i-1}.$$

Subjecting $A_{k,\xi}V$ as $\xi \to \infty$ to a similar procedure, one obtains

$$A_{m,x}(A_{k,\xi}V) = \sum_{\substack{0 \le j \le k+1 \\ 0 \le i \le m}} (-1)^i \frac{x^{j-i-1}}{j!} \varepsilon^i.$$

It is immediately evident that

$$A_{m,\xi}(A_{n,x}U) \equiv A_{n,x}(A_{m,\xi}V) \tag{0.8}$$

for all $m \ge 0$ and $n \ge 0$. This is the common part of the asymptotics (0.4) and (0.6). The equality (0.8) is an algebraic expression of the matching of the outer and inner expansions (0.4) and (0.6). The composite asymptotic expansion equals

$$f_n(x,\varepsilon) = A_{n,x}U + A_{n,\xi}V - A_{n,\xi}(A_{n,x}U); \tag{0.9}$$

i.e., it is constructed as the sum of the segments of the outer and inner expansions minus their common part. The function $f_n(x,\varepsilon)$ now approximates $f(x,\varepsilon)$ uniformly everywhere in $\overline{\Omega}$. Indeed, for $x \le \varepsilon^\alpha$, the function $f(x,\varepsilon)$ is close to $A_{n,\xi}V$, and the difference $A_{n,x}U - A_{n,\xi}(A_{n,x}U)$ is small. On the other hand, for $x \ge \varepsilon^\beta$, the function $f(x,\varepsilon)$ is close to $A_{n,x}U$ and the difference $A_{n,\xi}V - A_{n,\xi}(A_{n,x}U) = A_{n,\xi}V - A_{n,x}(A_{n,\xi}V)$ is small. We will not go into a detailed calculation, but the reader can show without difficulty that $|f_n(x,\varepsilon) - f(x,\varepsilon)| \le M_n \varepsilon^{n+1}$ everywhere for $0 \le x \le 1$. The graphs of $f(x,\varepsilon)$, $f_0(x,\varepsilon)$, $A_{0,x}U$, and $A_{-1,\xi}V$ for $\varepsilon = 1/10$ are shown in Figure 1.

The above asymptotic analysis performed on the function (0.3) does not seem to make much sense, to say the least. Indeed, the explicit form of the function (0.3) is in many respects simpler, more clear-cut and easy to grasp, than the asymptotic series (0.4), (0.6), not to mention cumbersome and frightening expression (0.9). However, one has to take into account that, as a rule, no explicit solution of the differential equation is available, while series of the types (0.1) and (0.2) can often be constructed. In constructing the series, one should strive to ensure that the matching conditions similar to (0.8) are satisfied. Then a uniform approximation of type (0.9) is quite acceptable, and, in a sense, provides an approximate solution to the boundary value problem for the differential equation under consideration. Some examples of this kind will be given in the following chapters. What we would like to note now is that even for such an elementary function as (0.3), the representation (0.9) can help to investigate the asymptotics of the function obtained on applying a nonlocal operator to $f(x,\varepsilon)$. Consider, for example, the operator $J(\varepsilon) = \int_0^1 f(x,\varepsilon)\,dx$. How can one find the asymptotics of this integral as $\varepsilon \to 0$? A straightforward expansion of $f(x,\varepsilon)$ in powers of ε results in diverging integrals and provides no answer. At the same time, the representation (0.9) yields an easy solution. Indeed, taking into account that

$f(x, \varepsilon) - f_n(x, \varepsilon) = O(\varepsilon^{n+1})$ uniformly on the closed interval $[0, 1]$, one has $J(\varepsilon) = J_n(\varepsilon) + O(\varepsilon^{n+1})$ where $J_n(\varepsilon) = \int_0^1 f_n(x, \varepsilon)\, dx$. The asymptotics of $J_n(\varepsilon)$ is computed without difficulty. Let, for example, $n = 1$. Then

$$J_1(\varepsilon) = \int_0^1 \{e^x(x^{-1} - \varepsilon x^{-2}) + (1 + \xi)^{-1}(\varepsilon^{-1} + \xi + 2^{-1}\varepsilon \xi^2)$$

$$- (\varepsilon^{-1} + \varepsilon^{-1}x + \varepsilon^{-1}2^{-1}x^2)(\varepsilon x^{-1} - \varepsilon^2 x^{-2})\}\, dx$$

$$= \int_0^1 [e^x - 1 - x - 2^{-1}x^2]x^{-1}\, dx - \varepsilon \int_0^1 \left[e^x - 1 - x - \frac{x^2}{2}\right]x^{-2}\, dx$$

$$+ \int_0^{\varepsilon^{-1}} (1 + \xi)^{-1}(1 + \varepsilon\xi + 2^{-1}\varepsilon^2 \xi^2)\, d\xi.$$

The last integral is found explicitly. Thus,

$$J(\varepsilon) = -\ln \varepsilon + a_0 + \varepsilon \ln \varepsilon - a_1 \varepsilon - 2^{-1}\varepsilon^2 \ln \varepsilon + O(\varepsilon^2),$$

where

$$a_0 = \int_0^1 [e^x - 1 - x - 2^{-1}x^2]x^{-1}dx + 5/4,$$

$$a_1 = \int_0^1 [e^x - 1 - x - 2^{-1}x^2]x^{-2}dx - 1/2.$$

Of course, this method of computing the asymptotics of the integral of $f(x, \varepsilon)$ is not new. It is, in fact, equivalent to the regularization procedure applied to the diverging integrals arising in the expansion of $f(x, \varepsilon)$ in powers of ε. Nevertheless, this example illustrates the use of a uniform asymptotic expansion (0.9) even in simplest situations.

We will now make some remarks on the notation to be used below. Unless otherwise stated, all series are assumed to be asymptotic either as the parameter (usually $\varepsilon > 0$) tends to zero, or the independent variable tends to zero or infinity.

If a series converges, this fact is always mentioned explicitly. The same applies to each case where a series is considered as a formal object with the obvious rules of addition, multiplication etc. In what follows, such series are called formal series.

Let us now make the definition of the operators $A_{\alpha, x}$ used in the previous example more precise. Let

$$U = \sum_{k=0}^{\infty} \nu_k(\varepsilon)u_k(x) \qquad (0.10)$$

be a formal series, where $\nu_k(\varepsilon)$ is a gauge sequence with the following property: for each natural n there is k_0 such that $\nu_{k_0}(\varepsilon) = o(\varepsilon^n)$ as $\varepsilon \to 0$ (and, consequently, $\nu_k(\varepsilon) = o(\varepsilon^n)$ as $\varepsilon \to 0$ for all $k \geq k_0$). By $A_{\alpha,x}U$ we always denote the partial sum of the series (0.10) consisting of those of its terms for which $\varepsilon^\beta = o(\nu_k(\varepsilon))$ for all $\beta > \alpha$. The meaning of the operator $A_{\alpha,x}$ is especially simple in the often encountered case where the gauge sequence $\nu_k(\varepsilon)$ consists of functions of the form $\varepsilon^m \ln^i \varepsilon$ where m and i are integers, and for each m there are only finitely many numbers i (either positive or negative). Then, for any integer p, we denote by $A_{p,x}U$ the partial sum of the series (0.10) containing all its terms for which the power of ε does not exceed p.

The subscript x in the notation for the operator $A_{\alpha,x}$ indicates that the coefficients of the series (0.10) are functions of the variable x. This subscript is important because, along with the independent variable x, there is another variable ξ depending on x and ε. The notation $A_{\alpha,\xi}$ means that one first has to write the function, to which the operator is to be applied, in the form of a series whose coefficients are functions of ξ and then apply the procedure described above. Let, for example, $x = \varepsilon\xi$ and $\mathscr{F}(x,\varepsilon) = e^x/x + \varepsilon \cos x/x^2 + \varepsilon^2/x^3$. Then

$$A_{1,x}\mathscr{F} = \frac{e^x}{x} + \varepsilon\frac{\cos x}{x^2}$$

and

$$A_{1,\xi}\mathscr{F} = A_{1,\xi}\left(\frac{e^{\varepsilon\xi}}{\varepsilon\xi} + \frac{\cos \varepsilon\xi}{\varepsilon\xi^3} + \frac{1}{\varepsilon\xi^3}\right) = \frac{1}{\varepsilon}\left(\frac{1}{\xi} + \frac{1}{\xi^2} + \frac{1}{\xi^3}\right) + 1 + \varepsilon\left(\frac{\xi}{2} - \frac{1}{6}\right).$$

The variable x may denote a point in a multidimensional space; the definition of the operator $A_{\alpha,x}$ remains unchanged.

The operator B_α, which will also be repeatedly used below, is simpler and also signifies a partial sum of a series. It is applied to a series of the form $H = \sum_{k=0}^\infty h_k(x)$ where $x = (x_1, \ldots, x_n)$ tends either to zero, or infinity, and the functions $h_k(x)$ are bounded by some gauge sequence. If $x \to 0$ then $B_\alpha H$ denotes the partial sum of all terms of the series H that are not $o(|x|^\alpha)$ as $x \to 0$. The definition of $B_\alpha H$ in the case $|x| \to \infty$ is formulated in the same way.

Each formula in the book is numbered by a pair of numbers. The first indicates the section number, and the second that of the formula. When referring to a formula in the same chapter, only the double number of the formula is given. For formulas of another chapter, the chapter number is added.

The symbol ∎ indicates the end of a proof. If a proof is evident, this symbol follows right after the statement.

CHAPTER I

Boundary Layer Functions of Exponential Type

This chapter is of an auxiliary nature. We consider a number of examples for which $u_k(x)$, i.e., the coefficients of the outer expansion (0.1), have no singularities in $\overline{\Omega}$. In this case the experience is that the behavior of the boundary layer functions $v_i(\xi)$, i.e., the coefficients of the series (0.2), is especially simple: the differences between the solution $u(x, \varepsilon)$ and partial sums of the series (0.1) decay exponentially with the increase of the distance from the set Γ. Therefore, one can (and it is more convenient to) seek the asymptotics of the solution in the form of the sum of the series (0.1) and the series $\sum_{i=0}^{\infty} \mu_i(\varepsilon) z_i(\xi)$.

The coefficients $z_i(\xi)$ tend to zero exponentially as $\xi \to \infty$. If one looks for the asymptotics in a neighborhood of Γ in the form (0.2) then, evidently, $v_i(\xi) = z_i(\xi) + \tilde{v}_i(\xi)$, where $\tilde{v}_i(\xi)$ are the functions obtained from the series (0.1) by re-expanding the coefficients $u_k(x)$ in Taylor series in a neighborhood of Γ.

As a rule, problems of this kind are much simpler than those constituting the main subject of this book. Their theory has been basically well developed about thirty years ago (see remarks and comments on bibliography at the end of this book). So the material of this chapter is of an introductory character, and its main goal is to make the reading of the main problems considered in Chapters II–VI independent of other (mainly, periodic) publications.

§1. Boundary value problems for ordinary differential equations

Ordinary differential equations are not a primary object of our study. We use them as simple examples to illustrate the most essential features of the asymptotic analysis which will later be used for partial differential equations. For that reason, in this section we consider only the technique of constructing the asymptotics without going into less essential details. We do not dwell on the justification of the procedure, which, if the reader so wishes, can be found in articles listed in the comments on bibliography.

EXAMPLE 1. Consider the following boundary value problem:

$$\varepsilon^2 u'' - q(x)u = f(x) \quad \text{for } 0 \leq x \leq 1, \tag{1.1}$$

$$u(0, \varepsilon) = R_0, \qquad (1.2)$$

$$u(1, \varepsilon) = R_1, \qquad (1.3)$$

where $q, f \in C^\infty[0, 1]$, $q(x) > \text{const} > 0$.

The outer expansion (0.1) is easily constructed, and is, evidently, of the form

$$U = \sum_{k=0}^{\infty} \varepsilon^{2k} u_{2k}(x), \qquad \varepsilon \to 0, \qquad (1.4)$$

where U is a formal series for now. Substituting U into equation (1.1) and equating the coefficients of the same powers of ε one obtains the recurrence system of equations

$$-q(x)u_0(x) = f(x), \qquad q(x)u_{2k}(x) = u''_{2k-2}(x) \quad \text{for } k \geq 1, \qquad (1.5)$$

whereby all $u_{2k}(x) \in C^\infty[0, 1]$ are defined uniquely.

Under a formal asymptotic solution (f.a.s.) of an equation as $\varepsilon \to 0$ we will mean a series whose partial sums satisfy the equation to within ε^m, where $m \to \infty$ together with the index of the partial sum. The f.a.s. of a boundary value problem is defined in the same way. In this case both the equation and the boundary conditions are assumed to be approximately satisfied.

Formulas (1.5) imply that the *series* (1.4) *is an f.a.s. to the equation* (1.1) *as* $\varepsilon \to 0$. However, it is not a solution to the whole boundary value problem because, in general, *it does not satisfy conditions* (1.2), (1.3) *even formally*. Although the coefficients of the series (1.4), i.e., the functions $u_{2k}(x)$, are smooth and normally do not make much trouble, there arises a *residual in the boundary conditions* (1.2), (1.3). In order to eliminate this residual one should introduce new, inner variables in neighborhoods of the points $x = 0$ and $x = 1$. This pair of points constitutes the singular set Γ in this problem. As both points are on an equal footing, let us consider just a neighborhood of the point $x = 0$. Under the assumption that the series (1.4) is close to the solution $u(x, \varepsilon)$ for the interior points of the closed interval $[0, 1]$, the solution $u(x, \varepsilon)$ undergoes an abrupt change from R_0 at $x = 0$ to $\approx u_0(0)$ for points close to zero. For a convenient description of this rapid change an inner "stretched" coordinate ξ should be introduced. The simplest substitution providing for the stretching of the coordinate is defined by the formula $x = \varepsilon^\alpha \xi$, where $\alpha > 0$. How should one choose α? One way is to set $q(x) \equiv \text{const} > 0$ in equation (1.1), find an explicit solution of the problem and verify that $\alpha = 1$. The same exponent should obviously be taken for the variable coefficient $q(x)$ as well.

We now consider another way of choosing α, which has a wider field of application. Denote $u(\varepsilon^\alpha \xi, \varepsilon) \equiv v(\xi, \varepsilon)$ and make the corresponding change of variable in the equation (1.1):

$$\varepsilon^{2-2\alpha} \frac{d^2 v}{d\xi^2} - q(\varepsilon^\alpha \xi)v = f(\varepsilon^\alpha \xi). \qquad (1.6)$$

Now one has to identify the principal terms of the equation for $\varepsilon \to 0$. If $\alpha < 1$, then the principal terms are $q(0)v(\xi, 0)$ and $f(0)$. Such a neglect of the second derivative has already led to the outer expansion (1.4) and is unable to yield anything new. If $\alpha > 1$, then the principal term is $\varepsilon^{2-2\alpha} d^2 v / d\xi^2$, and the first approximation equation is $d^2 v / d\xi^2 = 0$. Its solutions—linear functions—are also incapable of describing a smooth transition from the boundary condition $u(0, \varepsilon) = R_0$ to the outer expansion (1.4). For other problems the inner coordinate (or, in other words, the new scale) is chosen in such a way that the equation retains at least two principal terms. For partial differential equations, the choice of the latter is not so simple, and there is no general formal approach. In the subsequent chapters we shall face the problem of a correct choice of scale each time anew, and each time it will receive its own solution. In the above example, one evidently has to set $\alpha = 1$ in the equation (1.6). Then all terms of equation (1.6) acquire an equal status; one can represent $v(\xi, \varepsilon)$ by a series in the powers of ε, substitute this series into (1.6) and find the coefficients of the series.

However, as noted above, in this case it is more convenient (since $u_k(x)$ are smooth functions) to look for the f.a.s. to the problem (1.1), (1.2) in the form of the sum of the series (1.4) and the series

$$Z = \sum_{k=0}^{\infty} \varepsilon^k z_k(\xi), \qquad \varepsilon \to 0, \tag{1.7}$$

where $\xi = \varepsilon^{-1} x$. Since the series (1.4) is an f.a.s. of the nonhomogeneous equation (1.1), the series (1.7) must be an f.a.s. of the homogeneous equation which, in the variable ξ, is of the form

$$\frac{d^2 Z}{d\xi^2} - q(\varepsilon\xi)Z = 0. \tag{1.8}$$

The functions $z_k(\xi)$ must be defined on the segment $[0, \varepsilon^{-1}]$, but it is more convenient to assume that they are defined everywhere on $[0, \infty)$. The conditions imposed on the boundary layer functions $z_k(\xi)$ are as follows. They should not exert much influence for a fixed $x > 0$ and $\varepsilon \to 0$, which implies that

$$z_k(\xi) \to 0 \quad \text{as } \xi \to 0. \tag{1.9}$$

Their main mission is to eliminate the residual in the boundary condition (1.2). The sum of the series (1.4) and (1.7) must satisfy the condition $u(0, \varepsilon) = R_0$. This implies that

$$z_0(0) = R_0 - u_0(0), \qquad z_{2k}(0) = -u_{2k}(0) \quad \text{for } k > 0,$$
$$z_{2k+1}(0) = 0 \quad \text{for } k \geq 0. \tag{1.10}$$

The differential equations for $z_k(\xi)$ are obtained by substituting the series (1.7) into equation (1.8) and equating the coefficients of the same powers of ε. The coefficient $q(x)$ should first be expanded in a Taylor series

$q(\varepsilon\xi) = \sum_{k=0}^{\infty} q_k \varepsilon^k \xi^k$. Denoting $q_0 = b^2$, one obtains the recurrence system of equations for $z_k(\xi)$:

$$z_0'' - b^2 z_0 = 0,$$
$$z_1'' - b^2 z_1 = q_1 \xi z_0,$$
$$\cdots\cdots\cdots$$

$$z_k'' - b^2 z_k = \sum_{j=1}^{k} q_j \xi^j z_{k-j}(\xi). \tag{1.11}$$

The solutions to this system satisfying conditions (1.9), (1.10) are found easily:

$$z_0(\xi) = [R_0 - u_0(0)] \exp(-b\xi),$$
$$z_1(\xi) = q_1 [u_0(0) - R_0](2b)^{-2}(b\xi^2 + \xi) \exp(-b\xi),$$
$$\cdots\cdots\cdots$$

$$z_k(\xi) = P_{2k}(\xi) \exp(-b\xi).$$

Here P_{2k} are polynomials of degree $2k$ whose coefficients are calculated explicitly without much difficulty. The series (1.7) is, therefore, constructed. Substituting its partial sums $A_{n,\xi}Z$ into equation (1.8) one verifies that they satisfy it to within $O(\varepsilon^{n+1})$. Thus, the sum of the series (1.4) and (1.7) is an f.a.s. of the equation (1.1) and the boundary condition (1.2).

To eliminate the residual at the point $x = 1$ a similar series

$$W = \sum_{k=0}^{\infty} \varepsilon^k w_k(\eta), \qquad \varepsilon \to 0, \tag{1.12}$$

is constructed, where $\eta = \varepsilon^{-1}(1 - x)$. The functions $w_k(\eta)$ are solutions to the boundary value problems for $0 \leq \eta < \infty$ which are completely analogous to the problems for $z_k(\xi)$. The sum of the series (1.4), (1.7) and (1.12) is, by construction, an f.a.s. of equation (1.1). In addition, the sum of the series (1.4) and (1.7), satisfies condition (1.2) exactly while the sum of (1.4) and (1.12) satisfies condition (1.3) exactly. Therefore, the remaining residuals in the boundary conditions are given by the value of the series (1.7) for $x = 1$ and by that of the series (1.12) for $x = 0$. Since the coefficients $z_k(\xi)$ and $w_k(\eta)$ decay exponentially at infinity, these errors also decay exponentially. The construction of the asymptotics of the solution $u(x, \varepsilon)$ is thus completed: the sum of the series (1.4), (1.7) and (1.12) is an f.a.s. of the problem (1.1)–(1.3). Figure 2 shows the approximate graph of the solution $u(x, \varepsilon)$ and the graphs of the principal terms of the f.a.s.

In conclusion, let us examine the form the functions $v_i(\xi)$ take if one looks for an f.a.s. of the problem (1.1)–(1.3) in a neighborhood of the point $x = 0$ in the form of the series (0.2). Clearly, $\mu_i(\varepsilon) = \varepsilon^i$, and $v_i(\xi) = z_i(\xi) + \tilde{v}_i(\xi)$,

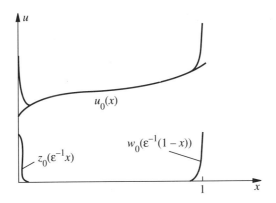

FIGURE 2

where $z_i(\xi)$ are coefficients of the series (1.7), and $\tilde{v}_i(\xi)$ are obtained from the series (1.4) in the following way: one has to expand each term of the series (1.4), i.e., each function $u_k(x) = u_k(\varepsilon\xi)$, into its Taylor series, and sum up all the coefficients of ε^i. Evidently, $\tilde{v}_i(\xi) = \sum_{j=0}^{i}(1/j!)(\partial^j u_{i-j}/\partial x^j)(0)\xi^j$. Thus, in terms of the inner variable ξ, each coefficient of the asymptotic expansion of the solution $u(x,\varepsilon)$ in the vicinity of the boundary is the sum of a polynomial and a function decaying exponentially at infinity.

EXAMPLE 2. Consider a slightly more difficult boundary value problem

$$\varepsilon\frac{d^4 u}{dx^4} + a(x)\frac{du}{dx} = f(x) \quad \text{for } 0 \leq x \leq 1, \ \varepsilon > 0, \tag{1.13}$$

$$u(0,\varepsilon) = \frac{du}{dx}(0,\varepsilon) = 0, \tag{1.14}$$

$$u(1,\varepsilon) = \frac{du}{dx}(1,\varepsilon) = 0, \tag{1.15}$$

where $a, f \in C^\infty[0,1]$, $a(x) \geq \text{const} > 0$. Here we are also looking for the asymptotics of the solution $u(x,\varepsilon)$ in the form of the sum of the outer expansion $U(x,\varepsilon)$ and the inner expansions $Z(\xi,\varepsilon)$, $W(\eta,\varepsilon)$ describing the behavior of the solution in the vicinity of the left and right endpoints of the closed interval, respectively. Then $\xi = \varepsilon^{-\alpha}x$, $\eta = \xi^{-\beta}(1-x)$, where $\alpha > 0$ and $\beta > 0$ are to be found. As in the preceding example, Z has to be an f.a.s. of the homogeneous equation which, in the variable ξ, is of the form

$$\varepsilon^{1-4\alpha}\frac{d^4 Z}{d\xi^4} + \varepsilon^{-\alpha}a(\varepsilon^\alpha\xi)\frac{dZ}{d\xi} = 0.$$

An argument similar to that used in Example 1 shows that one must choose α in such a way that $1 - 4\alpha = -\alpha$, whence $\alpha = 1/3$. Similarly, $\beta = 1/3$.

As will become clear from what follows, the relationship between the series U, Z, and W is now more complicated than in the preceding example. While in Example 1 the outer expansion U was constructed independently, now the inner and outer expansions affect each other. It is therefore clear that they must include the same powers of ε. Since the expansion of the coefficient $a(\varepsilon^{1/3}\xi)$ includes the powers of $\varepsilon^{1/3}$, so does the correct gauge sequence for all asymptotic expansions, viz., U, Z, and W.

Thus, we set

$$U(x,\varepsilon) = \sum_{k=0}^{\infty} \varepsilon^{k/3} u_k(x), \qquad Z(\xi,\varepsilon) = \sum_{k=0}^{\infty} \varepsilon^{k/3} z_k(\xi),$$

$$W(\eta,\varepsilon) = \sum_{k=0}^{\infty} \varepsilon^{k/3} w_k(\eta),$$

where $\xi = \varepsilon^{-1/3}x$, $\eta = \varepsilon^{-1/3}(1-x)$. As in Example 1, one obtains the recurrence system of equations for the functions $u_k(x)$:

$$a(x)u_0'(x) = f(x), \qquad u_1'(x) = 0, \qquad u_2'(x) = 0,$$
$$a(x)u_k'(x) = -u_{k-3}^{(4)}(x), \qquad k \geq 3. \tag{1.16}$$

However, the solution of this system is no longer unique: the *solution* of each equation *depends on one arbitrary parameter*. The presence of the parameters make it possible to satisfy at least one of the four boundary conditions (1.14), (1.15). But which one? In order to find the values of the parameters and, therefore, determine $u_k(x)$ uniquely, one has to take into account the behavior of the solution $u(x,\varepsilon)$ in the boundary layers at the endpoints of the closed interval. For that, an analysis of the auxiliary problems arising for the coefficients $z_k(\xi)$ and $w_k(\eta)$ is in order. Expand the coefficient $a(\varepsilon^{1/3}\xi)$ in the equation

$$\frac{d^4 Z}{d\xi^4} + a(\varepsilon^{1/3}\xi)\frac{dZ}{d\xi} = 0$$

into the series $a(\varepsilon^{1/3}\xi) = \sum_{k=0}^{\infty} a_k \varepsilon^{k/3}\xi^k$ and substitute the series Z into the equation. The resulting system of differential equations is of the form

$$\frac{d^4 z_0}{d\xi^4} + a(0)\frac{dz_0}{d\xi} = 0, \tag{1.17}$$

$$\frac{d^4 z_k}{d\xi^4} + a(0)\frac{dz_k}{d\xi} = -\sum_{j=1}^{k} a_j \xi^j \frac{dz_{k-j}}{d\xi}, \qquad k \geq 1. \tag{1.18}$$

Similarly,

$$\frac{d^4 w_0}{d\eta^4} - a(1)\frac{dw_0}{d\eta} = 0, \qquad \frac{d^4 w_k}{d\eta^4} - a(1)\frac{dw_k}{d\eta} = \sum_{j=1}^{k} b_j \eta^j \frac{dw_{k-j}}{d\eta}, \qquad k \geq 1, \tag{1.19}$$

where b_j are the coefficients of the Taylor expansion of the function $a(1-\varepsilon\eta)$. Assuming, as before, that both $z_k(\xi)$ and $w_k(\eta)$ decay exponentially at infinity, one arrives at the requirement that the sum of the series

U and Z must satisfy the boundary conditions (1.14), and the sum of U and W the boundary condition (1.15). This yields the boundary conditions for u_k, z_k, and w_k:

$$z_k(0) + u_k(0) = 0, \qquad k \geq 0, \tag{1.20}$$

$$z_0'(0) = 0, \quad z_k'(0) + u_{k-1}'(0) = 0, \qquad k \geq 1, \tag{1.21}$$

$$w_k(0) = -u_k(1), \quad k \geq 0, \qquad w_0'(0) = 0,$$
$$w_k'(0) - u_{k-1}'(1) = 0, \quad k \geq 1. \tag{1.22}$$

In addition, it is required that

$$z_k(\xi) \to 0 \quad \text{as } \xi \to \infty,$$
$$w_k(\eta) \to 0 \quad \text{as } \eta \to \infty. \tag{1.23}$$

Since $a(0) > 0$, the equation $d^4z/d\xi^4 + a(0)\,dz/d\xi = 0$ has just one linearly independent solution satisfying condition (1.23). It is the solution $z(\xi) = C\exp(-[a(0)]^{1/3}\xi)$. The equation $d^4w/d\eta^4 - a(1)\,dw/d\eta = 0$ has two linearly independent solutions satisfying condition (1.23), viz., $w(\eta) = \exp(-\gamma\eta)\{C_1\cos\gamma\sqrt{3}\eta + C_2\sin\gamma\sqrt{3}\eta\}$, where $\gamma = 2^{-1}[a(1)]^{1/3}$. Thus, at the right endpoint, the series W can, in general, *eliminate the residual in the two boundary conditions, but at the left endpoint this is impossible.* There the boundary layer functions $z_k(\xi)$ can help to satisfy only one of the boundary conditions. In order for the other boundary condition to be satisfied, one has to invoke the yet undetermined parameters in the coefficients of the outer expansion.

With that in mind, let us determine the coefficients u_k, z_k, and w_k as follows. The condition $z_0'(0) = 0$, condition (1.23) and equation (1.17) imply that $z_0(\xi) \equiv 0$. Therefore, for $k = 0$ condition (1.20) turns into the equality $u_0(0) = 0$. This together with equation (1.16) uniquely determine $u_0(x) = \int_0^x f(t)[a(t)]^{-1}dt$. Next, condition (1.21) $z_1'(0) = -u_0'(0)$ together with (1.23) and equation (1.18) for $k = 1$ uniquely determines $z_1(\xi) = f(0)[a(0)]^{-4/3}\exp\{-[a(0)]^{1/3}\xi\}$. It now follows from condition (1.20) $u_1(0) = -z_1(0)$ and equation (1.16) that $u_1(x) \equiv -z_1(0)$. The functions $z_1(\xi)$, $u_2(x)$, $z_3(\xi)$, etc. are found in exactly the same way. Once the functions $u_k(x)$ are constructed, equations (1.19) and conditions (1.22), (1.23) uniquely determine the functions $w_k(\eta)$. One can see that they decay exponentially as $\eta \to \infty$. The construction of the f.a.s. for the problem (1.13)–(1.15) is complete.

§2. Partial differential equations

We begin with an example of a boundary value problem in which the behavior of the solution is essentially the same as in Example 1.

EXAMPLE 3. Let Ω be a bounded domain in $\mathbf{R}^2$ with the boundary $S = \partial\Omega \in C^\infty$, and $u(x, \varepsilon)$ a solution of the boundary value problem

$$\mathscr{L}_\varepsilon u \equiv \varepsilon^2 \Delta u - q(x)u = f(x), \qquad x \in \Omega, \tag{2.1}$$

$$u(x, \varepsilon) = 0 \quad \text{for } x \in S. \tag{2.2}$$

Here $q, f \in C^\infty(\overline{\Omega})$, $q(x) > 0$ in $\overline{\Omega}$.

It is known (see, e.g., [61, Chapter 3]) that for any $\varepsilon > 0$ there exists a unique solution $u(x, \varepsilon)$ of the problem (2.1), (2.2).

The maximum principle yields the following bound for the solution, which does not depend on ε:

$$|u(x, \varepsilon)| \le \max_\Omega \left| [q(x)]^{-1} f(x) \right|. \tag{2.3}$$

Our task is to find the asymptotics to the solution $u(x, \varepsilon)$ for $\varepsilon \to 0$.

As in Example 1, we look for the outer expansion in the form

$$U(x, \varepsilon) = \sum_{k=0}^\infty \varepsilon^{2k} u_{2k}(x), \qquad \varepsilon \to 0.$$

The coefficients $u_{2k}(x)$ are found, as in Example 1, from the corresponding recurrence system of equations

$$u_0(x) = -f(x)[q(x)]^{-1}, \quad u_k(x) = [q(x)]^{-1}\Delta u_{k-1}, \qquad k \ge 1.$$

All $u_k(x) \in C^\infty(\overline{\Omega})$, but the series U fails to satisfy the boundary condition.

In order to eliminate the residual in the boundary condition, one has to construct a boundary layer function along the entire boundary S. Since S is smooth, one can introduce in its neighborhood the coordinate system s, y where s is a coordinate defined along the curve S, and y is the distance from a point $x \in \Omega$ to S (see Figure 3). The same argument as in Examples 1 and 2 suggests the change of variables $y = \varepsilon\zeta$ resulting in the inner expansion of the following form:

$$Z(s, \zeta, \varepsilon) = \sum_{k=0}^\infty \varepsilon^k z_k(s, \zeta). \tag{2.4}$$

The series Z has to be an f.a.s. of the homogeneous equation $\mathscr{L}_\varepsilon Z = 0$ which after the change of variables acquires the following form:

$$\mathscr{L}'_\varepsilon Z \equiv \frac{\partial^2 Z}{\partial \zeta^2} + \varepsilon L_1 \frac{\partial Z}{\partial \zeta} + \varepsilon^2 L_2 Z - q(x)Z = 0. \tag{2.5}$$

Here L_1 and L_2 are differential equations of the first and second order which include only the differentiations with respect to the variable s. The coefficients of L_1 and L_2 are smooth functions in s and y, i.e., s and $\varepsilon\zeta$. Suppose that $q(x)$ on the boundary equals $q_0(s)$. Taking the Taylor expansions with respect to $\varepsilon\zeta$ of all the coefficients of the equation (2.5) and substituting the series (2.4) into the same equation, one obtains the recurrence system of differential equations

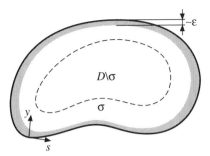

FIGURE 3

$$\frac{\partial^2 z_0}{\partial \zeta^2} - q_0(s)z_0 = 0, \qquad \frac{\partial^2 z_k}{\partial \zeta^2} - q_0(s)z_k = \mathcal{F}_k(s, \zeta), \qquad k \geq 1. \qquad (2.6)$$

Here the right-hand sides $F_k(s, \zeta)$ are linear in z_i and their derivatives for $i < k$, polynomial in ζ, and smooth in s. The system (2.6) is in fact the system of ordinary equations (1.11) of Example 1 but depending on s as a parameter. The boundary condition (2.2) implies the following requirement on $z_k(s, 0)$:

$$z_{2k}(s, 0) = -u_{2k}(x)|_{S(=\partial\Omega)}, \qquad z_{2k+1}(s, 0) = 0, \qquad k \geq 0. \qquad (2.7)$$

As in Example 1, there exist functions z_k satisfying equations (2.6), conditions (2.7), defined for $0 \leq \zeta < \infty$ and decaying exponentially as $\zeta \to \infty$. The only difference is that z_k are now also smooth functions in the variable s. The inner expansion (2.4) is constructed. The values of its coefficients are essential only for $0 \leq y \leq M\varepsilon$, i.e. in a narrow strip along the boundary whose width is of the same order of magnitude as ε (in Figure 3 this boundary layer is shaded). For $y = \varepsilon^\alpha$, where $\alpha < 1$, all terms of the series (2.4) decay exponentially, because for each of them one has

$$|z_k(s, \zeta)| \leq M_k \exp(-b\zeta), \qquad (2.8)$$

where $q_0(s) > b^2$. Here and in what follows we denote by M constants that depend neither on ε nor on the independent variables appearing in the functions being estimated. In general, these constants depend on the index of the asymptotic approximation, sometimes indicated by a subscript which, however, will mostly be omitted.

Thus, it is sufficient to consider boundary layer functions only in a narrow strip along the boundary. However, for the justification of the asymptotics it is convenient to consider them as defined everywhere in Ω after multiplying all $z_k(s, \varepsilon^{-1}y)$ by a fixed smooth function $\chi(x)$ (independent of ε). The function $\chi(x)$ equals 1 in a neighborhood of S and 0 in $\Omega\backslash\sigma$, where σ is a small neighborhood of the boundary S (see Figure 3). The use of such a truncating function is appropriate because the change of variables $x \leftrightarrow (s, y)$ is a diffeomorphism only in a small neighborhood of the boundary; equation (2.5) also makes sense in the same area. For that reason, in order to justify the asymptotics, we will consider, instead of the series (2.4), its product by the function $\chi(x)$ which, as we have just seen, describes an asymptotically equivalent situation.

Thus, let us show that $U(x, \varepsilon) + \chi(x)Z(s, \zeta, \varepsilon)$ is an asymptotic expansion for the solution of the problem (2.1), (2.2) uniformly everywhere in $\overline{\Omega}$. As above, denote the partial sums of the series as follows:

$$A_n U = \sum_{k=0}^{n} \varepsilon^{2k} u_{2k}(x), \qquad A_n Z = \sum_{k=0}^{n} \varepsilon^{k} z_k(s, \zeta).$$

By construction, the series U is an f.a.s. of the equation (2.1). Indeed, $\mathscr{L}_\varepsilon A_{2n} U - f(x) = \varepsilon^{2n+2} \Delta u_{2n} = O(\varepsilon^{2n+2})$. Similarly, the series Z is an f.a.s. of the equation (2.5): $\mathscr{L}_\varepsilon A_n Z = O(\varepsilon^{n+1} \exp(-b\zeta))$. Since the functions $z_k(s, \zeta)$ satisfy inequalities (2.8), one has

$$|\mathscr{L}_\varepsilon[A_{2n}(U + \chi Z)] - f(x)| \le M\varepsilon^{2n+1}$$

everywhere in $\overline{\Omega}$. Conditions (2.7) imply that $(U(x, \varepsilon) + \chi(x)Z(s, \zeta, \varepsilon))_S = 0$. Applying the maximum principle (see, for example, [61, Chapter 3]) to the boundary value problem for the operator $\mathscr{L}_\varepsilon$, one obtains from the last two relations the following bound:

$$|A_{2n}(U(x, \varepsilon) + \chi(x)Z(s, \zeta, \varepsilon)) - u(x, \varepsilon)| \le M\varepsilon^{2n+1} \quad \text{in } \overline{\Omega}. \qquad (2.9)$$

Take a strip σ_1 along the boundary such that $\chi(x) \equiv 1$ in σ_1 ($\sigma_1 \subset \sigma$). Then it follows from (2.8) and (2.9) that

$$|A_{2n} U - u(x, \varepsilon)| \le M\varepsilon^{2n+1} \quad \text{in } \overline{\Omega} \setminus \sigma_1,$$
$$|A_{2n}(U + Z) - u(x, \varepsilon)| \le M\varepsilon^{2n+1} \quad \text{in } \sigma_1.$$

Thus, the asymptotic expansion up to any power of ε which describes the solution $u(x, \varepsilon)$ uniformly in $\overline{\Omega}$ is constructed and justified. The Laplace operator in (2.1) can be replaced with any elliptic operator with smooth coefficients, which makes virtually no difference. The dimension of the domain Ω is also irrelevant in this example. If the dimension of the domain is $l > 2$, the only difference is that s becomes the coordinate of a smooth manifold of dimension $l - 1$.

The procedure becomes more complicated if the boundary is not smooth. Let us illustrate this type of a problem by an example.

EXAMPLE 4. Consider the problem (2.1), (2.2) in the square

$$\Omega = \{x_1, x_2 : 0 < x_1 < 1, 0 < x_2 < 1\}$$

where $q(x)$ and $f(x)$ satisfy the same conditions as in Example 3. For $\varepsilon > 0$ a solution of this problem exists, is continuous in $\overline{\Omega}$, satisfies the estimate (2.3) and is C^∞ everywhere in $\overline{\Omega}$ with the possible exception of the corners of the square. This information can be found, e.g., in the books [61], [82], but it is not particularly essential for understanding of what follows: our task is to find a uniform asymptotic expansion of the solution $u(x_1, x_2, \varepsilon)$ as $\varepsilon \to 0$.

It is natural to look for the asymptotics inside our domain in the same form as in the preceding example:

$$U(x, \varepsilon) = \sum_{k=0}^{\infty} \varepsilon^{2k} u_{2k}(x_1, x_2). \tag{2.10}$$

The functions $u_{2k}(x_1, x_2) \in C^{\infty}(\overline{\Omega})$ and are found in the same way as in Example 3. In the vicinity of the boundary the asymptotics is constructed as the sum of the series (2.10) and the boundary layer series. This procedure yields a separate series for each side of the square.

Consider, for example, the boundary layer series corresponding to the side $x_1 = 0$, $0 < x_2 < 1$ (see Figure 4, next page). It is, evidently, of the form

$$Z_1(y_1, x_2, \varepsilon) = \sum_{k=0}^{\infty} \varepsilon^k z_k^{(1)}(y_1, x_2), \qquad \varepsilon \to 0, \tag{2.11}$$

where $y_1 = \varepsilon^{-1} x_1$. Substituting this series into the homogeneous equation $\mathscr{L}_\varepsilon Z_1 = 0$ one obtains, as before, the recurrence system of equations for $z_k^{(1)}(y_1, x_2)$:

$$\frac{\partial^2 z_0^{(1)}}{\partial y_1^2} - q_0(x_2) z_0^{(1)} = 0,$$

$$\frac{\partial^2 z_1^{(1)}}{\partial y_1^2} - q_0(x_2) z_1^{(1)} = y_1 q_1(x_2) z_0^{(1)}, \tag{2.12}$$

$$\frac{\partial^2 z_k^{(1)}}{\partial y_1^2} - q_0(x_2) z_k^{(1)} = \sum_{j=1}^{k} y_1^j q_j(x_2) z_{k-j}^{(1)} - \frac{\partial^2 z_{k-2}^{(1)}}{\partial x^2}, \qquad k \geq 2.$$

Here $q_j(x_2)$ are the coefficients of the Taylor expansion of the function $q(x_1, x_2)$:

$$q(\varepsilon y_1, x_2) = \sum_{j=0}^{\infty} \varepsilon^j y_1^j q_j(x_2), \qquad \varepsilon \to 0.$$

The boundary condition (2.2) for the sum $U + Z_1$ on the side $x_1 = 0$ implies that

$$u_k(0, x_2) + z_k^{(1)}(0, x_2) = 0, \qquad k \geq 0. \tag{2.13}$$

By hypothesis, $q(x_1, x_2) > 0$ whence $q_0(x_2) > 0$ and, therefore, there are solutions of the equations (2.12) satisfying relations (2.13) and tending to zero as $y_1 \to \infty$:

$$|z_k^{(1)}(y_1, x_2)| \leq M_k e^{-\beta y_1}, \qquad \text{where } 0 < \beta < \min_{\overline{\Omega}}[q(x_1, x_2)]^{1/2}.$$

Evidently, in addition, $z_k^{(1)}(y_1, x_2)$ depends smoothly on x_2.

The boundary layer series arising along the remaining three sides of the square $\overline{\Omega}$ are similar to Z_1. By construction, the series U is an f.a.s. of the equation (2.1), and each of the boundary layer series of the form (2.4)

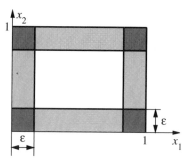

FIGURE 4

is an f.a.s. of the corresponding homogeneous equation. However, now the sum of U and the boundary layer series is no longer an f.a.s. of the whole problem (2.1.), (2.2) because it fails to satisfy condition (2.2). The source of the trouble lies in the vicinity of the corner points of the square. Indeed, the sum $U + Z_1$ vanishes for $x_1 = 0$. In the interior points of this section of the boundary the values of all the other boundary layer series decay exponentially, so that condition (2.2) is asymptotically satisfied. However, in the vicinity of the endpoints of this closed interval, i.e., near the points $(0, 0)$, $(0, 1)$, the values of the neighboring boundary layer functions become significant.

Consider, e.g., a neighborhood of the origin. The boundary layer series along the side $x_1 = 0$ is already constructed. The boundary layer series along the side $x_2 = 0$ is quite similar: $Z_2(x_1, y_2, \varepsilon) = \sum_{k=0}^{\infty} \varepsilon^k z_k^{(2)}(x_1, y_2)$, where $y_2 = \varepsilon^{-1} x_2$ and the properties of $z_k^{(2)}$ are the same as those of $z_k^{(1)}$ with x_2 replaced by x_1 and y_1 by y_2. Let us analyze the form of the residual arising in the vicinity of the origin. The values of the two remaining boundary layer series Z_3 and Z_4 in the vicinity of the origin decay exponentially, while the series $U + Z_1 + Z_2$ equals $\sum_{k=0}^{\infty} \varepsilon^k z_k^{(2)}(0, y_2)$ for $x_1 = 0$ and $\sum_{k=0}^{\infty} \varepsilon^k z_k^{(1)}(y_1, 0)$ for $x_2 = 0$. It is this residual that has newly arisen in the boundary condition. Note that it is continuous at the point $(0, 0)$ because $z_k^{(1)}(0, 0) = z_k^{(2)}(0, 0) = -u_k(0, 0)$. It is natural to expect that the residual affects only an ε-neighborhood of the origin (the doubly shaded domain in Figure 4). One should, therefore, introduce another , so-called *corner* boundary layer series. It is expressed in the variables $y_1 = \varepsilon^{-1} x_1$ and $y_2 = \varepsilon^{-1} x_2$. Denote this series by $W = \sum_{k=0}^{\infty} \varepsilon^k w_k(y_1, y_2)$. The recurrence system of equations for w_k is obtained in the standard way described above, and is of the form

$$\frac{\partial^2 w_0}{\partial y_1^2} + \frac{\partial^2 w_0}{\partial y_2^2} - \gamma^2 w_0 = 0,$$

$$\frac{\partial^2 w_k}{\partial y_1^2} + \frac{\partial^2 w_k}{\partial y_2^2} - \gamma^2 w_k = \sum_{j=1}^{k} P_j(y_1, y_2) w_{k-j}, \quad (2.14)$$

where

$$P_j(y_1, y_2) = \sum_{i=0}^{j} \frac{1}{i!(j-i)!} \frac{\partial^j q}{\partial x_1^i \partial x_2^{j-i}} y_1^i y_2^{j-i}$$

and $\gamma^2 = q(0, 0) > 0$. The boundary conditions for w_k must eliminate the earlier residuals:

$$\begin{aligned} w_k(y_1, 0) &= -z_k^{(1)}(y_1, 0), & 0 < y_1 < \infty, \\ w_k(0, y_2) &= -z_k^{(2)}(0, y_2), & 0 < y_2 < \infty. \end{aligned} \tag{2.15}$$

Thus, the problems (2.14), (2.15) are posed in the quadrant $E = \{y_1, y_2 : 0 < y_1 < \infty, 0 < y_2 < \infty\}$. These problems are known to have solutions decaying exponentially as $y_1 + y_2 \to \infty$.

This can be verified in various ways, but for this example we will prove this statement using the explicit form of the Green function for the equation (2.14) in the quadrant E. One can easily check that the function $\frac{1}{2\pi} K_0\left(\gamma\sqrt{y_1^2 + y_2^2}\right)$, where K_0 is Macdonald's function ([64, §5.3], [118, Chapter 7]) is the fundamental solution of the operator $\Delta - \gamma^2$ whence

$$\begin{aligned} G(y_1, y_2, \eta_1, \eta_2) = \frac{1}{2\pi} \Big\{ &K_0\left(\gamma\sqrt{(y_1 - \eta_1)^2 + (y_2 - \eta_2)^2}\right) \\ &- K_0\left(\gamma\sqrt{(y_1 - \eta_1)^2 + (y_2 + \eta_2)^2}\right) - K_0\left(\gamma\sqrt{(y_1 + \eta_1)^2 + (y_2 - \eta_2)^2}\right) \\ &+ K_0\left(\gamma\sqrt{(y_1 + \eta_1)^2 + (y_2 + \eta_2)^2}\right) \Big\} \end{aligned}$$

is the Green function of the first boundary value problem for the equation

$$\Delta w - \gamma^2 w = \varphi \quad \text{in } E. \tag{2.16}$$

Let $\varphi(y_1, y_2) \in C^\infty(\overline{E} \setminus (0, 0)) \cap C(\overline{E})$, $\psi_1(y_1)$, $\psi_2(y_2) \in C^\infty(0, \infty) \cap C[0, \infty)$ and suppose that the bounds

$$|\varphi(y_1, y_2)| < M \exp\{-\beta(y_1 + y_2)\},$$
$$|\psi_1(y_1)| < M \exp(-\beta y_1), \qquad |\psi_2(y_2)| < M \exp(-\beta y_2)$$

hold. Let also $\psi_1(0) = \psi_2(0)$. Then the function

$$\begin{aligned} w(y_1, y_2) = &\int_0^\infty \int_0^\infty G(y_1, y_2, \eta_1, \eta_2) \varphi(\eta_1, \eta_2) \, d\eta_1 \, d\eta_2 \\ &- \int_0^\infty \frac{\partial G}{\partial y_2}(y_1, y_2, \eta_1, 0) \psi_1(\eta_1) \, d\eta_1 - \int_0^\infty \frac{\partial G}{\partial y_1}(y_1, y_2, 0, \eta_2) \psi_2(\eta_2) \, d\eta_2 \end{aligned}$$

is continuous in $\overline{E}$, satisfies equation (2.16) in $\overline{E} \setminus (0, 0)$ and the boundary conditions $w(y_1, 0) = \psi_1(y_1)$, $w(0, y_2) = \psi_1(y_2)$. The explicit form of $w(y_1, y_2)$ and the properties of the function K_0 imply that $w(y_1, y_2) \in C^\infty(\overline{E} \setminus (0, 0))$ and

$$|w(y_1, y_2)| < M_1 \exp\{-\beta(y_1 + y_2)\}. \tag{2.17}$$

An iterated application of this statement to equations (2.14) and the boundary conditions (2.15) yields, by induction, that the solutions of these problems are continuous in

E and satisfy inequalities (2.17). The existence of the desired solutions $w_k(y_1, y_2)$ is proved.

Denote the resulting series by W_1, and construct the similar series W_2, W_3, W_4 in the vicinity of the remaining vertices of the square $\overline{\Omega}$. By virtue of inequalities (2.17) the coefficients of these series decay exponentially at the distance ε^α ($0 < \alpha < 1$) from the corresponding vertex. By construction, each of the series W_i is an f.a.s. of the equation $\mathscr{L}_\varepsilon W = 0$.

Thus, the series $U + \sum_{i=1}^{4} Z_i + \sum_{i=1}^{4} W_i$ is an f.a.s. not only of equation (2.1), but of the entire boundary value problem (2.1), (2.2). The correctness of this asymptotics is justified as in Example 3, and is in fact even simpler because the shape of the domain Ω makes the multiplication of the boundary layer function by a truncating function $\chi(x)$ unnecessary. The partial sum $Y_{2n}(x, \varepsilon) = A_{2n}\left(U + \sum_{i=1}^{4} Z_i + \sum_{i=1}^{4} W_i\right)$ where A_{2n} denotes the same operator as in the previous examples, satisfies the inequalities

$$|\mathscr{L}_\varepsilon(Y_{2n}(x, \varepsilon) - u(x, \varepsilon))| < M\varepsilon^{2n+1},$$

$$|Y_{2n}(x, \varepsilon)|_{\partial\Omega} < M\varepsilon^{2n+1}.$$

The maximum principle, together with these inequalities, implies that the bound $|Y_{2n}(x, \varepsilon) - u(x, \varepsilon)| < M_1\varepsilon^{2n+1}$ holds everywhere in $\overline{\Omega}$.

In Figure 4, the boundary layers along the sides of the square are shaded. The corner boundary layers are doubly shaded. One should realize that actually there is no specific curve serving as a border to a boundary layer. In this and the preceding examples we have conditionally depicted it as lying at the distance ε from the boundary of the square. The precise meaning is that at the distance $M\varepsilon$ from the boundary the values of the boundary layer functions are not small, while at the distance ε^α, where α is any number such that $0 < \alpha < 1$, they decay exponentially.

The constructions and proofs of Example 4 are transferred without serious modifications to an arbitrary plane domain with a piecewise smooth boundary. The Laplace operator can be replaced with any elliptic operator of the second order. These results can also be generalized without great difficulty to many domains of greater dimension (e.g., to a parallelepiped). Some modifications will be required for conic points of the boundary. However, the main condition $q(x) > \text{const} > 0$ ensures the exponential decay both of the boundary layer functions along smooth sections of the boundary, and those at the corners. The situation changes drastically if the coefficient $q(x)$ vanishes either at a point on the boundary or at an interior point of the domain. A short remark in that respect will be made in the next chapter where a similar situation for an ordinary differential equation is considered in detail.

CHAPTER II

Ordinary Differential Equations

In some sense this chapter is also of an introductory nature, but now with regard to the method of matched asymptotic expansions. The method will be demonstrated in sufficient detail using simple examples of ordinary differential equations. Only in passing shall we mention one boundary value problem for a partial differential equation. Other, more complicated problems for partial differential equations will be described in the following chapters.

We begin with an example, which on the surface looks very much like Example 1.

§1. A simple bisingular problem

EXAMPLE 5. Consider the boundary value problem

$$l_\varepsilon u \equiv \varepsilon^3 u'' - q(x)u = f(x), \qquad \varepsilon > 0, \, 0 \le x \le 1, \tag{1.1}$$

$$u(0, \varepsilon) = 0, \tag{1.2}$$

$$u(1, \varepsilon) = 0, \tag{1.3}$$

where $q, f \in C^\infty[0, 1]$ and $q(x) > 0$ for $x > 0$. Its difference from the problem of Example 1 is very small: ε^3 is written in place of ε^2 merely for convenience of notation later. The fact that nonhomogeneous boundary conditions are replaced with the homogeneous ones is also inessential. The main difference is that $q(x)$ is no longer positive on the whole closed interval $[0, 1]$, but only for $x > 0$. Suppose that

$$q(0) = 0, \qquad q'(0) = 1. \tag{1.4}$$

The requirement that $q'(0)$ be positive profoundly influences the structure of the asymptotics of the solution. The precise value of $q'(0)$ is unimportant: the unity is chosen just to simplify the notation.

As in Example 1, we are seeking the outer expansion in the form

$$U(x, \varepsilon) = \sum_{k=0}^{\infty} \varepsilon^{3k} u_{3k}(x), \qquad \varepsilon \to 0. \tag{1.5}$$

One obtains, as before, the recurrence system of equations

$$-q(x)u_0(x) = f(x), \qquad q(x)u_{3k}(x) = u''_{3(k-1)}(x), \qquad k \ge 1. \tag{1.6}$$

which defines all $u_{3k}(x) \in C^{\infty}(0, 1]$ uniquely. However, in general, all these functions have singularities at the origin. Denote by q_k and f_k the coefficients of the Taylor series for the functions $q(x)$ and $f(x)$, respectively. By (1.4),

$$q(x) = x + \sum_{k=2}^{\infty} q_k x^k, \quad f(x) = \sum_{k=0}^{\infty} f_k x^k, \quad x \to 0, \quad (1.7)$$

and both series can be infinitely differentiated term-by-term. Equalities (1.6), (1.7) now imply that

$$u_0(x) = -x^{-1} \left(1 + \sum_{k=2}^{\infty} q_k x^{k-1} \right)^{-1} \sum_{k=0}^{\infty} f_k x^k = x^{-1} \sum_{j=0}^{\infty} c_{0,j} x^j, \quad x \to 0.$$

This series can also be infinitely differentiated term-by-term. Hence, for $k = 1$, we have $u_3(x) = x^{-4} \sum_{j=0}^{\infty} c_{1,j} x^j$, $x \to 0$. By induction, one obtains

$$u_{3k}(x) = x^{-3k-1} \sum_{j=0}^{\infty} c_{k,j} x^j, \quad x \to 0. \quad (1.8)$$

Thus, *the problem* (1.1)–(1.3) *is a bisingular one; the coefficients of its outer expansion have increasing singularities at the origin.* In this problem, the set Γ consists of the points $x = 0$ and $x = 1$, but the behavior of the solution in a neighborhood of the point $x = 1$ is the same as in Example 1. The exponential boundary layer functions in a neighborhood of this point will be taken care of later, and now we turn to the point $x = 0$. In a neighborhood of this point the series (1.5) not only fails to approximate the solution $u(x, \varepsilon)$ but even loses its asymptotic properties. Indeed, for example, for $x \sim \varepsilon^2$ the functions $u_{3k}(x)$ are, according to (1.8), of the same order of magnitude as ε^{-6k-2}, whereby the ratio of each subsequent term of the series (1.5) to the preceding one is of the same order of magnitude as ε^{-3}.

For a correct description of the asymptotics of the solution $u(x, \varepsilon)$ in a neighborhood of zero, we introduce the stretching transformation $x = \varepsilon^2 \xi$ and denote $u(\varepsilon^{\alpha} \xi, \varepsilon)$ by $v(\xi, \varepsilon)$. After the change of variable equation (1.1) takes the form

$$\varepsilon^{3-2\alpha} \frac{d^2 v}{d\xi^2} - q(\varepsilon^{\alpha} \xi)v = f(\varepsilon^{\alpha} \xi).$$

In accordance with the argument in Example 1, our goal is to achieve that both terms on the left-hand side of the equation are of the same order of magnitude. Taking into account that $q(\varepsilon^{\alpha} \xi) \sim \varepsilon^{\alpha} \xi$, we have $3 - 2\alpha = \alpha \Rightarrow \alpha = 1$.

Thus, in the vicinity of the origin, the equation is of the form

$$\varepsilon \frac{d^2 v}{d\xi^2} - q(\varepsilon\xi)v = f(\varepsilon\xi), \quad (1.9)$$

and we look for the asymptotics of the solution $v(\xi, \varepsilon) \equiv u(\varepsilon\xi, \varepsilon)$ in the form of the inner expansion

$$V(\xi, \varepsilon) = \sum_{k=-1}^{\infty} \varepsilon^k v_k(\xi), \qquad (1.10)$$

where $\xi = \varepsilon^{-1}x$. This series starts with the term $\varepsilon^{-1}v_{-1}(\xi)$ because the principal term of the outer expansion is $u_0(x) \sim x^{-1} = \varepsilon^{-1}\xi^{-1}$, but, of course, this is only a heuristic argument. The formula will get its full justification only when the asymptotics will be rigorously validated. Then one has to insert the series (1.10) in the equation (1.9), expand the functions $q(\varepsilon\xi)$ and $f(\varepsilon\xi)$ in Taylor series, and equate the coefficients of the same powers of ε. The resulting system of equations is

$$v_{-1}'' - \xi v_{-1} = f_0,$$

$$v_k'' - \xi v_k = f_{k+1}\xi^{k+1} + \sum_{j=2}^{k+2} q_j\xi^j v_{k-j+1}, \qquad k \geq 0. \qquad (1.11)$$

Equality (1.2) yields the boundary condition

$$v_k(0) = 0, \qquad k \geq -1. \qquad (1.12)$$

THEOREM 1.1. *In the class of functions growing not faster than some power of ξ, the system (1.11) has a unique solution satisfying conditions (1.12). Each of the functions $v_k(\xi)$ has the following asymptotics at infinity:*

$$v_k(\xi) = \xi^k \sum_{j=0}^{\infty} h_{k,j}\xi^{-3j}, \qquad \xi \to \infty, k \geq -1. \qquad (1.13)$$

These statements concerning $v_k(\xi)$ are quite natural, because the coefficients of the series (1.13) are easily computed after the series are formally substituted into equation (1.11). Thus, for example, $h_{-1,0} = -f_0$, $h_{-1,1} = 2h_{-1,0}$, $h_{-1,2} = 20h_{-1,1}$, $\ldots$, $h_{0,0} = -f_1 + q_2f_0$, $h_{0,1} = -2q_2h_{-1,0}$, etc. As for the solutions of the homogeneous equation, one of them grows exponentially, and cannot, therefore, be used to solve the problems (1.11), (1.12), while the second decays exponentially and can help to satisfy the boundary conditions (1.12). We now present the proof of Theorem 1.1. The reader can, however, skip it if his principal interest is in the main, constructive side of the matter and he is prepared to take the statement for granted.

The Airy equation $v'' - \xi v = 0$ has, as is well known (see, e.g. [64, §5.17]), two linearly independent solutions:

$$Y_1 = \xi^{-1/4} \exp\left(\frac{2}{3}\xi^{3/2}\right)\left\{1 + \sum_{k=1}^{\infty} a_k\xi^{-3k/2}\right\}, \qquad \xi \to \infty.$$

$$Y_2 = \xi^{-1/4} \exp\left(-\frac{2}{3}\xi^{3/2}\right)\left\{1 + \sum_{k=1}^{\infty} b_k\xi^{-3k/2}\right\}, \qquad \xi \to \infty,$$

$$Y_2(\xi) > 0 \quad \text{for } \xi \geq 0.$$

If $\varphi(\xi) \in C[0, \infty)$ and $|\varphi(\xi)| \leq M|\xi|^{-N}$, then the explicit formula

$$\overline{v}(\xi) = Y_2(\xi) \int_0^{\xi} Y_1(s)\varphi(s)\, ds + Y_1(\xi) \int_{\xi}^{\infty} Y_2(s)\varphi(s)\, ds \qquad (1.14)$$

implies that $\overline{v}(\xi)$ is a solution of the equation $\overline{v}'' - \xi\overline{v} = -2\varphi(\xi)$, $\overline{v}(\xi) \in C^2[0, \infty)$ and that

$$|\overline{v}(\xi)| \leq M_1|\xi|^{-N}. \qquad (1.15)$$

LEMMA 1.1. *Let* $g(\xi) \in C^{\infty}[0, \infty)$. *Suppose that* $g(\xi)$ *can be expanded in an asymptotic series* $g(\xi) = \xi^p \sum_{j=0}^{\infty} q_j \xi^{-3j}$, $\xi \to \infty$, *and that this series can be infinitely differentiated term-by-term. Then, for* $0 \leq \xi < \infty$, *the equation*

$$lv \equiv v'' - \xi v = g(\xi), \qquad (1.16)$$

has a solution which can be expanded in an asymptotic series

$$V(\xi) = \xi^{p-1} \sum_{j=0}^{\infty} v_j \xi^{-3j}, \qquad \xi \to \infty. \qquad (1.17)$$

The series (1.17) *can be repeatedly differentiated term-by-term, and it is an f.a.s. of the equation* (1.16).

(An f.a.s. is defined as above but with respect to the variable ξ instead of the parameter ε. This means that the left-hand side of equation (1.16), after the insertion of the partial sum of the series (1.17) with a sufficiently large index N, differs from the right-hand side on $O\left(\xi^{-N_1}\right)$, where $N_1 \to \infty$ as $N \to \infty$.)

PROOF. We begin by formally constructing the series (1.17). Inserting it in the equation (1.16) and replacing the function $g(\xi)$ with its asymptotic series, one obtains a simple algebraic recurrence system of equations for v_j which uniquely determine $v_0 = -g_0$, $v_1 = -g_1 + (p-1)(p-2)v_0$, etc. Now take a partial sum $B_N V$ for N sufficiently large and multiply it by a function $\chi(\xi) \in C^{\infty}[0, \infty)$ vanishing for $\xi \leq 1$ and equal to 1 for $\xi \geq 2$. Denote $y_N(\xi) = \chi(\xi)B_N V$. By construction, $F_N(\xi) = g - ly_N = O\left(\xi^{-N}\right)$ for $\xi \to \infty$. Using formula (1.14), one can now construct the function $z_N(\xi)$ solving the equation $lz_N = F_N(\xi)$. Inequality (1.15) now implies that $z_N(\xi)$ decays rapidly as $\xi \to \infty$.

Set $v_N(\xi) = y_N(\xi) + z_N(\xi) + c_N Y_2(\xi)$, where the constant c_N is chosen in such a way that $v_N(0) = 0$. By construction, $lv_N = g(\xi)$, $v_N - B_N V = O\left(\xi^{-N}\right)$, and it only remains to show that the function v_N does not depend on N. Indeed, the difference $w_N(\xi) = v_N(\xi) - v_{N+1}(\xi)$ is a solution of the equation $lw_N = 0$, $w_N(0) = 0$, and the function $w_N(\xi)$ grows not faster than some power of ξ as $\xi \to \infty$. The explicit form of the solution of the homogeneous equation implies that $w_N \equiv 0$. The differentiability of the series (1.17) is a direct consequence of equation (1.16). ∎

A consecutive application of Lemma 1.1 to equations (1.11) yields the statements of Theorem 1.1 about the existence of solutions $v_k(\xi)$ and their asymptotics as $\xi \to \infty$. ∎

Now that both the outer (1.5) and inner (1.10) expansions are constructed, it only remains to find out whether they provide correct asymptotics for the solution $u(x, \varepsilon)$ of the problem (1.1)–(1.3), and if yes, in what regions. Since for $x \to 0$ one has $u_{3k}(x) \sim x^{-3k-1}$, the series (1.5) is of an asymptotic character (i.e., each subsequent term is less in the order of magnitude than the preceding one) for $x > \varepsilon^{\alpha}$ for any α such that $0 < \alpha < 1$. Similarly, since $v_k(\xi) \sim \xi^k$ for $\xi \to \infty$, the series (1.10) has an asymptotic character for $\varepsilon\xi < \varepsilon^{\beta}$, i.e., for $x < \varepsilon^{\beta}$ for any positive β. Note that the domains $x > \varepsilon^{\alpha}$ and $x < \varepsilon^{\beta}$ overlap for $\alpha > \beta$, and that these series are f.a.s. of equation (1.1). Indeed,

$$
\begin{aligned}
l_\varepsilon(A_{3n,x}U) &= f(x) + \varepsilon^{3n+3} u''_{3n}(x) = f(x) + O\left(\varepsilon^{3n+3} x^{-3n-3}\right) \\
&= f(x) + O\left(\varepsilon^{(3n+3)(1-\alpha)}\right) \quad \text{for } x > \varepsilon^{\alpha},
\end{aligned}
\tag{1.18}
$$

$$
\begin{aligned}
l_\varepsilon(A_{n,\xi}V) &= \varepsilon \frac{d^2 A_{n,\xi}V}{d\xi^2} - q(\varepsilon\xi)A_{n,\xi}V \\
&= \varepsilon \frac{d^2 A_{n,\xi}V}{d\xi^2} - \left\{A_{n+2,\xi}q(\varepsilon\xi)\right\}[A_{n,\xi}V] + O\left(\varepsilon^{n+3}\xi^{n+3}\left(\frac{1}{\varepsilon} + \varepsilon^n \xi^n\right)\right) \\
&= A_{n,\xi}f(\varepsilon\xi) - \sum_{\substack{i \leq n, j \leq n+2 \\ i+j \geq n+1}} q_j \xi^j \varepsilon^j v_i(\xi) \varepsilon^i + O\left(\varepsilon^{n+2}\xi^{n+3} + \varepsilon^{2n+3}\xi^{2n+3}\right) \\
&= f(x) + O\left(\varepsilon^{(n+1)\beta-1}\right), \quad \text{if } x < \varepsilon^{\beta}.
\end{aligned}
\tag{1.19}
$$

In deducing the last two equalities, equations (1.11) have been used, as well as the estimates of the functions $v_i(\xi)$ following from (1.13).

The series $V(\xi, \varepsilon)$ satisfies condition (1.2), but the series $U(x, \varepsilon)$ does not, in general, satisfy condition (1.3). To eliminate the residual in the boundary condition (1.3) one has to add to the series $U(x, \varepsilon)$ the series (1.12) of Chapter I, in exactly the same way as in Example 1 (with ε replaced by $\varepsilon^{3/2}$). Thus, the f.a.s. of the problem (1.1)–(1.3) is constructed on the whole interval $[0, 1]$. It is, therefore, quite natural to expect that the following theorem holds.

THEOREM 1.2. *Let α and β be numbers such that $0 < \beta < \alpha < 1$. Then the asymptotic series (1.10) constructed above is a uniform asymptotic expansion of the function $u(x, \varepsilon)$ solving the problem (1.1)–(1.3) on the closed interval $[0, \varepsilon^{\beta}]$, and the sum of the series (1.5) and (1.12) of Chapter I (with ε replaced by $\varepsilon^{3/2}$) is a uniform asymptotic expansion of the same solution on the closed interval $[\varepsilon^{\alpha}, 1]$.*

To prove this theorem, a uniform estimate of the operator inverse to the differential operator (1.1) subject to conditions (1.2), (1.3) is necessary.

LEMMA 1.2. *Let $w(x, \varepsilon) \in C^2[0, 1]$, and suppose that $w(x, \varepsilon)$ satisfies the equation $l_\varepsilon \equiv \varepsilon^3 w'' - q(x)w = \varphi(x)$, where $q(x) \in C[0, 1]$, $q(x) > 0$ for $x > 0$. Then* *the estimate*

$$|w(x, \varepsilon)| \le M \left(|w(0, \varepsilon)| + |w(1, \varepsilon)| + \varepsilon^{-3} \max_{x \in [0, 1]} |\varphi(x)| \right) \qquad (1.20)$$

holds.

PROOF. Inserting the expression $w = (2 - x^2)z$ one obtains the following equation for the function $z(x, \varepsilon)$:

$$\varepsilon^3 z'' - 2\varepsilon^3 x(2 - x^2)^{-1} z' - [q(x) + 2\varepsilon^3(2 - x^2)^{-1}]z = (2 - x^2)^{-1}\varphi(x).$$

It follows from the maximum principle that

$$|z(x, \varepsilon)| \le |z(0, \varepsilon)| + |z(1, \varepsilon)| + \varepsilon^{-3} \max_{x \in [0, 1]} |\varphi(x)|,$$

which yields the estimate (1.20). ∎

Lemma 1.2 implies, in particular, that for the solution of the problem (1.1)–(1.3) the estimate $|u(x, \varepsilon)| < M\varepsilon^{-3}$ holds. It is, however, a rough estimate. Although the functions $u(x, \varepsilon)$ are not uniformly bounded for $\varepsilon \to 0$, the asymptotic expansion (1.10) implies that $|u(x, \varepsilon)| \le \varepsilon^{-1} \max_\xi |v_{-1}(\xi)| + M$. After Theorem 1.2 is proved, it will be clear that this estimate does indeed hold and that it is sharp.

Now one has to verify that the outer expansion (1.5) and the inner expansion (1.10) are matched to each other as shown in the Introduction for the example of an elementary function. We have to prove that the equality

$$A_{m,\xi} A_{n,x} U = A_{n,x} A_{m,\xi} V \qquad (1.21)$$

holds $\forall m > 0$ and $\forall n > 0$. It is *the focal point of the matching procedure applied to any bisingular problem.* Here, as in all the problems considered below, the left- and right-hand sides of equalities (1.21) are linear combinations of elementary functions in x or ξ with coefficients depending on ε. In the example mentioned, these were powers of x or ξ, and the coefficients were powers of ε. When both left- and right-hand sides are expressed in the same variable (either x or ξ), the verification of (1.21) is reduced to the comparison of the corresponding coefficients. E.g., in this example, the left-hand side of relation (1.21) equals, by (1.5), (1.8), $\varepsilon^{-1} \sum_{3k \le n} \xi^{-3k-1} \sum_{j \le m+1} c_{k,j} \varepsilon^j \xi^j$, while, by virtue of (1.10), (1.13) the right-hand side equals $\sum_{k=-1}^{m} x^k \sum_{3j \le n} h_{k,j} x^{-3j} \varepsilon^{3j}$. It is now clear that equality (1.21) is equivalent to the sequence of equalities $h_{k,j} = c_{j,k+1}$ for $k \ge -1$, $j \ge 0$. They can be verified without difficulty with the use of the recurrence systems of equations (1.6) and (1.11), but we will rather present another proof of relation (1.21) which is also applicable to more complicated situations.

The proof is based on the fact that the series U and V are the f.a.s. of the same equation but in different, although overlapping domains. Let, for simplicity, $n = 3N$. The asymptotics (1.8) implies that

$$A_{m,\xi} A_{3N,x} U = \sum_{l=-1}^{m} \varepsilon^l \xi^l B_{3N} V_l, \qquad (1.22)$$

where the V_l are series of the form

$$\sum_{j=0}^{\infty} c_{j,l+1}\xi^{-3j}, \tag{1.23}$$

and B_p, as usual, denotes the partial sum of the series containing the terms ξ^{-s} for $s \leq p$. Inserting $A_{3N,x}U$ into the left-hand side of equation (1.1), one has

$$l_\varepsilon(A_{3N,x}U) = f(x) + \varepsilon^{3N+3}u_{3N}''(x). \tag{1.24}$$

Applying the operator $A_{m+1,\xi}$ to both sides of this equality, we obtain:

$$A_{m+1,\xi}\left(\varepsilon\frac{d^2}{d\xi^2}A_{3N,x}U - q(\varepsilon\xi)A_{3N,x}U\right)$$

$$= A_{m+1,\xi}f(x) + A_{m+1,\xi}\left(\varepsilon^{3N+1}\frac{d^2u}{d\xi^2}(\varepsilon\xi,\varepsilon)\right).$$

Taking into account the asymptotic expansion (1.8) and relation (1.22), one can rewrite this equality in the following form:

$$\varepsilon\sum_{l=-1}^{m}\varepsilon^l\frac{d^2}{d\xi^2}(\xi^l B_{3N}V_l) - A_{m+1,\xi}\left\{\left(\sum_{l=-1}^{\infty}\varepsilon^l\xi^l B_{3N}V_l\right)\left(\sum_{j=1}^{\infty}\varepsilon^j\xi^j q_j\right)\right\}$$

$$= A_{m+1,\xi}f(\varepsilon\xi) + A_{m+1,\xi}\frac{d^2}{d\xi^2}\left(\xi^{-3N-1}\sum_{j=0}^{\infty}c_{3N,j}\varepsilon^j\xi^j\right).$$

No infinite series appears in this relation. Both the left- and right-hand parts of it are finite sums of terms each of which is a product of a power of ξ by a power of ε. This provides a lawful ground for equating (not just formally, as before) the terms on both sides containing the same powers of ε. One can easily note that this procedure is, in fact, hardly different from the one used to write out equation (1.11). Indeed, equation (1.24) differs from (1.9) only in the term $\varepsilon^{3N+3}u_{3N}''(x)$ so that the resulting equations for $\xi^l B_{3N}V_l$ differ only in the terms corresponding to it:

$$\frac{d^2}{d\xi^2}(\xi^{-1}B_{3N}V_{-1}) - \xi(\xi^{-1}B_{3N}V_{-1}) = f_0 + O(\xi^{-3N-3}),$$

$$\frac{d^2}{d\xi^2}(B_{3N}V_0) - \xi(B_{3N}V_0) = f_1\xi + q_2\xi^2(\xi^{-1}B_{3N}V_{-1}) + O(\xi^{-3N-2}),$$

etc. This implies that the functions $\xi^k V_k$, where V_k are the series (1.23), are the f.a.s. of equations (1.11) but now as $\xi \to \infty$. In each of equations (1.11) the coefficients of the asymptotics as $\xi \to \infty$ are found uniquely whereby $v_k(\xi) = \xi^k V_k$, $\xi \to \infty$, and relations (1.21) are satisfied.

TABLE 1

U \ V	$\varepsilon^{-1}v_1(\xi)$	$v_0(\xi)$	$\varepsilon v_1(\xi)$	$\varepsilon^2 v_2(\xi)$	$\varepsilon^3 v_3(\xi)$	$\cdots$
$u_0(x)$	$\varepsilon^{-1}h_{-1,0}\xi^{-1}$ $c_{0,0}x^{-1}$	$h_{0,0}$ $c_{0,1}$	$\varepsilon h_{1,0}\xi$ $c_{0,2}x$	$\varepsilon^2 h_{2,0}\xi^2$ $c_{0,3}x^2$	$\varepsilon^3 h_{3,0}\xi^3$ $c_{0,4}x^3$	$\cdots$
$\varepsilon^3 u_3(x)$	$\varepsilon^{-1}h_{-1,1}\xi^{-4}$ $\varepsilon^3 c_{1,0}x^{-4}$	$h_{0,1}\xi^{-3}$ $\varepsilon^3 c_{1,1}x^{-3}$	$\varepsilon h_{1,1}\xi^{-2}$ $\varepsilon^3 c_{1,2}x^{-2}$	$\varepsilon^2 h_{2,1}\xi^{-1}$ $\varepsilon^3 c_{1,3}x^{-1}$	$\varepsilon^3 h_{3,1}$ $\varepsilon^3 c_{1,4}$	$\cdots$
$\varepsilon^6 u_6(x)$	$\varepsilon^{-1}h_{1,2}\xi^{-7}$ $\varepsilon^6 c_{2,0}x^{-7}$	$h_{0,2}\xi^{-6}$ $\varepsilon^6 c_{2,1}x^{-6}$	$\varepsilon h_{1,2}\xi^{-5}$ $\varepsilon^6 c_{2,2}x^{-5}$	$\varepsilon^2 h_{2,2}\xi^{-4}$ $\varepsilon^6 c_{2,3}x^{-4}$	$\varepsilon^3 h_{3,2}\xi^{-3}$ $\varepsilon^6 c_{2,4}x^{-3}$	$\cdots$
$\varepsilon^9 u_9(x)$	$\varepsilon^{-1}h_{-1,3}\xi^{-10}$ $\varepsilon^9 c_{3,0}x^{-10}$	$h_{0,3}\xi^{-9}$ $\varepsilon^9 c_{3,1}x^{-9}$	$\varepsilon h_{1,3}\xi^{-8}$ $\varepsilon^9 c_{3,2}x^{-8}$	$\varepsilon^2 h_{2,3}\xi^{-7}$ $\varepsilon^9 c_{3,3}x^{-7}$	$\varepsilon^3 h_{3,3}\xi^{-6}$ $\varepsilon^9 c_{3,4}x^{-6}$	$\cdots$
$\cdots$	$\cdots$	$\cdots$	$\cdots$	$\cdots$	$\cdots$	$\cdots$

A clear and practical way of presenting the basic equality (1.21) will be repeatedly used below. Let us write the terms constituting the left- and right-hand sides of (1.21) in the form of Table 1.

Each row of the table presents the asymptotic expansion as $x \to 0$ of the term $\varepsilon^{3k}u_{3k}(x)$ of the outer expansion (1.5), the bottom half of each square of this row includes, in accordance with (1.8), the expression $\varepsilon^{3k}c_{k,j}x^{j+3k-1}$. Each column of the table presents the asymptotic expansion as $\xi \to \infty$ of the term $\varepsilon^i v_i(\xi)$ of the inner expansion (1.10), the top half of each square of this column includes, in accordance with (1.13), the expression $\varepsilon^i h_{i,l}\xi^{i-3l}$. The matching procedure for the asymptotic expansions signifies that one and the same function appears in both the top and bottom halves of each square! This implies equality (1.21), and, in our particular case, the converse statement holds as well. In more complicated cases (e.g., those considered in Chapter III, §3, and Chapter IV, §2) the functions in the different halves of a same square do not coincide, the change of variables moves some of the terms to neighboring squares. This phenomenon is mostly related to the presence of logarithmic terms in the asymptotics. Nevertheless, in the expanding system of rectangular portions of the table the sum of the terms appearing in the top halves of the square is equal to the sum of the terms in the bottom halves, and this is what is meant by equality (1.21).

PROOF OF THEOREM 1.2. It follows from the construction of $A_{m,\xi}A_{3N,x}U$ and the asymptotic expansions (1.8), (1.13) that the estimate

$$|A_{m,\xi}A_{3N,x}U - A_{3N,x}U| \le M\sum_{k=0}^{N}\varepsilon^{3k}x^{m+1-3k-i} \le Mx^m(1 + \varepsilon^{3N}x^{-3N})$$

holds, as well as a similar estimate for the difference between the second derivatives of these functions. Similarly,

$$|A_{3N,x}A_{m,\xi}V - A_{m,\xi}V| \le M\sum_{i=-1}^{m}\varepsilon^i\xi^{-3N+i} \le M\xi^{-3N}(\varepsilon^{-1}\xi^{-1} + (\varepsilon\xi)^m),$$

$$\left|\frac{d^2}{d\xi^2}(A_{3N,x}A_{m,\xi}V - A_{m,\xi}V)\right| \le M\xi^{-3N-2}(\varepsilon^{-1}\xi^{-1} + \varepsilon^m\xi^m).$$

We now construct a composite asymptotic expansion which is almost the same as the function (0.9) in the Introduction:

$$Y_N(x,\varepsilon) = A_{3N,x}U + A_{3N,\xi}V - A_{3N,\xi}A_{3N,x}U + A_{3N}W,$$

where W is the boundary layer series (1.12) of Chapter I in the vicinity of the right-hand endpoint of the closed interval. It follows from the inequalities we have just obtained, estimates (1.18), (1.19), and the basic equality (1.21) that

$$|l_\varepsilon Y_N(x,\varepsilon) - f| < M\varepsilon^{(3N+3)(1-\gamma)-1} \quad \text{for } x \ge \varepsilon^\gamma,$$

$$|l_\varepsilon Y_N(x,\varepsilon) - f| < M\varepsilon^{(3N+1)\gamma-1} \quad \text{for } x \le \varepsilon^\gamma.$$

Taking into account that $Y_N(1,\varepsilon) = O\left(\varepsilon^{3N+1}\right)$, $Y_N(0,\varepsilon) = O(e^{-\mu/\varepsilon})$, $\mu > 0$, and comparing $Y_N(x,\varepsilon)$ with $u(x,\varepsilon)$, one concludes, using Lemma 1.2, that

$$|Y_N(x,\varepsilon) - u(x,\varepsilon)| < M\varepsilon^{(3N+1)\gamma_1-1}, \qquad \gamma_1 > 0. \quad \blacksquare$$

Figure 5 illustrates the problem (1.1)–(1.3) for $q(x) = \sin x$, $f(x) = -\cos x$, $\varepsilon = 1/20$ depicting the graphs of the following functions:

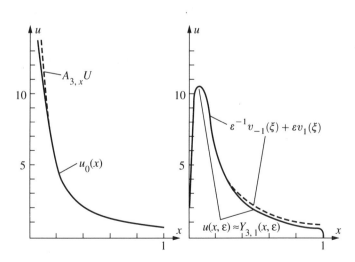

FIGURE 5

$u_0(x) = \cot x$, $A_{3,x}U = \cot x + 2\varepsilon^3 \cot x(\sin x)^{-3}$, $A_{1,\xi}V = \varepsilon^{-1}v_{-1}(\xi) + \varepsilon v_1(\xi)$, $Y_{3,1}(x,\varepsilon) = A_{3,x}U + A_{1,\xi}V - A_{3,x}A_{1,\xi}V + A_{3,\eta}W$. The explicit solutions of the problems (1.11), (1.12) are easily written out:

$$v_{-1}(\xi) = \pi \left\{ \mathrm{Bi}(\xi) \int_\xi^\infty \mathrm{Ai}(t)\,dt + \mathrm{Ai}(\xi) \left[\int_0^\xi \mathrm{Bi}(t)\,dt - \sqrt{3} \int_0^\infty \mathrm{Ai}(t)\,dt \right] \right\},$$

$$v_1(\xi) = \pi \left\{ \mathrm{Bi}(\xi) \int_\xi^\infty \mathrm{Ai}(t)\,dt \left[\frac{t^3}{6}v_{-1}(t) - \frac{t^2}{2} \right] dt \right.$$

$$+ \mathrm{Ai}(\xi) \int_0^\xi \mathrm{Bi}(t) \left[\frac{t^3}{6}v_{-1}(t) - \frac{t^2}{2} \right] dt$$

$$\left. - \mathrm{Ai}(\xi)\sqrt{3} \int_0^\infty \mathrm{Ai}(t) \left[\frac{t^3}{6}v_{-1}(t) - \frac{t^2}{2} \right] dt \right\}.$$

Theorem 1.2 implies that the estimate $|Y_{3,1}(x,\varepsilon) - u(x,\varepsilon)| < M\varepsilon^3$ holds everywhere on the closed interval $[0,1]$. It is quite interesting to compare these graphs with those of the example given in the Introduction (see Figure 1).

REMARK. Similarly to Example 5, the problem (1.1)–(1.3) can also be analyzed in the case where the coefficient $q(x)$ is positive everywhere except an interior point x_0 at which $q(x)$ has a zero of finite order (the simplest case is that of a zero of second order). Then there is an exponential boundary layer in the vicinity of both endpoints, and an inner expansion arises in a neighborhood of the point x_0. Its coefficients are functions of the inner variable ξ ranging from $-\infty$ to ∞. In the inner variable, the differential operator is of the form $d^2/d\xi^2 - \xi^{2m}$, where $2m$ is the order of the zero of $q(x)$ at x_0.

No significant modifications, as compared with Example 5, arise in the analysis of, for example, the following boundary value problem for a partial differential equation:

$$\varepsilon^4 \Delta u - q(x)u = f(x), \qquad x \in \Omega \subset \mathbf{R}^m,$$
$$u(x,\varepsilon) = 0 \quad \text{for } x \in \partial\Omega,$$

where Ω is a bounded domain, $\partial\Omega \in C^\infty$, $q(x) = |x|^2 h(x)$, $h \in C^\infty(\overline{\Omega})$, $h(x) > 0$ in $\overline{\Omega}$. The outer expansion for this problem is constructed as easily, as in all the preceding examples.

If the origin lies inside the domain Ω, an exponential boundary layer arises along the boundary of Ω, the same one as in Example 3, and the inner expansion

$$W = \sum_{i=-2}^\infty \varepsilon^i w_i(\xi), \tag{1.25}$$

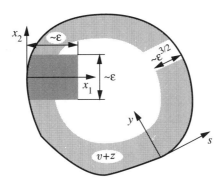

FIGURE 6

where $\xi = \varepsilon^{-1}x$, is valid in a neighborhood of the origin. Its coefficients satisfy the system of equations

$$\Delta_\xi w_{-2} - h(0)w_{-2} = f(0),$$

$$\Delta_\xi w_i - h(0)w_i = Q_{i+2}(\xi) + \sum_{j=3}^{i+4} P_j(\xi)w_{i-j+2}, \qquad i \geq -1, \tag{1.26}$$

where P_j and Q_j are polynomials of degree j obtained by expanding $q(x)$ and $f(x)$ in Taylor series. One can prove the existence of solutions for equations (1.26) in the whole space and find the asymptotics of $w_i(\xi)$ as $\xi \to \infty$. The series (1.25) turns out to be matched, as in Example 5, to the outer expansion so that a theorem completely analogous to Theorem 1.2 holds.

If the origin $O \in \partial\Omega$, then equations (1.26) have to be studied in a half-space so that the inner boundary layer in the vicinity of the point O overlaps with the exponential boundary layer. The analysis is only slightly more complicated, but basically the asymptotics is justified as in Example 5. Figure 6 shows a domain Ω with $O \in \partial\Omega$. The boundary layer along the whole boundary $\partial\Omega$ is shaded. The coefficients of the expansion in this boundary layer have singularities at the origin. The corresponding series of type (2.4) Chapter I, as well as the outer expansion loses its asymptotic properties in the vicinity of the point O. The area of the inner boundary layer, where the asymptotic expansion (1.25) holds, is double shaded. Of course, the comment on depicting the borders of the boundary layers made at the end of Example 4 with regard to Figure 4 applies to Figure 5 and all the following figures of this kind.

In fact, neither Example 5 nor other similar boundary value problems just described included the process of matching outer to inner expansions. Their coefficients are uniquely determined, and the proof is reduced to the verification of the fact that the asymptotic expansions are indeed matched. In the next section we consider an example where the coefficients of the asymptotic expansions are no longer a priori uniquely determined and have to be chosen by a matching process. Furthermore, the sequence of gauge functions of ε in this simple example turns out to be not as simple as before.

§2. Matching procedure for asymptotic expansions

EXAMPLE 6. Consider the boundary value problem

$$l_\varepsilon u \equiv \varepsilon^2 u'' - b(x)u' - q(x)u = f(x) \quad \text{for } 0 \le x \le 1, \tag{2.1}$$

$$u(0, \varepsilon) = 0, \tag{2.2}$$

$$u(1, \varepsilon) = 0, \tag{2.3}$$

where $b(x) = xa(x)$, a, q, $f \in C^\infty[0, 1]$, $q(x) > 0$, $a(x) > 0$ for $0 \le x \le 1$, $a(0) = 1$, $q(0) = \mu > 0$.

As before, we shall seek the outer expansion in the form

$$U = \sum_{k=0}^\infty \varepsilon^{2k} u_{2k}(x). \tag{2.4}$$

The equations for $u_{2k}(x)$ are

$$\begin{aligned} b(x)u_0' + q(x)u_0 &= -f(x), \\ b(x)u_{2k}' + q(x)u_{2k} &= u_{2k-2}'', \qquad k \ge 1. \end{aligned} \tag{2.5}$$

Now, as in Example 2, $u_{2k}(x)$ are not determined by equations (2.5) uniquely. It is not reasonable to require that U satisfies the boundary condition (2.3) because one can easily eliminate the residual at the point $x = 1$ with the help of an exponential boundary layer (as in Examples 1 and 2). At the other endpoint, i.e., $x = 0$, the boundary condition (2.2) cannot be satisfied because, in general, the solutions of equations (2.5) have singularities. For that reason, we start by investigating the asymptotics of the solutions of (2.5) at the origin in order to see to what extent these solutions are undetermined. This is easily achieved with the use of the explicit formula giving the solution of the equation $b(x)u' + q(x)u = F(x)$:

$$u(x) = E(x) \left\{ \int_1^x (b(\eta)E(\eta))^{-1} F(\eta)\, d\eta + C \right\}, \tag{2.6}$$

where $E(x) = \exp \int_x^1 [b(\theta)]^{-1} q(\theta)\, d\theta$ and C is an arbitrary constant. For a smooth $F(x)$ the solution (2.6), in general, has a singularity at the origin—it behaves like $x^{-\mu}$. However, it turns out that a special choice of the constant C ensures that $u(x)$ is a smooth function.

In this example, we denote by $s_i(x)$ functions in $C^\infty[0, 1]$ and their Taylor series for $x \to 0$, omitting the subscript wherever this does not lead to any misunderstanding. The explicit form of the solution $E(x)$ of the homogeneous equation implies that

$$E(x) = \exp \int_x^1 \left[\frac{\mu}{\theta} + s(\theta) \right] d\eta = x^{-\mu} s_1(x),$$

and $[E(x)]^{-1} = x^{\mu} s_2(x)$. We now find the asymptotics of the function (2.6) assuming that $F(x) \in C^{\infty}[0, 1]$:

$$\int_1^x [b(\eta)E(\eta)]^{-1} F(\eta) \, d\eta = \int_1^x \eta^{\mu-1} s(\eta) \, d\eta = x^{\mu} s_3(x) - c_0.$$

(The last equality can be verified, for example, by integrating by parts; see, for example, [107, §9].) Therefore, if the constant C in formula (2.6) equals c_0 then $u(x) \in C^{\infty}[0, 1]$.

By consecutively applying this statement to equations (2.5) and each time choosing a smooth solution, one obtains $u_{2k}(x) \in C^{\infty}[0, 1]$ and hereby the series (2.4). This series is evidently an f.a.s. of the equation (2.1), but, in general, it does not satisfy conditions (2.2) and (2.3). We have already mentioned the vicinity of the point $x = 1$, but the situation near the point $x = 0$ is much more complicated. First, we make the change of variables in a neighborhood of this endpoint of the closed interval: $x = \varepsilon^{\alpha} \xi$. Since this procedure does not change the order of the second term in equation (2.1) one clearly has to take $\alpha = 1$, which ensures that all three terms in the left-hand side of the equation are of the same order.

Thus, letting $x = \varepsilon \xi$, we write the inner expansion in the form

$$V = \sum_{i=0}^{\infty} \varepsilon^i v_i(\xi). \tag{2.7}$$

One could look for the asymptotics of the whole problem (2.1)–(2.3) near the origin in the form (2.7). However, since a smooth f.a.s. of equation (2.1) is already found in the form (2.4), it is simpler, as in Examples 1 and 2, to seek the solution in the form of the sum of the series (2.4) and (2.7). Then V must be an f.a.s. of the homogeneous equation. But this is where the analogy with Examples 1 and 2 ends. It turns out that the functions $v_i(\xi)$ do not decay exponentially as $\xi \to \infty$. Moreover, as we shall see, (2.4) is not a correct asymptotic expansion of the solution $u(x, \varepsilon)$ inside the closed interval. And if $\mu < 2$ then only $u_0(x)$ is the principal term of the asymptotics of the solution $u(x, \varepsilon)$ inside the closed interval $[0, 1]$. The following term is no longer of the form $\varepsilon^2 u_2(x)$ but quite different. The true asymptotics of the solution inside the closed interval $[0, 1]$ is the sum of the series (2.4) and another series the coefficients of which have growing singularities at the point $x = 0$. Therefore, in spite of the fact that the functions $u_{2k}(x)$ are smooth, the problem (2.1)–(2.3) is bisingular.

We now go back to the series V. The homogeneous equation which this series has to satisfy is of the form

$$\frac{d^2V}{d\xi^2} - \xi a(\varepsilon\xi)\frac{dV}{d\xi} - q(\varepsilon\xi)V = 0. \tag{2.8}$$

Expanding the coefficients in Taylor series

$$a(x) = 1 + \sum_{j=1}^{\infty} a_j x^j, \qquad q(x) = \mu + \sum_{j=1}^{\infty} q_j x^j, \qquad x \to 0$$

one obtains, as before, the system of differential equations

$$lv \equiv v_0'' - \xi v_0' - \mu v_0 = 0, \tag{2.9}$$

$$lv_i = \xi \sum_{j=1}^{i} a_j \xi^j v_{i-j}' + \sum_{j=1}^{i} q_j \xi^j v_{i-j}, \qquad i \geq 1. \tag{2.10}$$

The boundary condition (2.2) leads to the boundary conditions for $v_k(\xi)$:

$$v_{2i}(0) = -u_{2i}(0),$$
$$v_{2i+1}(0) = 0, \qquad i \geq 0. \tag{2.11}$$

From now on until the end of our treatment of problem (2.1)–(2.3), we denote by $\sigma_j(\xi)$ the asymptotic series of the form $\sum_{p=0}^{\infty} c_p \xi^{-p}$ as $\xi \to \infty$ which can be infinitely differentiated term-by-term. The subscript in σ_j will often be omitted. The following theorem holds.

THEOREM 2.1. *There is a solution of the system* (2.9), (2.10) *satisfying conditions* (2.11) *such that* $v_i(\xi) \in C^{\infty}[0, \infty)$ *and*

$$v_i(\xi) = \xi^{-\mu} \sum_{j=0}^{i} \sigma_j(\xi) \xi^{1-j} \ln^j \xi, \qquad \xi \to \infty. \tag{2.12}$$

We begin by writing out the solution of the equation $lv = \varphi$ and finding its estimate. As is known, the homogeneous equation (2.9) has two linearly independent solutions which we denote by $Y(\xi)$ and $Y_1(\xi)$. These solutions satisfy the relations

$$Y(\xi) = \xi^{-\mu} \left(1 + \sum_{j=1}^{\infty} c_j \xi^{-2j} \right), \qquad \xi \to \infty, \, Y(\xi) > 0,$$

$$Y_1(\xi) = e^{\xi^2/2} \xi^{\mu-1} \cdot \left(1 + \sum_{j=1}^{\infty} \bar{c}_j \xi^{-2j} \right), \qquad \xi \to \infty.$$

This asymptotics easily follows from general properties of solutions of ordinary linear differential equations (see, for example, [28, Chapter 2, §6]), but a simpler way is to express the solutions of equation (2.9) through parabolic cylinder functions for which both the integral representations and the asymptotics are known (see, for example, [64, §10.2]). One can directly verify that the function

$$v(\xi) = \int_{\xi}^{\infty} [Y(\xi)Y_1(\theta) - Y_1(\xi)Y(\theta)] \exp(-\theta^2/2) \varphi(\theta) \, d\theta$$

is a solution of the equation $lv = \varphi$ if $\varphi(\theta) \in C^{\infty}[0, 1)$ and $|\varphi(\theta)| < M|\theta|^{-N}$, $N > \mu$. This formula shows that

$$|v(\xi)| < M|\xi|^{-N}. \tag{2.13}$$

We will now prove a lemma similar to Lemma 1.1.

LEMMA 2.1. *Let* $g(\xi) \in C^{\infty}[0, \infty)$ *and* $g(\xi) = \xi^{-\mu+p} \ln^i \xi$ *for* $\xi > 1$, *where* i *and* p *are nonnegative integers. Then, for* $0 \leq \xi < \infty$, *there exists a solution of the equation* $lv \equiv v'' - \xi v' - \mu v = g(\xi)$ *which can be expanded in an asymptotic series*

$$V(\xi) = \xi^{-\mu+p} \sum_{j=0}^{i+1} \sigma_j(\xi) \ln^j \xi. \tag{2.14}$$

This series can be infinitely differentiated term-by-term.

The proof almost repeats that of Lemma 1.1: it is sufficient to construct an f.a.s. in the form (2.14). Then one has to use the estimate (2.13) for the remainder, and the fact that the solution $Y(\xi)$ is positive (it is the only solution of the homogeneous equation which is not fast-growing as $\xi \to \infty$). The formal construction of the series (2.14) is easily obtained from the obvious relations

$$l(\xi^{-\mu+p} \ln^i \xi)$$
$$= \xi^{-\mu+p}[-p \ln^i \xi + c_1 \ln^{i-1} \xi + c_2 \xi^{-2} \ln^i \xi + c_3 \xi^{-2} \ln^{i-1} \xi + c_4 \xi^{-2} \ln^{i-2} \xi],$$
$$l(\xi^{-\mu} \ln^i \xi) = \xi^{-\mu}[i \ln^{i-1} \xi + \bar{c}_1 \xi^{-2} \ln^i \xi + \bar{c}_2 \xi^{-2} \ln^{i-1} \xi + \bar{c}_3 \xi^{-2} \ln^{i-2} \xi]. \blacksquare$$

PROOF OF THEOREM 2.1. Evidently, $v_0(\xi) = -u_0(0)[Y(0)]^{-1} Y(\xi)$. Then the equation for $v_1(\xi)$ is of the form $lv_1 = \xi^{-\mu+1} \sigma(\xi)$. Now Lemma 2.1 has to be applied to each term of the asymptotic series $\sigma(\xi)$ thus yielding the f.a.s. of the equation for v_1. Using the estimate (2.13) for the remainder, one has the solution $\tilde{v}_1(\xi) = \xi^{-\mu+1} \sum_{j=0}^{1} \sigma_j(\xi) \ln^j \xi$. The addition of $cY(\xi)$ to it and the appropriate choice of the constant c results in satisfying conditions (2.11). The existence of the remaining $v_i(\xi)$ and their asymptotics (2.12) is proved by induction. $\blacksquare$

REMARK. In fact, taking into account the parity of the powers of ξ in the series $\sigma(\xi)$, one can note that the powers of the logarithmic terms are smaller than stated in Lemma 2.1 and Theorem 2.1. Indeed, the asymptotics of the functions $v_i(\xi)$ contains the terms $\ln^j \xi$ only for $2j < i$. We have written the asymptotics of the functions $v_i(\xi)$ in a rougher but simpler form in order to make our notation less cumbersome.

We now proceed to the construction of the part of the outer expansion which has to be matched to the series V. The form of this expansion follows from the matching condition: one has to take the series V and formally rewrite it in terms of the outer variable x. The resulting series is of the form $\sum_{m=0}^{\infty} \nu_m(\varepsilon) Z_m(x)$, where $\nu_m(\varepsilon)$ is the gauge sequence of functions, and $Z_m(x)$ are formal series as $x \to 0$. It is easy to see that in our example $\nu_m(\varepsilon)$ are the products of the form $\varepsilon^{\mu+k} \ln^l \varepsilon$, $l \leq k$, and that the outer expansion is given by the formula

$$Z = \varepsilon^{\mu} \sum_{k=0}^{\infty} \varepsilon^k \sum_{l=0}^{k} z_{k,l}(x) \ln^l \varepsilon. \tag{2.15}$$

It remains to find the functions $z_{k,l}(x)$. The equations for them are obtained directly from equation (2.1):

$$b(x) z'_{k,l}(x) + q(x) z_{k,l}(x) = z''_{k-2,l}(x), \tag{2.16}$$

where, according to (2.15), $l \leq k$ and the functions having at least one negative index are identically equal to zero. Moreover, the matching condition yields asymptotic series for the functions $z_{k,l}(x)$ as $x \to 0$. Indeed, the asymptotic series (2.12) imply that for any $n \geq 0$, $m \geq 0$ the equalities

$$A_{m,x}A_{n,\xi}V = A_{n,\xi}A_{m,x}Z^* \qquad (2.17)$$

hold, where Z^* is the same series as (2.15) with $z_{k,l}(x)$ replaced by

$$Z_{k,l} = x^{-k}\sum_{j=0}^{k-l}x^j s_j(x)\ln^j x, \qquad x \to 0. \qquad (2.18)$$

These relations are conveniently illustrated by Table 2 which is quite similar to Table 1, but of a slightly more complicated form. Each column of Table 2 contains the asymptotic expansion of the function $\varepsilon^i \tilde{v}_i(\xi) = \varepsilon^i \xi^\mu v_i(\xi)$ (the factor ξ^μ is singled out only to simplify the notation) as $\xi \to \infty$. Individual terms of this asymptotics appear in the top half of each square of the table. (According to the remark following the proof of Lemma 2.1 and Theorem 2.1 the table does not include those terms of the asymptotic series (2.12) of functions $v_i(\xi)$ the coefficients of which vanish. This is why the terms $\tilde{z}_{2,2}$, $\tilde{z}_{3,2}$, $\tilde{z}_{3,3}$, etc. are missing.) The rows of the table contain the asymptotic series for $\tilde{z}_{k,l} = \varepsilon^{-\mu}x^\mu$ with individual terms of the corresponding expansions appearing in the bottom halves of the squares. We see that in this case the top expressions do not necessarily coincide with the bottom ones. The change of variables $\ln\xi = \ln x - \ln\varepsilon$ moves a part of the expression into the same square, and another part into the neighboring ones. However, the coefficients of the series $Z_{k,l}$ are constructed in such a way that relations (2.17) are satisfied.

The construction of the asymptotics for the solution of the problem (2.1)–(2.3) is virtually completed. It remains to construct the exponential boundary layer functions near the right endpoint of the closed interval. This series is of the same form as that in Example 1. The only difference is that it has to eliminate the residuals due both to the series (2.4) and (2.15). Thus

$$W(\eta, \varepsilon) = \sum_{k=0}^{\infty}\varepsilon^{2k}w_k(\eta) + \varepsilon^\mu\sum_{k=0}^{\infty}\varepsilon^k\sum_{l=0}^{k}w_{k,l}(\eta)\ln^l\varepsilon, \qquad (2.19)$$

where $\eta = \varepsilon^{-1}(1-x)$. The construction of the functions $w_k(\eta)$ and $w_{k,l}(\eta)$ is obvious and is therefore omitted. The following theorems hold.

THEOREM 2.2. *There are functions $z_{k,l}(x) \in C^\infty(0, 1]$, $2l \leq k$, satisfying (2.16) and having asymptotic expansions $z_{k,l}$ as $x \to 0$ defined by formulas (2.18). Thus, the matching conditions (2.17), where Z^* is replaced with Z, are satisfied for the series (2.7) and (2.15).*

THEOREM 2.3. *Let α and β be two numbers such that $0 < \beta < \alpha < 1$. Then the sum of the series (2.4) and (2.7) is a uniform asymptotic expansion of the function $u(x, \varepsilon)$ solving the problem (2.1)–(2.3) on the closed interval*

TABLE 2

$\tilde{Z}$ \ $\tilde{V}$	$\tilde{v}_0$	$\varepsilon\tilde{v}_1$	$\varepsilon^2\tilde{v}_2$	$\varepsilon^3\tilde{v}_3$	$\varepsilon^4\tilde{v}_4$	$\cdots$
$\tilde{z}_{0,0}$	c_0 c_0	$\varepsilon c_{1,-1}\xi$ $c_{1,-1}x$	$\varepsilon^2 c_{2,-2}\xi^2$ $c_{2,-2}x^2$	$\varepsilon^3 c_{3,-3}\xi^3$ $c_{3,-3}x^3$	$\varepsilon^4 c_{4,-4}\xi^4$ $c_{4,-4}x^4$	$\cdots$
$\varepsilon\tilde{z}_{1,0}$		$\varepsilon c_{1,0}$ $\varepsilon c_{1,0}$	$\varepsilon^2 c_{2,-1}\xi$ $\varepsilon c_{2,-1}x$	$\varepsilon^3 c_{3,-2}\xi^2$ $\varepsilon c_{3,-2}x^2$	$\varepsilon^4 c_{4,-3}\xi^3$ $\varepsilon c_{4,-3}x^3$	$\cdots$
$\varepsilon^2\ln\varepsilon\,\tilde{z}_{2,1}$			$\varepsilon^2 d_{2,0}\ln\xi$ $-\varepsilon^2 d_{2,0}\ln\varepsilon$	$\varepsilon^3 d_{3,-1}\xi\ln\xi$ $-\varepsilon^3\ln\varepsilon d_{3,-1}x$	$\varepsilon^4 d_{4,-2}\xi^2\ln\xi$ $-\varepsilon^2\ln\varepsilon d_{4,-2}x^2$	$\cdots$
$\varepsilon^2\tilde{z}_{2,0}$	$c_{0,2}\xi^{-2}$ $\varepsilon^2 c_{0,2}x^{-2}$	$\varepsilon c_{1,1}\xi^{-1}$ $\varepsilon^2 c_{1,1}x^{-1}$	$\varepsilon^2 c_{2,0}$ $\varepsilon^2(c_{2,0}+d_{2,0}\ln x)$	$\varepsilon^3 c_{3,-1}\xi$ $\varepsilon^2 x(c_{3,-1}+d_{3,-1}\ln x)$	$\varepsilon^4 c_{4,-2}\xi^2$ $\varepsilon^2 x^2(c_{4,-2}+d_{4,-2}\ln x)$	$\cdots$
$\varepsilon^3\ln\varepsilon\,\tilde{z}_{3,1}$				$\varepsilon^3 d_{3,0}\ln\xi$ $-\varepsilon^3\ln\varepsilon d_{3,0}$	$\varepsilon^4 d_{4,-1}\xi\ln\xi$ $-\varepsilon^3\ln\varepsilon d_{4,-1}x$	$\cdots$
$\varepsilon^3\tilde{z}_{3,0}$		$\varepsilon c_{1,2}\xi^{-2}$ $\varepsilon^3 c_{1,2}x^{-2}$	$\varepsilon^2 c_{2,1}\xi^{-1}$ $\varepsilon^3 c_{2,1}x^{-1}$	$\varepsilon^3 c_{3,0}$ $\varepsilon^3(c_{3,0}+d_{3,0}\ln x)$	$\varepsilon^4 c_{4,-1}\xi$ $\varepsilon^3 x(c_{4,-1}+d_{4,-1}\ln x)$	$\cdots$
$\varepsilon^4\ln^2\varepsilon\,\tilde{z}_{4,2}$					$\varepsilon^4 e_{4,0}\ln^2\xi$ $\varepsilon^4\ln^2\varepsilon\cdot e_{4,0}$	$\cdots$
$\varepsilon^4\ln\varepsilon\,\tilde{z}_{4,1}$			$\varepsilon^2 d_{2,2}\xi^{-2}\ln\xi$ $-\varepsilon^4\ln\varepsilon d_{2,2}x^{-2}$	$\varepsilon^3 d_{3,1}\xi^{-1}\ln\xi$ $-\varepsilon^4\ln x\,x^{-1}$	$\varepsilon^4 d_{4,0}\ln\xi$ $-\varepsilon^4\ln\varepsilon(d_{4,0}+2e_{4,0}\ln x)$	$\cdots$
$\varepsilon^4\tilde{z}_{4,0}$	$c_{0,4}\xi^{-4}$ $\varepsilon^4 c_{0,4}x^{-4}$	$\varepsilon c_{1,3}\xi^{-3}$ $\varepsilon^4 c_{1,3}x^{-3}$	$\varepsilon^2 c_{2,2}\xi^{-2}$ $\varepsilon^4 x^{-2}(c_{2,2}+d_{2,2}\ln x)$	$\varepsilon^3 c_{3,1}\xi^{-1}$ $\varepsilon^4 x^{-1}(c_{3,1}+d_{3,1}\ln x)$	$\varepsilon^4 c_{4,0}$ $\varepsilon^4(c_{4,0}+d_{4,0}\ln x+e_{4,0}\ln^2 x)$	$\cdots$
$\cdots$	$\cdots$	$\cdots$	$\cdots$	$\cdots$	$\cdots$	$\cdots$

$[0,\varepsilon^\beta]$. *The sum of the series* (2.4), (2.15) *and* (2.19) *is a uniform asymptotic expansion of the same solution on the closed interval* $[\varepsilon^\beta,1]$.

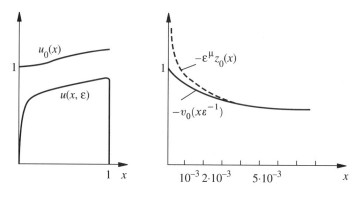

FIGURE 7

For the composite asymptotic expansion the estimate

$$|A_{N,x}U + A_{N,x}Z + A_{N,\xi}V - A_{N,x}A_{N,\xi}V + A_{N,\eta}W - u(x,\varepsilon)| < M\varepsilon^{N_1}, \quad (2.20)$$

where $N_1 \to \infty$ *as* $N \to \infty$, *holds. Here* U, Z, V *and* W *are the series* (2.4), (2.15), (2.7), *and* (2.19), *respectively.*

Figure 7 shows the graph of the solution of the boundary value problem

$$l_\varepsilon u = \varepsilon^2 u'' - \tan x \cdot u' - \mu u = -\mu(1+x),$$
$$u(0,\varepsilon) = u(1,\varepsilon) = 0$$

for $\varepsilon = 10^{-3}$, $\mu = 1/5$ together with the graphs of some terms of the asymptotic approximations of $u_0(x)$, $z_0(x)$, and $v_0(\xi)$. Here $u_0(x) = 1 + x - (\sin x)^{-\mu} \int_0^x (\sin t)^\mu \, dt$, $z_0(x) = -2^{-\mu/2}C(\sin x)^{-\mu}$, $v_0(\xi) = -CH_{-\mu}(\xi/\sqrt{2})$, $C = 2^\mu \Gamma((1+\mu)/2)/\Gamma(1/2)$, $H_\nu(y)$ is Hermite's function (see [64, §10.2]).

The second term of the series (2.4) evidently plays no role of any importance while the first terms of the series (2.7) and (2.15) are quite significant. It is of interest that the coefficient of the leading term in the equation is six orders (!) lower than other coefficients. Nevertheless, if one considers only the series (2.4), then even far from the boundary, at the point $x = 1/2$, the error is more than 20%. Only the correct matching of the outer to inner expansion yields a satisfactory result. Indeed, denote

$$Y_\mu(x,\varepsilon) = u_0(x) + \varepsilon^\mu z_0(x) + v_0(\xi) + 2^{-\mu/2}C\xi^{-\mu}.$$

Then Theorem 2.3 and a closer examination of the structure of the terms of the series (2.4), (2.7), and (2.15) imply the estimate

$$|Y_\mu(x,\varepsilon) - u(x,\varepsilon)| < M\varepsilon^{\mu+2}|\ln\varepsilon|$$

for $x < 1 < \delta$ ($\delta > 0$). (The uniform asymptotics in the vicinity of the endpoint $x = 1$ is obtained by the addition of the series (2.19) and is of no interest.) Applying the operator l_ε to $Y_\mu(x, \varepsilon)$ and estimating the result for the above values of μ and ε, one has (even after a rough estimate) $|Y_\mu(x, \varepsilon) - u(x, \varepsilon)| < 10^{-5}$ for $0 \le x \le 0.99$. Thus, the graphs of $u(x, \varepsilon)$ and $Y_\mu(x, \varepsilon)$ in Figure 7 are practically indistinguishable.

PROOF OF THEOREM 2.2. We first show that the series (2.18) are the f.a.s. of the system (2.16). Summing up equations (2.9), (2.10) multiplied by ε^i, one has

$$\frac{d^2}{d\xi^2}(A_{n,\xi}V) - \xi a(\varepsilon\xi)\frac{d}{d\xi}(A_{n,\xi}V) - q(\varepsilon\xi)A_{n,\xi}V$$

$$= \xi \sum_{i=0}^{n} \varepsilon^i v_i'(\xi)\left(\sum_{j=0}^{n-i}\varepsilon^j\xi^j a_j - a(x)\right) + \sum_{i=0}^{n}\varepsilon^i v_i(\xi)\left(\sum_{j=0}^{n-i}\varepsilon^j\xi^j q_j - q(x)\right),$$

or, in terms of x,

$$\varepsilon^2\frac{d^2}{dx^2}(A_{n,\xi}V) - xa(x)\frac{d}{dx}A_{n,\xi}V - q(x)A_{n,\xi}V$$

$$= x\sum_{i=0}^{n}\varepsilon^i\frac{d}{dx}v_i\left(\frac{x}{\varepsilon}\right)\left[\sum_{j=0}^{n-i}a_j x^j - a(x)\right] + \sum_{i=0}^{n}\varepsilon^i v_i\left(\frac{x}{\varepsilon}\right)\left[\sum_{j=0}^{n-i}q_j x^j - q(x)\right].$$

$$(2.21)$$

Now apply the operator $A_{m+\mu,x}$ to both parts of this equality, and take into account that $A_{m+\mu,x}A_{n,\xi}V = \varepsilon^\mu \sum_{l=0}^{m}\varepsilon^l \sum_{j=0}^{l}B_{n-l-\mu}Z_{l,j}\ln^j\varepsilon$. The resulting equation includes only finite sums of terms of the form $\varepsilon^{\mu+l}\ln^j\varepsilon\,\varphi_{l,j}(x)$. Equating the coefficients of same powers $\varepsilon^{\mu+l}\ln^j\varepsilon$ yields a system of equations for $B_{n-l-\mu}Z_{l,j}$ which is only slightly different from (2.16), viz., the right-hand sides differ by functions that decay rapidly as $x \to 0$:

$$b(x)(B_{n-\mu}Z_{0,0})'_x + q(x)B_{n-\mu}Z_{0,0} = O(x^{-\mu+n}), \qquad x \to 0,$$

$$b(x)(B_{n-1-\mu}Z_{1,0})'_x + q(x)B_{n-1-\mu}Z_{1,0} = O(x^{-\mu+n-1}), \qquad x \to 0,$$

$$b(x)(B_{n-2-\mu}Z_{2,0})'_x + q(x)B_{n-2-\mu}Z_{2,0}$$
$$= (B_{n-\mu}Z_{0,0})''_x + O(x^{-\mu+n-2}), \qquad x \to 0,$$

etc. This implies that the series $Z_{k,l}(x)$ are the f.a.s. of equations (2.16) as $x \to 0$.

LEMMA 2.2. *Let $F(x) \in C^\infty(0, 1]$. Suppose that $F(x)$ can be expanded in an asymptotic series*

$$F(x) = x^{-\mu-l}\sum_{k=0}^{\infty}x^k\sum_{j=0}^{k}a_{k,j}\ln^j x, \qquad x \to 0;$$

this series can be repeatedly differentiated term-by-term, and the series

$$Z(x) = x^{-\mu-l}\sum_{k=0}^{\infty}x^k\sum_{j=0}^{k}b_{k,j}\ln^j x, \qquad x \to 0; \qquad (2.22)$$

is an f.a.s. of the equation

$$L_1 Z \equiv b(x)Z' + q(x)Z = F(x) \qquad (2.23)$$

as $x \to 0$. *Then there exists a solution* $z(x)$ *of this equation which can be expanded in the asymptotic series* (2.22). *This series can also be infinitely differentiated term-by-term.*

PROOF. Consider a partial sum $z_N(x) = B_N Z$. By definition, it satisfies the equality $L_1 B_N Z = F(x) + \varphi_N(x)$, where $\varphi_N(x) = O(x^{N_1})$, $x \to 0$ and $N_1 \to \infty$ for $N \to \infty$. According to formula (2.6), we construct the function $y_N(x)$, which is the solution of the equation $L_1 y_N(x) = \varphi_N(x)$:

$$y_N(x) = E(x) \int_0^x [b(\eta) E(\eta)]^{-1} \varphi_n(\eta) \, d\eta,$$

where $E(x) = \exp \int_x^1 [b(\theta)]^{-1} q(\theta) \, d\theta$. The estimate on the function $\varphi_N(x)$ implies that $|y_N(x)| < M x^{N_1}$. Let $w_N(x) = z_N(x) - y_N(x)$. By construction, the function $w_N(x)$ satisfies (2.23). It only remains to show that $w_N(x)$ does not depend on N. Indeed, $\sigma_N(x) = w_{N+1}(x) - w_N(x)$ is a solution of the equation $L_1 \sigma_N = 0$ for which the estimate $|\sigma_N(x)| < M x^{N_1}$ holds. Hence $\sigma_N(x) = c E(x)$, and the asymptotics of $E(x)$ implies that $c = 0$. The termwise differentiation of the series (2.22) is a direct consequence of equation (2.23). ∎

The statement of Theorem 2.2 is now obtained by consecutively applying Lemma 2.4 to equations (2.16) and taking into account that $Z_{k,l}$ are the f.a.s of this system of equations. ∎

The proof of Theorem 2.3 follows the same scheme as that of Theorem 1.2. One has to check that

(a) the series (2.4) is an f.a.s. of the equation (2.1) uniformly on the closed interval $[0, 1]$;
(b) the series (2.7) is an f.a.s. of the equation (2.8) for $x \ll 1$;
(c) the series (2.15) is an f.a.s. of the equation $l_\varepsilon Z = 0$ for $x/\varepsilon \gg 1$;
(d) the series (2.19) is an f.a.s. of the equation $l_\varepsilon W = 0$ on the closed interval $[0, 1]$;
(e) the matching conditions for the series Z and V are satisfied;
(f) the series $U + V$ satisfies the boundary condition (2.2), and the series $U + W$ satisfies condition (2.3).

Statement (a) follows from the relation $l_\varepsilon A_{2n,x} U = f(x) + \varepsilon^{2n+2} u_{2n}''$, (b) is a consequence of (2.21): $|l_\varepsilon A_{n,\xi} V| \le M x^n$, and (c) is implied by (2.16), (2.18):

$$|l_\varepsilon A_{n+\mu} Z| = \left| \varepsilon^{\mu+n+2} \sum_{l=0}^n \frac{d^2}{dx^2} z_{n,l} \right| \le M \varepsilon^{\mu+n+2} [1 + x^{-\mu-n-2} |\ln^n x|].$$

Statement (e) is the condition (2.17), and the construction of the series Z was based on that relation. Statements (d) and (f) are obvious: they can be verified in the same manner as in the preceding examples. The proof of the estimate (2.20) almost repeats that of Theorem 1.2 and takes into account the fact that $q(x) \ge q_0 > 0$ whereby the following estimate holds for the solution of the equation $l_\varepsilon \psi = \varphi$:

$$|\psi(x, \varepsilon)| \le |\psi(0, \varepsilon)| + |\psi(1, \varepsilon)| + q_0^{-1} \max_{0 \le x \le 1} |\varphi(x)|. \quad \blacksquare$$

§3. Nonlinear equation. Intermediate boundary layer

This last section of Chapter II has some peculiar features. By content it stands somewhat aside from the course of the main exposition. This is the only place in the book where we consider the Cauchy problem for an ordinary differential equation with a small parameter at the highest derivative, a subject well covered in numerous literature and therefore considered nowhere else in this book. There are at least two reasons for such a deviation. First, we would like to show, by means of a relatively simple example, that the method can be successfully applied to a nonlinear problem. Second, it demonstrates an interesting phenomenon of the intermediate boundary layer. From the technical point of view, this section is rather complicated and can be skipped at the first reading: no reference is made to it in the following chapters.

Consider the Cauchy problem for the differential equation

$$\varepsilon \frac{du}{dx} = f(x, u) \tag{3.1}$$

with the initial condition

$$u(0, \varepsilon) = R_0, \tag{3.2}$$

where $x \in [0, d]$, $\varepsilon > 0$ is a small parameter, and R_0 is a positive constant. Let $f(x, u)$ be an infinitely differentiable function for $x \in [0, d]$ and all values of u such that

$$f(0, 0) = \frac{\partial f}{\partial u}(0, 0) = 0, \quad \frac{\partial f}{\partial x}(0, 0) = 1, \quad \frac{\partial^2 f}{\partial u^2}(0, 0) = -2, \tag{3.3}$$

$$f(x, 0) > 0 \quad \text{for } x > 0,$$

$$\frac{\partial^2 f}{\partial u^2}(x, u) < \text{const} < 0 \quad \text{for } 0 \le x \le d, u \ge 0. \tag{3.4}$$

The meaning of these numerous conditions is fairly simple: (3.3) implies that

$$f(x, u) = x - u^2 + O(x^2 + |xu| + |u^3|)$$

in a neighborhood of the origin, while relations (3.4) provide sufficient conditions for the existence and uniqueness of a nonnegative function $u_0(x)$ such that $f(x, u_0(x)) \equiv 0$ and $(\partial f / \partial u)(x, u_0(x)) < 0$ for $x > 0$.

One can naturally expect that it is the function $u_0(x)$ that provides the limit of the solution of the problem (3.1), (3.2) as $\varepsilon \to 0$ provided such a solution $u(x, \varepsilon)$ exists. The proof of the existence of the solution $u(x, \varepsilon)$ will be given later, together with the investigation of its asymptotics. Meanwhile, we formally write the outer expansion of the solution $u(x, \varepsilon)$ in the same form as in the preceding sections:

$$U = \sum_{k=0}^{\infty} \varepsilon^k u_k(x). \tag{3.5}$$

Inserting this series into equation (3.1), expanding the function $f(x, U)$ in a Taylor series at the point $(x, u_0(x))$, and equating the coefficients of same powers of ε, one obtains the recurrence system of equations

$$f(x, u_0(x)) = 0, \qquad (3.6)$$

$$\frac{du_0}{dx} = \frac{\partial f}{\partial u}(x, u_0(x))u_1,$$

$$\frac{du_1}{dx} = \frac{\partial f}{\partial u}(x, u_0(x))u_2 + \frac{1}{2}\frac{\partial^2 f}{\partial u^2}(x, u_0(x))u_1^2,$$

$$\cdot \quad \cdot \quad \cdot \quad \cdot \quad \cdot \quad \cdot \qquad (3.7)$$

$$\frac{du_{k-1}}{dx} = \frac{\partial f}{\partial u}(x, u_0(x))u_k$$

$$+ \sum_{j=2}^{k} \frac{1}{j!}\frac{\partial^j f}{\partial u^j}(x, u_0(x)) \sum_{p_1+\ldots+p_j=k} \prod_{l=1}^{j} u_{p_l}, \qquad k \geq 2.$$

As we have already mentioned above, conditions (3.4) imply the existence of the function $u_0(x)$, i.e., the solution of equation (3.6). The same conditions together with conditions (3.3) imply that $u_0(0) = 0$, $u_0(x) > 0$ for $x > 0$ and $(\partial f/\partial u)(x, u_0(x)) < 0$ for $x > 0$. Hence $u_0(x) \in C^\infty(0, d]$, and equations (3.7) consecutively determine $u_k(x) \in C^\infty(0, d]$ for $k \geq 1$. This defines all coefficients of the outer expansion (3.5).

It can be shown without difficulty that the asymptotic expansion

$$u_0(x) = \sqrt{x} + \sum_{j=2}^{\infty} a_{j,0}x^{j/2}, \qquad x \to 0, \qquad (3.8)$$

is valid for the function $u_0(x)$ as $x \to 0$. This relation can be infinitely differentiated term-by-term. In order not to interrupt the construction of the asymptotics of the solution $u(x, \varepsilon)$ the proof of relation (3.8) is postponed until the end of the present section. Instead, using (3.8), we will analyze the asymptotic behavior of all the other coefficients $u_k(x)$ for $x \to 0$. Denote by $s_j(x)$ the series of the form $\sum_{l=0}^{\infty} c_l x^{l/2}$, $x \to 0$, omitting the subscripts in s_j wherever it does not cause any misunderstanding. Expanding $f(x, u)$ in a Taylor series at the origin, and using (3.3), (3.8), one obtains

$$\frac{\partial f}{\partial u}(x, u_0(x)) = -2\sqrt{x} + xs(x), \qquad x \to 0,$$

$$\frac{\partial^j f}{\partial u^j}(x, u_0(x)) = s_j(x), \qquad x \to 0, j \geq 2.$$

One can easily deduce, by induction, from these equalities and equations (3.7) that

$$u_k(x) = x^{(1-3k)/2}s_k(x), \qquad x \to 0, \qquad (3.9)$$

and that this equality allows repeated term-by-term differentiation.

Thus, the coefficients $u_k(x)$ have growing singularities as $x \to 0$, i.e., the problem (3.1), (3.2) is bisingular. We now proceed to the construction of the outer expansion in a neighborhood of the point $x = 0$ according to our general principle. Since in the vicinity of this point one has $u(x, \varepsilon) \approx R_0 > 0$, the natural change of variable $x = \varepsilon \xi$ ensures that both left- and right-hand sides of equation (3.1) are of the same order. Introducing the notation $u(\varepsilon \xi) \equiv v(\xi, \varepsilon)$, one obtains the following equation for v:

$$\frac{dv}{d\xi} = f(\varepsilon \xi, v). \tag{3.10}$$

The inner expansion is of the form

$$V = \sum_{i=0}^{\infty} \varepsilon^i v_i(\xi), \qquad \varepsilon \to 0. \tag{3.11}$$

The system of equations for $v_i(\xi)$ is obtained in the usual way by expanding the function $f(\varepsilon \xi, V)$ in the equation (3.10) into the series in the powers of ε:

$$\frac{dv_0}{d\xi} = f(0, v_0), \tag{3.12}$$

$$\frac{dv_1}{d\xi} = \frac{\partial f}{\partial u}(0, v_0)v_1 + \frac{\partial f}{\partial x}(0, v_0)\xi,$$

$$\cdot \quad \cdot \quad \cdot \quad \cdot \quad \cdot \quad \cdot \quad \cdot$$

$$\frac{dv_i}{d\xi} = \frac{\partial f}{\partial u}(0, v_0)v_i + \frac{1}{i!}\frac{\partial^i f}{\partial x^i}(0, v_0)\xi^i \tag{3.13}$$

$$+ \sum_{\substack{2 \le q+m \le i \\ m \ge 1}} \frac{1}{q!m!}\frac{\partial^{q+m} f}{\partial x^q \partial u^m}(0, v_0)\xi^q \sum_{p_1+\cdots+p_m=i-q} \prod_{l=1}^{m} v_{p_l}, \qquad i \ge 2.$$

The initial conditions also evidently follow from (3.2):

$$v_0(0) = R_0, \qquad v_k(0) = 0 \quad \text{for } k > 0. \tag{3.14}$$

The solution of equation (3.12) with the initial condition (3.14) can be found in quadratures:

$$\xi = \int_{R_0}^{v_0} \frac{dv}{f(0, v)}. \tag{3.15}$$

Conditions (3.3), (3.4) imply that $f(0, v) < 0$ for $0 < u \le R_0$ whereby condition (3.15) defines a positive, monotonically decreasing function $v_0(\xi)$ for $\xi \ge 0$. It follows from the condition $(\partial f/\partial u)(0, 0) = 0$ that the function $v_0(\xi)$ is defined for all $\xi \ge 0$ and $v_0(\xi) \to 0$ as $\xi \to \infty$.

Our next task is to investigate the asymptotics of $v_0(\xi)$ as $\xi \to \infty$. It turns out that

$$v_0 = \xi^{-1} \sum_{j=0}^{\infty} \xi^{-j} \sum_{l=0}^{j} c_{j,l}(\ln \xi)^l, \qquad \xi \to \infty, c_{0,0} = 1. \tag{3.16}$$

We now verify this asymptotic expansion. Let $(\partial^3 f / \partial u^3)(0, 0) = 6\gamma$. Then

$$f(0, v) = -v^2 + \gamma v^3 + \sum_{k=4}^{N+4} c_k v^k + O\left(|v|^{N+5}\right), \qquad v \to 0 \qquad (3.17)$$

for any positive integer N. Therefore,

$$\frac{1}{f(0, v)} = -\frac{1}{v^2} - \frac{\gamma}{v} + \sum_{k=0}^{N} \overline{c}_k v^k + O\left(|v|^{N+1}\right), \qquad v \to 0.$$

Inserting this expression into (3.15), one has

$$\xi = \frac{1}{v_0} + \gamma \ln v_0 + \sum_{k=0}^{N+1} d_k v_0^k + O\left(|v_0|^{N+2}\right), \qquad v_0 \to 0, \qquad (3.18)$$

whence

$$v_0 = \frac{1}{\xi} \left(1 - \frac{\gamma \ln v_0}{\xi} - \sum_{k=0}^{N+1} d_k \frac{v_0^k}{\xi} + O\left(\frac{|v_0|^{N+2}}{\xi}\right) \right)^{-1}. \qquad (3.19)$$

The iteration approach can now be used. It follows from (3.18) that $v_0(\xi) = O(\xi^{-1})$ and $[v_0(\xi)]^{-1} = O(\xi)$. Hence $\ln v_0 = O(\ln \xi)$. Therefore, one concludes from (3.19) that $v_0(\xi) = \xi^{-1} + O(\xi^{-2} \ln \xi)$. Substituting this expression for $v_0(\xi)$ into the right-hand side of (3.19), one has

$$v_0(\xi) = \xi^{-1} - \gamma \xi^{-2} \ln \xi + d_0 \xi^{-2} + O(\xi^{-3} \ln^2 \xi).$$

Relation (3.16) can now be proved by induction. The explicit formulas show that, for $\gamma \neq 0$, the asymptotic expansion includes infinitely many terms containing powers of $\ln \xi$.

The existence of the solutions of equations (3.13) satisfying conditions (3.14) is evident. For each of the functions v_i the equation is linear provided that $v_j(\xi)$ are determined for $j < i$. Thus, $v_i(\xi) \in C^\infty[0, \infty)$, the solution is expressed in quadratures, and its explicit form yields the asymptotics of the functions $v_i(\xi)$ for $\xi \to \infty$. It can be shown to be of the following form:

$$v_i(\xi) = \xi^{3i-1} \sum_{j=0}^{\infty} \xi^{-j} \sum_{l=0}^{j} c_{i, l, j} (\ln \xi)^l, \qquad \xi \to \infty. \qquad (3.20)$$

The proof of relation (3.20) is relatively simple, but requires some accuracy. Denote by $\sigma_i(\xi)$ the asymptotic series of the form $\sum_{j=0}^{\infty} \xi^{-j} \sum_{l=0}^{j} a_{j, l} (\ln \xi)^l$, $\xi \to \infty$. (The subscript in σ_i will again be sometimes omitted.) Formula (3.16) means that $v_0(\xi) = \xi^{-1} \sigma(\xi)$, $\xi \to \infty$, and the principal term of the asymptotics is ξ^{-1} so that $[v_0(\xi)]^{-1} = \xi \sigma(\xi)$. However, this representation of $[v_0(\xi)]^{-1}$ is too rough. If it is used, extra terms containing too high degrees of $\ln \xi$ appear in the asymptotic expansion of $v_i(\xi)$. We will, therefore, make the asymptotics of $[v_0(\xi)]^{-1}$ more precise using relation (3.18). It follows immediately from (3.16) and (3.18) that

$$[v_0(\xi)]^{-1} = \xi + \gamma \ln \xi + \sigma(\xi), \qquad \xi \to \infty. \qquad (3.21)$$

Now consider equation (3.13) for $v_i(\xi)$ where $i > 0$. It is of the form

$$\frac{dv_i}{d\xi} = \frac{\partial f}{\partial u}(0, v_0(\xi))v_i + G_i(\xi),$$

where $G_i(\xi)$ is a known function belonging to $C^\infty[0, \infty)$ if all $v_i(\xi)$ for $j < i$ are determined. Equation (3.12) implies that the function $Z(\xi) = dv_0/d\xi$ is a solution of the homogeneous equation $dZ/d\xi = (\partial f/\partial u)(0, v_0(\xi))Z$. Therefore, for the solution of the equation (3.13) vanishing at $\xi = 0$, there is the explicit formula

$$v_i(\xi) = Z(\xi) \int_0^\zeta [Z(\theta)]^{-1} G_i(\theta)\, d\theta. \tag{3.22}$$

It remains to find the asymptotics of the functions $Z(\xi), [Z(\theta)]^{-1}$, $G_i(\theta)$ and integrate the resulting asymptotic series. Equation (3.12) and asymptotics (3.16), (3.17) imply that

$$Z(\xi) = \frac{dv_0}{d\xi} = f(0, v_0(\xi)) = \xi^{-2}\sigma(\xi), \qquad \xi \to \infty.$$

For the computation of the asymptotics $[Z(\xi)]^{-1}$ relation (3.21) can be used:

$$[Z(\xi)]^{-1} = [f(0, v_0(\xi)]^{-1} = -[v_0(\xi)]^{-2}\left(1 - \gamma v_0(\xi) - \sum_{k=4}^\infty c_k v_0^{k-2}(\xi)\right)^{-1}$$

$$= -[\xi + \gamma\ln\xi + \sigma(\xi)]^2[1 + \xi^{-1}\sigma(\xi)] = -[\xi^2 + 2\gamma\xi\ln\xi + \xi\sigma(\xi)], \quad \xi \to \infty.$$

For the first of the equations (3.13) the right-hand side is

$$G_1(\xi) = \frac{\partial f}{\partial x}(0, v_0(\xi))\xi = \left(1 + \sum_{k=1}^\infty c_k'[v_0(\xi)]^k\right)\xi = \xi + \sigma(\xi), \qquad \xi \to \infty.$$

Hence

$$[Z(\theta)]^{-1}G_1(\theta) = -\theta^3 - 2\gamma\theta^2\ln\theta + \theta^2\sigma(\theta), \qquad \theta \to \infty.$$

Integrating the resulting asymptotic series in (3.22), one obtains

$$v_1(\xi) = Z(\xi)\left[-\frac{1}{4}\xi^4 - \frac{2}{3}\gamma\xi^3\ln\xi + \xi^3\sigma(\xi) + c\ln^4\xi\right] = \xi^2\sigma_1(\xi).$$

Relation (3.20) is hereby proved for $i = 1$. The proof now proceeds by induction. One has to take into account that the principal term in the right-hand side of (3.13) is given by the expression

$$\frac{1}{2}\frac{\partial^2 f}{\partial u^2}(0, v_0(\xi))\sum_{j=1}^{i-1} v_j(\xi)v_{i-j}(\xi),$$

whereby the principal terms in the integrand of (3.22) are expressions of the form

$$[Z(\theta)]^{-1}\frac{\partial^2 f}{\partial u^2}(0, v_0(\theta))v_j(\theta)v_{i-j}(\theta)$$

$$= Z(\theta)\frac{\partial^2 f}{\partial u^2}(0, v_0(\theta))\tilde{v}_j(\theta)\tilde{v}_{i-j}(\theta) = v_0''(\theta)\tilde{v}_j(\theta)\tilde{v}_{i-j}(\theta),$$

where $v_i(\theta) = Z(\theta)\tilde{v}_i(\theta)$, and $\tilde{v}_i(\theta)$ is an expression of a special form, similar to the one obtained above for $\tilde{v}_1(\theta)$.

Thus, for the solution $u(x, \varepsilon)$ of the problem (3.1), (3.2), we have constructed the outer expansion (3.5), the inner expansion (3.11), and investigated the asymptotics of the coefficients as $x \to 0$ and $\xi \to \infty$, respectively. Following the prescriptions given above, it remains to verify the matching condition for the series U and V and construct the composite asymptotic expansion. Alas, it turns out that *the series* (3.5) *and* (3.11) *are not matched*. This is, for example, evident from the fact that the asymptotics (3.20) include the terms $(\ln \xi)^l$ which, expressed through x, turn into $(\ln x - \ln \varepsilon)^l$, while there is no $\ln x$ in the asymptotic (3.9). Moreover, the principal term in $\varepsilon^i v_i(\xi)$ as $\xi \to \infty$ equals $\varepsilon^i \xi^{3i-1}$ and, expressed in terms of x, turns into $\varepsilon^{1-2i} x^{3i-1}$, causing the appearance of negative and growing in absolute value powers of ε. Not only there is no domain in which both series U and V are asymptotic, but, e.g., for $x = \varepsilon^{2/3}$, both series forfeit their asymptotic character, and, therefore, neither of them can serve as an asymptotic expansion for the solution $u(x, \varepsilon)$.

This leads to the idea that *there should exist another scale and another asymptotic expansion of the solution* $u(x, \varepsilon)$ *in the domain intermediate with respect to the extreme values of* x *and extreme values of* ξ. And that is actually the case. In order to find the correct scales of the new asymptotic expansion, one has to take into account that equation (3.1) is nonlinear so that not only the change of scale of the independent variable is essential, but also that of the unknown function u. *The intermediate boundary layer* must, evidently, be located in the vicinity of the point $x = 0$. The solution $u_0(x)$ vanishes at this point, and $v_0(\xi)$ is also small for large ξ. One can assume, therefore, that in the intermediate layer the solution is small. One can, therefore, write equation (3.1) approximately, by replacing the function $f(x, u)$ with the principal terms of its Taylor series:

$$\varepsilon \frac{du}{dx} \approx x - u^2. \tag{3.23}$$

After the change of variables $u = \varepsilon^\alpha w$, $x = \varepsilon^\beta \eta$, one obtains the equality

$$\varepsilon^{1+\alpha-\beta} \frac{dw}{d\eta} \approx \varepsilon^\beta \eta - \varepsilon^{2\alpha} w^2.$$

We now have to equate the orders of the terms in this equality. Note that for the outer expansion (3.5) the principal terms in equation (3.23) are two terms in the right-hand side, while the inner expansion (3.11) meant equating the orders of $\varepsilon(du/dx)$ and u^2. The only remaining possibility is that $\varepsilon(du/dx)$ and x should be principal terms, i.e., $1+\alpha-\beta = \beta$, while the term u^2 must be of the same or lesser order, i.e., $2\alpha \geq \beta \Rightarrow \alpha \geq 1/3$. Practice shows that when there is arbitrariness in the choice of scale one should take the extreme values of the parameters. In this case we have $\beta = 2/3$, $\alpha = 1/3$, which corresponds to equality of orders of all three terms in (3.23).

Thus, we consider the change of variables

$$x = \varepsilon^{2/3}\eta, \qquad u(x, \varepsilon) \equiv \varepsilon^{1/3}w(\eta, \varepsilon). \tag{3.24}$$

The series for $w(\eta, \varepsilon)$ must have a more complicated form than a power series in $\varepsilon^{1/3}$. If we wish to satisfy the condition of matching the intermediate expansion to the inner expansion (3.11), then, taking into account the asymptotics (3.20), the powers of $\ln \varepsilon$ have to be included into the intermediate expansion. As usual, the gauge functions of the asymptotic expansion in the variable η are those functions in ε which appear in the expression $A_{m,\eta}A_{n,\xi}V$, where V is the series (3.11), and ξ and η are related by the change of variables (3.24) and $x = \varepsilon\xi$, i.e., $\eta = \varepsilon^{1/3}\xi$. It is now evident that the gauge functions are of the form $\varepsilon^{k/3}(\ln \varepsilon)^l$ where $k \geq 0$, $0 \leq l \leq k$ and the intermediate expansion has to be sought in the form

$$W = \sum_{k=0}^{\infty} \varepsilon^{k/3} \sum_{l=0}^{k} w_{k,l}(\eta)(\ln \varepsilon)^l. \tag{3.25}$$

Equations for $w_{k,l}(\eta)$ are obtained from the equation for w:

$$\varepsilon^{2/3}\frac{dw}{d\eta} = f(\varepsilon^{2/3}\eta, \varepsilon^{1/3}w)$$

on expanding the function $f(x, u)$ in a Taylor series:

$$\frac{dw_{0,0}}{d\eta} - \eta + w_{0,0}^2 = 0, \tag{3.26}$$

$$\frac{dw_{1,1}}{d\eta} + 2w_{0,0}w_{1,1} = 0,$$

$$\frac{dw_{1,0}}{d\eta} + 2w_{0,0}w_{1,0} - \eta\frac{\partial^2 f(0,0)}{\partial x\partial u}w_{0,0} - \frac{1}{3!}\frac{\partial^3 f(0,0)}{\partial u^3}w_{0,0}^3 = 0, \tag{3.27}$$

etc.

The equations for $w_{k,l}$ for $k > 1$ are simple:

$$\frac{dw_{k,l}}{d\eta} + 2w_{0,0}w_{k,l} - F_{k,l}(\eta) = 0, \tag{3.28}$$

where the function $F_{k,l}(\eta)$ depends on $w_{i,s}$ with lesser subscripts and is of a rather cumbersome form

$$F_{k,l}(\eta) = -\sum_{\substack{1 \leq j \leq k-1 \\ 0 \leq l \leq s}} w_{j,s}w_{k-j,l-s} + \sum_{q=3}^{\infty}\frac{1}{q!}\frac{\partial^q f}{\partial u^q}(0,0)\sum_{\substack{\sum i_j = k+2-q \\ \sum s_j = l}}\prod_{j=1}^{q}w_{i_j,s_j}$$

$$+ \sum_{\substack{p \geq 1, q \geq 0 \\ p+q \geq 2}}\frac{1}{p!q!}\frac{\partial^{p+q} f}{\partial x^p\partial u^q}(0,0)\eta^p\sum_{\substack{\sum i_j = k+2-q-2p \\ \sum s_j = l}}\prod_{j=1}^{q}w_{i_j,s_j}.$$

The functions $w_{k,l}(\eta)$ have to be considered for $0 < \eta < \infty$, and they should be chosen in such a way that the series (3.25) for $\eta \to 0$ is matched to the series (3.11) for $\xi \to \infty$, while for $\eta \to \infty$ it must be matched to the series (3.5) for $x \to 0$.

We begin, as always, with the first equation (3.26). On the change of variable

$$w_{0,0}(\eta) = \frac{g'(\eta)}{g(\eta)}$$

it reduces to the Airy equation (see, for example, [64, §5.17]) we have already met in §1. If $g(\eta) \neq 0$ then equation (3.26) is equivalent to the equation

$$\frac{d^2 g}{d\eta^2} - \eta g = 0. \tag{3.29}$$

Now we have to find the condition the function $w_{0,0}(\eta)$ must satisfy for $\eta \to 0$. The principal term of the series (3.11) is $v_0(\xi)$, while the principal term of the asymptotics of $v_0(\xi)$ as $\xi \to \infty$ is $\xi^{-1} = \varepsilon^{1/3}\eta^{-1}$. Therefore, the relation to be satisfied is

$$w_{0,0}(\eta) \sim \eta^{-1} \quad \text{for } \eta \to 0.$$

Clearly, it is sufficient to choose a solution of the Airy equation (3.29) such that $g(0) = 0$, $g'(0) = 1$. It follows from equation (3.29) that $g(\eta)$ is an increasing function so that $g(\eta) > 0$ for $\eta > 0$. Equation (3.29) also implies that the Taylor series for the function $g(\eta)$ is of the form $g(\eta) = \eta + \sum_{k=1}^{\infty} g_k \eta^{3k+1}$, $\eta \to 0$. Thus, the function is constructed. Evidently, it has the asymptotic expansion

$$w_{0,0}(\eta) = \eta^{-1}\left(1 + \sum_{j=1}^{\infty} b_{0,0,j}\eta^{3j}\right), \qquad \eta \to 0. \tag{3.30}$$

The remaining equations (3.26)–(3.28) are linear, and their solutions are defined to within the additive term $C \exp\left(-\int 2w_{0,0}(\eta)\,d\eta\right) = C[g(\eta)]^{-2}$. All these constants are determined by the matching condition for the series (3.25) and (3.11). The matching process repeats that of the preceding sections. Each term of the series (3.11) has to be replaced by its asymptotics (3.20) as $\xi \to \infty$, then the change of variable η according to formula (3.24) made, and the terms containing the same powers of ε and $\ln \varepsilon$ regrouped. The resulting equality is of the form

$$V = \varepsilon^{1/3} \sum_{k=0}^{\infty} \varepsilon^{k/3} \sum_{l=0}^{k} W_{k,l}(\eta)(\ln \varepsilon)^l,$$

where $W_{k,l}(\eta)$ are the following formal series:

$$W_{k,l}(\eta) = \eta^{-1-k} \sum_{j=0}^{\infty} \eta^{3j} \sum_{s=0}^{l} b_{k,l,j,s}(\ln \eta)^s. \tag{3.31}$$

The most convenient way of verifying these equalities is by arranging the coefficients of the asymptotic expansions into a table similar to Table 2. We do not give this table here. The following theorem holds.

THEOREM 3.1. *There is a solution of the system* (3.26)–(3.28) *such that each of the functions* $w_{k,l}(\eta)$ *for* $\eta \to 0$ *can be expanded in the asymptotic series* $W_{k,l}(\eta)$ *given by formula* (3.31). *Therefore, the matching conditions*

$$A_{m,\eta} A_{n,\xi} V = A_{n,\xi} A_{m,\eta} W$$

are satisfied for the series (3.25) *and* (3.11).

The proof of this theorem is achieved along the same lines as that of Theorem 2.3 but is more cumbersome because equation (3.1) is nonlinear. First, we prove two lemmas.

LEMMA 3.1. *Let* $w(\eta)$ *be a solution of the equation*

$$\frac{dw}{d\eta} = \eta - w^2 + \varphi(\eta), \tag{3.32}$$

where $\varphi(\eta) \in C^\infty(0, 1]$, $\varphi(\eta) = O(\eta^N)$, *and* $w(\eta) = \eta^{-1} + O(\eta^2)$ *as* $\eta \to +0$, $N > 0$. *Then* $w(\eta) - w_{0,0}(\eta) = O(\eta^{N+1})$ *as* $\eta \to +0$, *where* $w_{0,0}(\eta)$ *is the function constructed above.*

PROOF. Denote $z(\eta) = w(\eta) - w_{0,0}(\eta)$. It follows from equations (3.26) and (3.32) that $dz/d\eta + z(w_{0,0} + w) = \varphi(\eta)$. By the hypothesis, $w_{0,0}(\eta) + w(\eta) = 2\eta^{-1} + \varphi_1(\eta)$ where $\varphi_1(\eta) = O(\eta^2)$. Denote $\mu(\eta) = \exp \int_0^\eta \varphi_1(\theta)\,d\theta$. Multiplying the equation for z by the integrating factor $\eta^2 \mu(\eta)$, one has $(d/d\eta)(\eta^2\mu(\eta)z) = \eta^2\mu(\eta)\varphi(\eta)$ whence $\eta^2\mu(\eta)z(\eta) = \int_0^\eta \theta^2\mu(\theta)\varphi(\theta)\,d\theta + c$. Taking into account that $z(\eta) \to 0$ as $\eta \to 0$, we have $c = 0$ which implies the statement of the lemma. ∎

LEMMA 3.2. *Suppose that* $\overline{w}(\eta)$ *satisfies the equation* $d\overline{w}/d\eta + 2y(\eta)\overline{w} = \overline{F}(\eta)$, *where* $y(\eta)$, $\overline{F}(\eta) \in C^\infty(0, 1]$. *Also let* $F(\eta) \in C^\infty(0, 1]$, $\overline{F}(\eta) - F(\eta) = O(\eta^N)$, $\overline{w}(\eta) = O(\eta^{-k})$, $y(\eta) - w_{0,0}(\eta) = O(\eta^N)$ *as* $\eta \to +0$, $N > k+2 > 0$, *where* $w_{0,0}(\eta)$ *is the function constructed above. Then there exists a solution* $w(\eta) \in C^\infty(0, 1]$ *of the equation*

$$\frac{dw}{d\eta} + 2w_{0,0}(\eta)w = F(\eta)$$

such that $w(\eta) - \overline{w}(\eta) = O(\eta^{N-k+1})$ *as* $\eta \to +0$.

PROOF. Let $z(\eta) = w(\eta) - \overline{w}(\eta)$. This difference satisfies the equation

$$\frac{dz}{d\eta} + 2w_{0,0}(\eta)z + \varphi(\eta)\overline{w}(\eta) = \varphi_1(\eta),$$

where $\varphi(\eta) = 2w_{0,0}(\eta) - 2y(\eta) = O(\eta^N)$, $\varphi_1(\eta) = F(\eta) - \overline{F}(\eta) = O(\eta^N)$. Hence

$$\frac{dz}{d\eta} + 2w_{0,0}(\eta)z = \psi(\eta) = O(\eta^{N-k}).$$

Multiplying this equation by $\exp(-2\int_{\eta}^{1} w_{0,0}(\theta)\,d\theta)$ and integrating the resulting equality, one obtains

$$
z(\eta)\exp\left(-2\int_{\eta}^{1} w_{0,0}(\theta)\,d\theta\right) = \int_{0}^{\eta}\psi(\xi)\exp\left(-2\int_{\xi}^{1} w_{0,0}(\theta)\,d\theta\right)d\xi + c.
$$

The conclusion of the lemma now follows by letting $c = 0$ and taking into account the asymptotics (3.30) for $w_{0,0}$. ∎

PROOF OF THEOREM 3.1. First, we prove that the partial sums of the series (3.11) satisfy equation (3.1) approximately, i.e., that (3.11) is an f.a.s. of equation (3.1), or, which is the same, of equation (3.10) for $x \ll \varepsilon^{2/3}$. Denote the partial sum $A_{n,\xi}V$ by $v_0(\xi) + z_n(\xi, \varepsilon)$ and insert it into (3.10). The resulting equality is

$$
\frac{d}{d\xi}(A_{n,\xi}V) - f(\varepsilon\xi, A_{n,\xi}V) = R_n(\xi, \varepsilon), \tag{3.33}
$$

where one has to estimate the right-hand side

$$
R_n(\xi, \varepsilon) = \sum_{i=0}^{n}\varepsilon^{i}\frac{dv_i}{d\xi} - f(\varepsilon\xi, v_0(\xi) + z_n(\xi, \varepsilon)).
$$

Expanding the function $f(\varepsilon\xi, v_0(\xi) + z_n(\xi, \varepsilon))$ in a Taylor series at $(0, v_0(\xi))$ with the remainder written in the integral form, one can represent $R_n(\xi, \varepsilon)$ as the sum

$$
R_n(\xi, \varepsilon) = G_n(\xi, \varepsilon) + \rho_n(\xi, \varepsilon),
$$

where

$$
G_n(\xi, \varepsilon) = \sum_{i=0}^{n}\varepsilon^{i}\frac{dv_i}{d\xi} - \sum_{k=0}^{n}\frac{1}{k!}\varepsilon^{k}\xi^{k}\sum_{j=0}^{n-k}\frac{1}{j!}[z_n(\xi, \varepsilon)]^{j}\frac{\partial^{k+j}}{\partial x^{k}\partial u^{j}}f(0, v_0(\xi)),
$$

$$
\rho_n(\xi, \varepsilon) = -\frac{1}{n!}\varepsilon^{n+1}\xi^{n+1}\int_{0}^{1}(1-\theta)^{n}\frac{\partial^{n+1}f}{\partial x^{n+1}}(\theta\varepsilon\xi, v_0(\xi) + z_n(\xi, \varepsilon))\,d\theta
$$

$$
-\sum_{k=0}^{n}\frac{\varepsilon^{k}\xi^{k}}{k!}\frac{[z_n(\xi, \varepsilon)]^{n-k+1}}{(n-k)!}
$$

$$
\times\int_{0}^{1}(1-\theta)^{n-k}\frac{\partial^{n+1}}{\partial x^{k}\partial u^{n-k+1}}f(0, v_0 + \theta z_n(\xi, \varepsilon))\,d\theta.
$$

By virtue of the system (3.12), (3.13) all the coefficients of ε^{k} for $k \le n$ in $G_n(\xi, \varepsilon)$ vanish. (Actually, equations (3.26)–(3.28) have been obtained precisely by equating the coefficients at ε^{k} after expanding $f(\varepsilon\xi, V)$ in a Taylor series. The expression $R_n(\xi, \varepsilon)$ includes, instead of the series, partial sums up to but excluding the terms with ε^{n+1} and on.) Thus, $G_n(\xi, \varepsilon)$ is a linear combination of terms of the form

$$
(\varepsilon\xi)^{i}\prod_{s}\varepsilon^{j_s}v_{j_s}(\xi)\frac{\partial^{p}f}{\partial x^{p_1}\partial u^{p_2}}(0, v_0(\xi)), \tag{3.34}
$$

where $i + \sum_s j_s \geq n+1$. Since the asymptotics (3.20) for $v_j(\xi)$ implies the estimate $|v_j(\xi)| \leq M(1+\xi^{3j-1})$, and the function $v_0(\xi)$ is bounded, one obtains the following estimate for the products (3.34) and the entire sum $G_n(\xi, \varepsilon)$:

$$|G_n(\xi, \varepsilon)| \leq M\varepsilon^{n+1}\left(1 + \xi^{3(n+1)-1}\right).$$

If $\varepsilon\xi^3$ is bounded, so is $v_0(\xi) + \theta z_n(\xi, \varepsilon)$. Therefore, the same estimate holds for $\rho_n(\xi, \varepsilon)$ as for $G_n(\xi, \varepsilon)$. Thus

$$|R_n(\xi, \varepsilon)| \leq M\varepsilon^{n+1}(1 + \xi^{3(n+1)-1}) \quad \text{for } \varepsilon\xi^3 < \text{const} \tag{3.35}$$

and, therefore, (3.33) implies that (3.11) is an f.a.s. of equation (3.10) for $x \ll \varepsilon^{2/3}$.

We now have to verify that the series $W_{k,l}(\eta)$ constructed above (formula (3.31)) are f.a.s. as $\eta \to 0$ of equations (3.26)–(3.28). The construction of $W_{k,l}(\eta)$ implies that

$$A_{m,\eta}A_{n\xi}V = \varepsilon^{1/3}\sum_{k=0}^{3m-1}\varepsilon^{k/3}\sum_{l=0}^{k}B_{3n-k-1}W_{k,l}(\eta)(\ln\varepsilon)^l,$$

where B_jW denotes, as before, the partial sum of the series up to and including η^j (cf. the similar formula in the proof of Theorem 2.2). Applying the operator $A_{m,\eta}$ to both sides of (3.33), one has

$$A_{m,\eta}\left(\frac{d}{d\xi}A_{n,\xi}V\right) = \varepsilon^{2/3}\sum_{k=0}^{3m-1}\varepsilon^{k/3}\sum_{l=0}^{k}(\ln\varepsilon)^l\frac{d}{d\eta}B_{3n-k-1}W_{k,l}(\eta).$$

The result of applying $A_{m,\eta}$ to $f(\varepsilon\xi, A_{n,\xi}V)$ is slightly more complicated to compute. One has to expand the function $f(\varepsilon\xi, A_{n,\xi}V) = f(\varepsilon^{2/3}\eta, A_{n,\xi}V)$ in an asymptotic series for $\varepsilon \to 0$ and constant η. It is, however, clear that this process coincides with the formal procedure leading to equations (3.26)–(3.28). Thus $A_{mn}((d/d\xi)A_{n,\xi}V - f(\varepsilon\xi, A_{n,\xi}V))$ is a linear combination of the terms $\varepsilon^{k/3}\ln^l\varepsilon$, where $k \leq 3m$, and the coefficients of these are given by the left-hand sides of equations (3.26)–(3.28) with $w_{k,l}(\eta)$ replaced by $B_{3n-k-1}W_{k,l}$.

We now have to apply the operator $A_{m,\eta}$ to the sum $R_n(\xi, \varepsilon) = G_n(\xi, \varepsilon) + \rho_n(\xi, \varepsilon)$. The function $G_n(\xi, \varepsilon)$ is a linear combination of the expressions (3.34), where each of the terms has to be subjected to the change of variable $\xi = \varepsilon^{-1/3}\eta$, then expanded in a series as $\varepsilon \to 0$ omitting all terms except $\varepsilon^{k/3}\ln^l\varepsilon$ for $k \leq 3m$. The asymptotics (3.20) imply that

$$A_{m,\eta}G_n(\xi, \varepsilon) = \varepsilon^{1/3}\sum_{k=1}^{3m-1}\varepsilon^{k/3}\sum_{l=0}^{k}\varphi_{k,l}(\eta)(\ln\varepsilon)^l,$$

where $\varphi_{k,l}(\eta) = O(\eta^{3n-1-k})$. One can easily see that $A_{m,\eta}\rho(\xi, \varepsilon)$ is of the same form. Equating in the resulting equality the coefficients of the same powers ε^k and $\ln^l\varepsilon$ one obtains the system of equations similar to (3.26)–(3.28). The only difference is that $w_{k,l}(\eta)$ in the left-hand side is replaced with $B_{3n-k-1}W_{k,l}(\eta)$, and the zeros in the right-hand side are replaced by $O(\eta^{3n-k-1})$. The conclusion of Theorem 3.1 is now obtained by consecutively applying Lemmas 3.1 and 3.2 to these equations. ∎

The coefficients of the series (3.25), i.e., the functions $w_{k,l}(\eta)$, are thus constructed for all $\eta \geq 0$. It remains to verify that the series (3.25) for

$\eta \to \infty$ is matched to the series (3.5) for $x \to 0$. First, one has to investigate the behavior of the functions $w_{k,l}(\eta)$ as $\eta \to \infty$.

THEOREM 3.2. *The functions* $w_{k,0}(\eta)$ *constructed above and solving the system* (3.26)–(3.28) *admit the following asymptotic expansions*

$$w_{k,0}(\eta) = \eta^{(k+1)/2} \sum_{j=0}^{\infty} h_{k,j} \eta^{-3j/2}, \qquad \eta \to \infty, \, h_{0,0} = 1, \tag{3.36}$$

and these series can be repeatedly differentiated term-by-term.

For $l > 0$, $s \geq 0$, *one has*

$$\left| \frac{d^s}{d\eta^s} w_{k,l}(\eta) \right| \leq M \exp(-\eta^{3/2}). \tag{3.37}$$

PROOF. For $k = 0$, the asymptotics (3.36) follows from the explicit representation of the function $w_{0,0}(\eta) = g'(\eta)[g(\eta)]^{-1}$, where $g(\eta)$ is the monotonically increasing Airy function. Its asymptotics as $\eta \to \infty$ is known (it is of the same form as the asymptotics of the function $Y_1(\xi)$ in Example 5, §1):

$$g(\eta) = C\eta^{-1/4} \exp\left(\frac{2}{3}\eta^{3/2}\right) \left\{ 1 + \sum_{j=1}^{\infty} a_j \eta^{-3j/2} \right\}, \qquad \eta \to \infty,$$

which immediately implies the asymptotics for the function $w_{0,0}(\eta)$.

The function $w_{1,1}(\eta)$ is expressed explicitly from the first equation (3.27):

$$w_{1,1}(\eta) = C \exp\left\{ -2 \int g'(\theta)[g(\eta)]^{-1} d\theta \right\} = C_1 [g(\eta)]^{-2}$$

$$= \eta^{1/2} O\left(\exp\left(-\frac{4}{3}\eta^{3/2} \right) \right), \qquad \eta \to \infty.$$

Each of the functions $w_{k,l}(\eta)$ for $k \geq 1$ satisfies equation (3.28), where $F_{k,l}(\eta)$ is a polynomial in η and $w_{i,j}$ for $i < k$. If $l > 0$ then each of the terms $F_{k,l}(\eta)$ includes at least one factor $w_{i,j}(\eta)$ for $j > 0$. The estimate (3.37) now follows by induction. The asymptotics (3.36) for the functions $w_{k,0}(\eta)$ also follows by induction from the explicit formula for the solution of equation (3.28). ■

THEOREM 3.3. *The series* (3.5) *and* (3.25) *satisfy the matching condition*

$$A_{n,x} A_{m,\eta} \varepsilon^{1/3} W = A_{m,\eta} A_{n,x} U \tag{3.38}$$

for all positive integers m *and* n.

The proof of this theorem follows the same scheme as that of Theorem 3.1. First, two lemmas, almost repeating Lemmas 3.1 and 3.2, are proved. The only difference is that, instead of the asymptotics as $\eta \to 0$, one considers the asymptotics as $\eta \to \infty$, and the words "there exists a solution $w(\eta)$ ... such that" in the statement of Lemma 3.2 must be replaced with the words "for any solution $w(\eta)$ ... one has" because the solution of the homogeneous equation decays exponentially at infinity.

Next one proves that the partial sums of the series (3.5) satisfy equation (3.1) approximately, viz.,

$$\varepsilon \frac{d}{dx} A_{n,x} U - f(x, A_{n,x}U) = O(\varepsilon^{n+1} x^{-(3n+1)/2}). \tag{3.39}$$

It is simultaneously verified that the formal series for $w_{k,0}(\eta)$ for $\eta \to \infty$ obtained from the sums $A_{n,x}U$ are the f.a.s. of the system (3.26)–(3.28) for $l = 0$ under the assumption that $w_{i,s} \equiv 0$ for $s < 0$. This together with the lemmas mentioned above imply the matching condition (3.38) because, for $l > 0$, the functions $w_{k,l}(\eta)$ decay exponentially at infinity. The detailed elaboration of this proof is a good exercise for the reader interested in justification of asymptotic expansions. ∎

For a convenient overview we now write the series U, V, and W constructed above, as well as the asymptotics of their coefficients. We denote by $P_j(z)$ polynomials in z whose degree does not exceed j, by $s(z)$ asymptotic series of the form $\sum_{k=0}^{\infty} a_k z^k$, $z \to 0$, and by $\bar{s}(z)$ asymptotic series of the form $\sum_{k=0}^{\infty} a_k z^k P_k(\ln z)$, $z \to 0$.

Thus,

$$U = \sum_{k=0}^{\infty} \varepsilon^k u_k(x), \qquad u_k(x) = x^{(1-3k)/2} s_k(x), \qquad x \to 0,$$

$$V = \sum_{k=0}^{\infty} \varepsilon^k v_k(\xi), \qquad v_k(\xi) = \xi^{3k-1} \bar{s}_k(\xi), \qquad \xi \to \infty,$$

$$W = \sum_{k=0}^{\infty} \varepsilon^{k/3} \sum_{l=0}^{k} w_{k,l}(\eta) \ln^l \varepsilon,$$

$$w_{k,l}(\eta) = \eta^{-1-k} \sum_{j=0}^{l} s_{k,l,j}(\eta^3)(\ln \eta)^j, \qquad \eta \to 0,$$

$$w_{k,l}(\eta) = \eta^{(k+1)/2} s_k(\eta^{-3/2}), \qquad \eta \to \infty,$$

where $\eta = \varepsilon^{1/3}\xi = \varepsilon^{-2/3}x$.

THEOREM 3.4. *For all sufficiently small $\varepsilon > 0$ there exists a solution $u(x, \varepsilon)$ of the problem (3.1), (3.2). The series U, $\varepsilon^{1/3}W$, and V are asymptotic expansions of the solution $u(x, \varepsilon)$ as $\varepsilon \to 0$ in the domains where these series preserve their asymptotic character, i.e., for $x \gg \varepsilon^{2/3}$, for $\varepsilon \ll x \ll 1$, and for $x \ll \varepsilon^{2/3}$, respectively.*

Figure 8 (next page) shows the graphs of the functions $u(x, \varepsilon)$, $\varepsilon^{1/3} w_{0,0}(\eta)$ and $v_0(\xi)$.

THEOREM 3.5. *The following estimate holds:*

$$|u(x, \varepsilon) - z_N(x, \varepsilon)| \le M \varepsilon^{\gamma N} \qquad \forall x \in [0, d], \tag{3.40}$$

where $u(x, \varepsilon)$ is a solution of the problem (3.1), (3.2), γ a positive number, and

$$z_N(x, \varepsilon) = A_{N,x}U + A_{N,\xi}V + A_{N,\eta}(\varepsilon^{1/3}W) - A_{N,\eta}(A_{N,x}U) - A_{N,\eta}(A_{N,\xi}V).$$

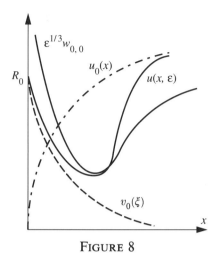

FIGURE 8

Theorem 3.4 follows immediately from Theorem 3.5 taking into account the matching condition for the series U, V, $\varepsilon^{1/3}W$ and the asymptotics of their coefficients.

PROOF OF THEOREM 3.5. First we have to verify that the composite asymptotic expansion $z_N(x, \varepsilon)$ approximately satisfies equation (3.1) in the closed interval $[0, d]$. To achieve that, one has to consider the function $z_N(x, \varepsilon)$ in three sections of the closed interval: for $x \le \varepsilon^{3/4}$, $\varepsilon^{3/4} \le x \le \varepsilon^{1/2}$, and $\varepsilon^{1/2} \le x \le 1$. In each of these sections $z_N(x, \varepsilon)$ approximately equals $A_{N,\xi}V$, $A_{N,\eta}(\varepsilon^{1/3}W)$, and $A_{N,x}U$, respectively, while, by virtue of the matching conditions (Theorems 3.1 and 3.3), the sum of the remaining terms is small. (The numbers 3/4 and 1/2 can be replaced by any pair of numbers α and β provided $2/3 < \alpha < 5/6$, $0 < \beta < 2/3$.) On the other hand, the construction procedure for the series and the estimates of their coefficients imply that the partial sums of these series approximately satisfy equation (3.1) (see (3.33), (3.35), (3.38)) so that for sufficiently large N one has

$$\varepsilon \frac{dz_N}{dx} - f(x, z_N) = \varphi_N(x, \varepsilon), \quad \text{where } |\varphi_N(x, \varepsilon)| < M\varepsilon^{\gamma_1 N}, \; \gamma_1 > 0. \quad (3.41)$$

The precise value of γ_1 can be established without difficulty but is totally irrelevant for what follows. Let $y_N(x, \varepsilon) = z_N(x, \varepsilon) - u(x, \varepsilon)$. As the existence of a solution for the problem (3.1), (3.2) on the entire closed interval $[0, d]$ has not yet been proved, the function $y_N(x, \varepsilon)$ is defined only for those values of x for which the solution $u(x, \varepsilon)$ exists. It follows from the general theorem on the existence of a solution for a Cauchy problem that the function $u(x, \varepsilon)$ exists for $x \in [0, \lambda]$, where the constant $\lambda > 0$, in general, depends on ε. The solution $u(x, \varepsilon)$ can be extended to the whole closed interval $[0, d]$ if it is uniformly bounded. Let us consider the function $y_N(x, \varepsilon)$ in the closed interval $[0, \lambda(\varepsilon)]$ and prove that $y_N(x, \varepsilon)$ is bounded (and even small). This immediately implies that the solution $u(x, \varepsilon)$ can be extended to the whole closed interval $[0, d]$. Subtracting, term-by-term, equation (3.1) from equation (3.41), we obtain

$$\varepsilon \frac{dy_N}{dx} - y_N \frac{\partial f}{\partial u}(x, z_N - \theta y_N) = \varphi_N(x, \varepsilon), \qquad 0 < \theta < 1.$$

Since, by construction, $A_{N,\xi}V|_{\xi=0} = u(0,\varepsilon) = R_0$ one has $y_N(0,\varepsilon) = O(\varepsilon^{\gamma N})$. Integrating the resulting equation for y_N, and setting

$$\beta_N(x,\varepsilon) = \frac{\partial f}{\partial u}(x, z_N(x,\varepsilon) - \theta y_N(x,\varepsilon)),$$

one obtains the following formula for y_N:

$$y_N(x,\varepsilon) = \varepsilon^{-1} \int_0^x \varphi_N(\theta,\varepsilon) \exp\left(\varepsilon^{-1} \int_\theta^x \beta_N(\zeta,\varepsilon)\,d\zeta\right) d\theta + O(\varepsilon^{\gamma N}). \qquad (3.42)$$

The conclusion of the theorem is a direct consequence of this formula and the estimate for $\varphi_N(x,\varepsilon)$ if only $\beta_N(x,\varepsilon) \le 0$.

One has to realize, however, that $\beta_N(x,\varepsilon)$ depends, in its turn, on the functions $z_N(x,\varepsilon)$ and $y_N(x,\varepsilon)$ whereby the verification of the fact that β_N is nonpositive requires some effort. First, we have to bound the function $z_N(x,\varepsilon)$ from below. For $x \le \varepsilon^{3/4}$, one has

$$A_{N,\xi}V \ge (\xi + M)^{-1} - M\sum_{k=1}^N \varepsilon^k(1 + \xi^{3k-1}) \ge M\varepsilon^{1/4}.$$

(As usual, we denote by M positive constants depending neither on x nor on ε, and assume that ε is sufficiently small.)

For $\varepsilon^{3/4} \le x \le \varepsilon^{1/2}$, i.e., for $\varepsilon^{1/12} \le \eta \le \varepsilon^{-1/6}$, one has

$$A_{N,\eta}(\varepsilon^{1/3}W) \ge \varepsilon^{1/3}\left\{ w_0(\eta) - M\sum_{k=1}^{3N-1} \varepsilon^{k/3}|\ln\varepsilon|^k(\eta^{-1-k} + \eta^{(k+1)/2})\right\} \ge M\varepsilon^{1/3}.$$

Finally, for $x \ge \varepsilon^{1/2}$, we have

$$A_{N,x}U \ge u_0(x) - M\sum_{k=1}^N \varepsilon^k x^{(1-3k)/2} \ge u_0(x) - M\varepsilon^{1/2}.$$

Hence, $z_N(x,\varepsilon) \ge M\varepsilon^{1/4}$ for $x \le \varepsilon^{1/2}$ and $z_N(x,\varepsilon) \ge u_0(x) - M\varepsilon^{1/2}$ for $x \ge \varepsilon^{1/2}$. These inequalities together with conditions (3.3), (3.4) imply that if $|z_N(x,\varepsilon) - u(x,\varepsilon)| < M\varepsilon$ then

$$\frac{\partial f}{\partial u}(x, z_N(x,\varepsilon) - \theta(z_N(x\varepsilon) - u(x,\varepsilon))) \le 0$$

for $0 < \theta < 1$. Therefore, for those x_0 for which $|y_N(x,\varepsilon)| \le M\varepsilon$ holds for $0 \le x \le x_0$, relation (3.42) yields $|y_N(x,\varepsilon)| \le M\varepsilon^{\gamma_1 N-1}$. For sufficiently large N, this inequality implies (3.40). ∎

Since the estimate (3.40) holds for all sufficiently large N, it follows from the explicit form of the series U, V and W that in this inequality N can be any positive integer, and one can set $\gamma = 1/4$.

We conclude with the lemma describing the asymptotics of the function $u_0(x)$ as $x \to 0$ promised at the beginning of this section.

LEMMA 3.3. *Let conditions* (3.3), (3.4) *be satisfied. Then the asymptotics* (3.6) *is valid for the nonnegative solution $u_0(x)$ of the equation $f(x, u_0(x)) = 0$.*

PROOF. Make the following change of the unknown function and the independent variable: $u_0(x) \equiv yw(y)$, $y = \sqrt{x}$. Then the equation for $u_0(x)$ is equivalent to the equation $y^{-2}f(y^2, yw(y)) = 0$. Denote the left-hand side of this equality by $F(y, w(y))$. It follows from the Taylor expansion for $f(x, u)$ that $F(y, w)$ is infinitely differentiable for $w \in \mathbf{R}^1$, $y \in [0, d^{1/2}]$. We also have $F(0, 1) = 0$, and $(\partial F/\partial w)(0, 1) = -2$. The implicit function theorem now implies that $w(y) \in C^\infty[0, d^{1/2}]$. Returning to the variables x, $u_0(x)$ one obtains the conclusion of the lemma. ■

Singular Perturbations of the Domain Boundary in Elliptic Boundary Value Problems

Let Ω be a bounded domain in $\mathbf{R}^m$ in which a linear elliptic differential operator $\mathscr{L}$ with smooth coefficients is defined. If the boundary $\partial\Omega$ is smooth and undergoes a smooth deformation, then, in general, the solution of the boundary value problem for the operator $\mathscr{L}$ in Ω also changes in a smooth manner. However, other deformations of the boundary, so-called *singular* ones, are also of interest. Suppose, for example, that a component of the boundary $\partial\Omega$ is an $(m-1)$-dimensional sphere of radius ε. What is the behavior of the solution as $\varepsilon \to 0$, i.e., in the process of disappearance of this component of the boundary? The sphere may, of course, be replaced with any other surface having a small diameter. Another example: suppose that the boundary of the limit domain Ω_0 has singularities such as corners, edges, cuts, conic points, etc., and the domain Ω_ε is obtained by smoothing these singularities. In all these cases, the boundary value problems for elliptic equations are bisingular.

This chapter considers only the first boundary value problem for a second order operator in the domain obtained by deleting a "small" subdomain from a fixed domain Ω.

Thus, Ω is a bounded domain in $\mathbf{R}^m$ with smooth boundary $\partial\Omega$. Denote by Ω_ε the domain $\Omega\backslash\overline{\omega}_\varepsilon$, where ω_ε is an interior subdomain which, as $\varepsilon \to 0$, contracts to a set σ of measure zero (e.g., a point, a smooth curve, etc.). Our goal is to find the asymptotics as $\varepsilon \to 0$ of the solution $u(x, \varepsilon)$ of the following problem:

$$\mathscr{L}u = f(x), \qquad x \in \Omega_\varepsilon, \tag{0.1}$$

$$u(x, \varepsilon) = \varphi(x) \quad \text{for } x \in \partial\Omega_\varepsilon. \tag{0.2}$$

Here $\mathscr{L}$ is a linear second order elliptic operator, $f(x) \in C^\infty(\overline{\Omega})$, and the function $\varphi(x)$ is assumed to be smooth everywhere in $\overline{\Omega}$. The boundary $\partial\Omega_\varepsilon$ evidently consists of two parts: $\partial\Omega$ and $\partial\omega_\varepsilon$.

One can naturally expect that $u_0(x) = \lim_{\varepsilon\to 0} u(x, \varepsilon)$ is the solution of equation (0.1) satisfying the condition $u|_{x\in\partial\Omega} = \varphi(x)$. Whether after the limit transition the boundary condition (0.2) holds on σ or not, depends on the dimension k of the set σ. For example, it is known (see [82, Chapter 4]) that for $m = 2$ no boundary condition can be imposed if σ is a point, but it is perfectly possible if σ is a straight line segment.

Thus, it is reasonable to assume that $u_0(x)$ equals $\varphi(x)$ on σ whenever the corresponding problem has a solution ($m - k = 1$). Otherwise ($m - k \geq 2$) one can expect that $u_0(x)$ is a smooth function in Ω whose restriction to σ has no relation to $\varphi(x)$. This statement is indeed valid if the solution of the limit problem $\mathscr{L} u_0 = f(x)$ with the given boundary conditions exists. Leaving the other possibility aside for the time being, i.e., that there is no such limit solution $u_0(x)$ (this alternative will be discussed in §4), we will look for a sharper approximation to the solution $u(x, \varepsilon)$. An attempt to find the next (after $u_0(x) = \lim_{\varepsilon \to 0} u(x, \varepsilon)$) terms of the asymptotic expansion of the function leads to a situation similar to that considered in the foregoing chapter. The following terms of the asymptotic expansion on any compact set in $\overline{\Omega} \backslash \sigma$ are functions of the form $\varepsilon^k u_k(x)$. The functions $u_k(x)$ are defined and smooth everywhere in $\overline{\Omega} \backslash \sigma$, but, in general, have singularities on σ. The orders of these singularities grow with k so that the problem is again bisingular.

The uniform asymptotics of the solution $u(x, \varepsilon)$ is constructed with the help of this outer expansion and the inner expansion valid in a neighborhood of σ. Here the method of matched asymptotic expansions works to its full capacity: the coefficients of both inner and outer expansions are not defined from their respective boundary value problems uniquely, and only the matching procedure makes it possible to completely define both asymptotic expansions. Various typical examples of these problems are considered in §§1–4. Brief remarks on higher order equations and other generalizations can be found at the end of the book.

§1. Three-dimensional problem in a domain with a small cavity

1. The Laplace equation. Consider a bounded domain $\Omega \subset \mathbf{R}^3$ with smooth boundary $\partial \Omega$. Suppose that Ω contains the origin O. The aim of this section is to study the asymptotics of the solution of the Laplace equation $\Delta u = 0$ in the domain obtained by deleting from Ω a small neighborhood of the point O (see Figure 9). A more precise description of this neighborhood can be given as follows. Let ω be a bounded domain with smooth boundary containing the origin and suppose that the complement of ω is connected (e.g., ω is a unit ball, or a torus). Denote by ω_ε the domain obtained from ω by a contraction with coefficient ε^{-1}. In other words, $x \in \omega_\varepsilon \Leftrightarrow \varepsilon^{-1} x \in \omega$. We assume that $\varepsilon > 0$ is a small parameter and denote $\Omega_\varepsilon = \Omega \backslash \overline{\omega}_\varepsilon$ whereby $\partial \Omega_\varepsilon = \partial \Omega \cup \partial \omega_\varepsilon$. Thus, $u(x, \varepsilon) \in C(\overline{\Omega}_\varepsilon)$ is the solution of the equation

$$\Delta u = 0 \quad \text{for } x \in \Omega_\varepsilon, \tag{1.1}$$

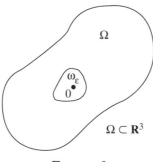

FIGURE 9

satisfying the boundary conditions

$$u(x, \varepsilon) = \varphi(x) \quad \text{for } x \in \partial\Omega, \tag{1.2}$$

$$u(x, \varepsilon) = 0 \quad \text{for } x \in \partial\omega_\varepsilon, \tag{1.3}$$

$$x = (x_1, x_2, x_3), \qquad \varphi(x) \in C^\infty(\partial\Omega).$$

We will look for the asymptotics of the solution $u(x, \varepsilon)$ as $\varepsilon \to 0$ using the method of matching the outer and inner asymptotic expansions. It is natural to look for the outer expansion in the form

$$U = \sum_{k=0}^{\infty} \varepsilon^k u_k(x). \tag{1.4}$$

Inserting this series into equation (1.1) and condition (1.2) shows that $u_0(x)$ is the solution of the limit problem (1.1), (1.2) defined not only in Ω_ε but everywhere in $\overline{\Omega}$ so that $u_0(x) \in C^\infty(\Omega) \cap C(\overline{\Omega})$. The remaining functions $u_k(x)$ are harmonic and vanish on $\partial\Omega$. If a function of this kind is smooth everywhere in Ω, it vanishes identically. Therefore, for $k > 0$, *the functions $u_k(x)$ must have singularities at the origin O.* What is the form of these singularities? A priori this is by no means clear. The behavior of the functions $u_k(x)$ at the origin, and, therefore, the definition of the functions $u_k(x)$ can be analyzed only after matching the series (1.4) to the inner expansion. Thus,

$$\Delta u_k(x) = 0 \quad \text{for } x \in \Omega \backslash O, \qquad u_k(x) = 0 \quad \text{for } x \in \partial\Omega. \tag{1.5}$$

The form of the inner variables is clear. They have to be chosen in such a way that, in the new variables, the boundary $\partial\omega_\varepsilon$ does not depend on ε. Therefore, $x = \varepsilon\xi$, and one can look for the inner expansion in the form

$$V = \sum_{i=0}^{\infty} \varepsilon^i v_i(\xi). \tag{1.6}$$

We assume that the functions $v_i(\xi)$ are defined not just in the domain obtained by stretching Ω_ε, but everywhere in $\mathbf{R}^3 \backslash \omega$. Equation (1.1) and condition (1.3) are then transformed into the equations

$$\Delta v_i = 0 \quad \text{for } \xi \in \mathbf{R}^3 \backslash \omega \tag{1.7}$$

and conditions

$$v_i(\xi) = 0 \quad \text{for } \xi \in \partial\omega. \tag{1.8}$$

Evidently, *the functions $v_i(\xi)$ are not defined uniquely,* and the question now arises how to choose them. Thus, for the problem (1.1)–(1.3) *there is arbitrariness in the choice of the coefficients of both the outer and the inner asymptotic expansions.*

In order to determine the functions $u_k(x)$ and $v_i(\xi)$, one has to find the general form of solutions of the problems (1.5) and (1.7),(1.8), and examine the degree of arbitrariness in these solutions. The matching conditions for the series (1.4) and (1.6) imply that the singularities of the functions $u_k(x)$ at the origin and those of the functions $v_i(\xi)$ at infinity can be of power type only. To be more precise,

$$|u_k(x)| < M_k r^{-k}, \tag{1.9}$$

$$|v_i(\xi)| < M_i \rho^i. \tag{1.10}$$

(Here and in what follows the notation $r = |x|$, $\rho = |\xi|$ is used.) If inequality (1.9) were not satisfied for some k, the function $\varepsilon^k u_k(x)$, after the change of variable $x = \varepsilon\xi$, would not be bounded for bounded ξ, and, therefore, the series (1.4) would not be matched to the series (1.5). A similar argument yields inequalities (1.10). Harmonic functions satisfying conditions (1.9), (1.10) are well known. The following statements hold.

(a) Suppose that the function $u(x)$ is harmonic in the domain $\Omega\backslash O$ and satisfies the estimate (1.9) for some $k > 0$. Then $u(x) = \tilde{u}(x) + y(x)$, where $\tilde{u}(x)$ is a harmonic function which is smooth everywhere in Ω, and $y(x)$ is a linear combination of the function r^{-1} and its derivatives up to the order $k - 1$.

(b) Suppose that the function $v(\xi)$ is harmonic in the domain $\mathbf{R}^3\backslash\omega$ and satisfies estimate (1.10) for some $i \geq 0$. Then $v(\xi) = \tilde{v}(\xi) + Y(\xi)$, where $\tilde{v}(\xi)$ is a harmonic function in $\mathbf{R}^3\backslash\omega$, $\tilde{v}(\xi) \to 0$ as $\rho \to \infty$, and $Y(\xi)$ is a harmonic polynomial (i.e., $\Delta Y = 0$) the degree of which does not exceed i.

The statements (a) and (b) provide a useful insight into the structure of increasing harmonic functions but are not required for our immediate purposes. To construct the asymptotics of the solution of the problem (1.1)–(1.3) reverse statements are needed. Before formulating these statements we note that r^{-1} is a harmonic function, and its derivative of order l is of the form $X_l(x)r^{-2l-1}$, where $X_l(x)$ is a homogeneous harmonic polynomial of degree l (i.e., $\Delta X_l = 0$, $X_l(\alpha x) = \alpha^l X_l(x)$ $\forall\alpha$). The fact that X_l is a homogeneous polynomial is an easy consequence of differentiation formulas, while the fact that it is harmonic can be deduced, for example, with the use of the inversion transformation: $x = \overline{x}|\overline{x}|^{-2}$. Since $g(x) = X_l(x)r^{-2l-1}$ is a harmonic function, it is known (see, for example, [118, Chapter 4, §1]) that $|\overline{x}|^{-1}g(\overline{x}|\overline{x}|^{-2}) = |\overline{x}|^{-1}X_l(\overline{x}|\overline{x}|^{-2}) \cdot |\overline{x}|^{2l+1} = X_l(\overline{x})$ is also a harmonic function. Conversely, for each homogeneous harmonic polynomial $X_l(x)$ of degree l the function $X_l(x)r^{-2l-1}$ is a harmonic one.

In this section we denote by $X_l(x)$, $Y_l(x)$, $Z_l(x)$, $W_l(x)$, or $X_{l,j}(x)$, $Y_{l,j}(x)$ etc. homogeneous polynomials of degree l.

In what follows the following statements are needed.

LEMMA 1.1. *Suppose that* $z(x)$ *is a linear combination of functions of the form* $X_l(x)r^{-2l-1}$. *Then there exists a function* $u(x) = \tilde{u}(x) + z(x)$ *such that* $\tilde{u}(x) \in C^{\infty}(\Omega) \cap C(\overline{\Omega})$, $\Delta u(x) = 0$ *for* $x \in \Omega \backslash O$, $u(x) = 0$ *for* $x \in \partial\Omega$.

LEMMA 1.2. *Suppose that* $v(\xi)$ *is a harmonic function in a neighborhood of infinity* ($\xi \in \mathbf{R}^3$), *and* $v(\xi) \to 0$ *as* $\rho \to \infty$. *Then the asymptotic expansion*

$$v(\xi) = \sum_{j=0}^{\infty} X_j(\xi)\rho^{-2j-1} \qquad (1.11)$$

holds.

LEMMA 1.3. *Let* $Y(\xi)$ *be a harmonic polynomial. Then there exists a function* $v(\xi) = \tilde{v}(\xi) + Y(\xi)$ *such that* $v(\xi) \in C^{\infty}(\mathbf{R}^3\backslash\overline{\omega}) \cap C(\mathbf{R}^3\backslash\omega)$, $\Delta v(\xi) = 0$ *for* $\xi \in \mathbf{R}^3\backslash\omega$, $v(\xi) = 0$ *for* $\xi \in \partial\omega$, *and for the function* $\tilde{v}(\xi)$ *the asymptotic expansion* (1.11) *holds.*

PROOF OF LEMMA 1.1 reduces to the construction of a harmonic function $\tilde{u}(x)$ because $z(x)$ is harmonic by the hypothesis. For $\tilde{u}(x)$ one evidently has to take the solution of the boundary value problem $\Delta\tilde{u}(x) = 0$ for $x \in \Omega$, $\tilde{u}(x) = -z(x)$ for $x \in \partial\Omega$, $\tilde{u}(x) \in C(\overline{\Omega})$. Such a solution (see [100, Chapter 3]) is known to exist. ∎

PROOF OF LEMMA 1.2. The asymptotic expansion (1.11) can, for example, be obtained by representing $v(\xi)$ in the form of a surface-distribution potential. It is, however, simpler to use the inversion transformation. The function $g(\overline{\xi}) = |\overline{\xi}|^{-1}v(\overline{\xi}|\overline{\xi}|^{-2})$ is smooth and harmonic in the neighborhood of the origin; here $\overline{\xi} = \xi\rho^{-2}$. By expanding this function in a Taylor series, one obtains the relation $g(\overline{\xi}) = \sum_{j=0}^{\infty} X_j(\overline{\xi})$. Going back to the variable ξ, one has (1.11). ∎

PROOF OF LEMMA 1.3. It is sufficient to construct a harmonic function $\tilde{v}(\xi)$ such that $\tilde{v}(\xi) \to 0$ as $\rho \to \infty$, $\tilde{v}(\xi) = -Y(\xi)$ for $\xi \in \partial\omega$. The problem is known (see [118], [61]) to have a solution, and $\tilde{v}(\xi) \in C^{\infty}(\mathbf{R}^3\backslash\overline{\omega}) \cap C(\mathbf{R}^3\backslash\omega)$. ∎

One can now proceed to the actual construction of the asymptotic expansions (1.4) and (1.6). Thus, $u_0(x)$ is the solution of the boundary value problem

$$\Delta u_0(x) = 0 \quad \text{for } x \in \Omega, \qquad u_0(x) = \varphi(x) \quad \text{for } x \in \partial\Omega.$$

By expanding the function $u_0(x)$ in a Taylor series, one obtains the asymptotic expansion

$$u_0(x) = \sum_{j=0}^{\infty} X_{j,0}(x), \qquad x \to 0.$$

For $k > 0$ define the functions $u_k(x)$ in accordance with Lemma 1.1, by the formula $u_k(x) = \tilde{u}_k(x) + z_k(x)$, where $z_k(x) = \sum_{j=0}^{k-1} Z_{j,x}(x) r^{-2j-1}$. Here $\Delta \tilde{u}_k = 0$ for $x \in \Omega$, $u_k(x) = 0$ for $x \in \partial\Omega$. The harmonic functions $\tilde{u}_k(x)$ can also be expanded in Taylor series at the origin. The resulting formulas are

$$u_k(x) = \sum_{j=0}^{k-1} Z_{j,k}(x) r^{-2j-1} + \sum_{j=0}^{\infty} X_{j,k}(x), \qquad x \to 0. \tag{1.12}$$

We reiterate that $Z_{j,k}(x)$ are as yet *arbitrarily chosen harmonic polynomials of degree j*, while for a fixed k the harmonic polynomials $X_{j,k}(x)$ are *determined uniquely* provided $Z_{j,k}(x)$ are already chosen for all j such that $0 \le j \le k - 1$.

Similarly, in accordance with Lemma 1.3, one constructs (nonuniquely for now) functions $v_i(\xi)$:

$$\Delta v_i(\xi) = 0 \quad \text{for } \xi \in \mathbf{R}^3 \backslash \omega, \qquad v_i(\xi) = 0 \quad \text{for } \xi \in \partial\omega,$$

$$v_i(\xi) = \sum_{j=0}^{i} Y_{j,i}(\xi) + \sum_{j=0}^{\infty} W_{j,i}(\xi) \rho^{-2j-1}, \qquad \rho \to \infty. \tag{1.13}$$

Here $Y_{j,i}(\xi)$ are again arbitrary harmonic polynomials of degree j, and the harmonic polynomials $W_{j,i}(\xi)$ are determined by $Y_{j,i}(\xi)$ uniquely.

We now insert the asymptotic expansions (1.12) and (1.13) into the series (1.4), (1.6) and apply the matching condition

$$A_{N,\xi} A_{N,x} U = A_{N,x} A_{N,\xi} V,$$

taking into account the change of variable $x = \varepsilon\xi$.

We have

$$A_{N,\xi} A_{N,x} U = \sum_{k=0}^{N} \varepsilon^k \left(\sum_{j=0}^{k-1} \varepsilon^{-j-1} Z_{j,k}(\xi) \rho^{-2j-1} + \sum_{j=0}^{N-k} \varepsilon^j X_{j,k}(\xi) \right),$$

$$A_{N,x} A_{N,\xi} V = \sum_{i=0}^{N} \varepsilon^i \left(\sum_{j=0}^{i} \varepsilon^{-j} Y_{j,i}(x) + \sum_{j=0}^{N-i-1} \varepsilon^{j+1} W_{j,i}(x) r^{-2j-1} \right),$$

whereby

$$Y_{j,k}(\xi) = X_{j,k-j}(\xi) \quad \text{for } k \ge 0, 0 \le j \le k, \tag{1.14}$$

$$Z_{j,k}(\xi) = W_{j,k-j-1}(\xi) \quad \text{for } k \ge 0, 0 \le j \le k - 1. \tag{1.15}$$

Now all the polynomials can be found in succession. Since $X_{j,0}(x)$ are defined uniquely, relation (1.14) determines all $Y_{j,j}(\xi)$. Next, starting from $Y_{0,0}$, relation (1.13) determines all $W_{j,0}(\xi)$, then (1.15) yields $Z_{0,1}$, etc. It is convenient to follow the chain of definitions for the functions $u_k(x)$ and $v_i(\xi)$ using Table 3 matching the series (1.4) and (1.6).

TABLE 3

V / U	$v_0(\xi)$	$\varepsilon v_1(\xi)$	$\varepsilon^2 v_2(\xi)$	$\cdots$
$u_0(x)$	$Y_{0,0}$ $-----$ $X_{0,0}$	$\varepsilon Y_{1,1}(\xi)$ $-----$ $X_{1,0}(x)$	$\varepsilon^2 Y_{2,2}(\xi)$ $-----$ $X_{2,0}(x)$	$\cdots$
$\varepsilon u_1(x)$	$W_{0,0}\rho^{-1}$ $-----$ $\varepsilon Z_{0,1}r^{-1}$	$\varepsilon Y_{0,1}$ $-----$ $\varepsilon X_{0,1}$	$\varepsilon^2 Y_{1,2}(\xi)$ $-----$ $\varepsilon X_{1,1}(x)$	$\cdots$
$\varepsilon^2 u_2(x)$	$W_{1,0}(\xi)\rho^{-3}$ $-----$ $\varepsilon^2 Z_{1,2}(x)r^{-3}$	$\varepsilon W_{0,1}\rho^{-1}$ $-----$ $\varepsilon^2 Z_{0,2}r^{-1}$	$\varepsilon^2 Y_{0,2}$ $-----$ $\varepsilon^2 X_{0,2}$	$\cdots$
$\cdots$	$\cdots$	$\cdots$	$\cdots$	$\cdots$

This table is organized in the same way as the preceding ones. The lower halves of the squares in each row include the terms of the asymptotic expansion of the function $\varepsilon^k u_k(x)$ as $x \to 0$. The upper halves of the squares in each column contain the terms of the asymptotic expansion of the function $\varepsilon^i v_i(\xi)$ as $\rho \to \infty$. The pair of equalities (1.14), (1.15) is equivalent to the coincidence of the terms in the upper and lower halves of each square.

Thus, one begins by determining the first row, then (from $Y_{0,0}$) the first column, then, from $Z_{0,1}$, the second row. The second column is found from $Y_{1,1}$ and $Y_{0,0}$, etc. The construction of the functions $u_k(x)$ and $v_i(\xi)$, and hereby of the series U and V is therefore complete.

THEOREM 1.1. *The estimate*

$$|A_{N,x}U + A_{N,\xi}V - A_{N,x}A_{N,\xi}V - u(x,\varepsilon)| < M\varepsilon^{N+1} \qquad (1.16)$$

holds for all natural N everywhere in $\overline{\Omega}_\varepsilon$, where $u(x,\varepsilon)$ is the solution of the problem (1.1)–(1.3), and U and V are the series (1.4), (1.6) constructed above.

Proof. Let $T_N(x, \varepsilon) = A_{N,x} U + A_{N,\xi} V - A_{N,x} A_{N,\xi} V - u(x, \varepsilon)$ and estimate the values of $T_N(x, \varepsilon)$ on $\partial \Omega_\varepsilon$. By construction, the function $A_{N,\xi} V$ vanishes on $\partial \omega_\varepsilon$, while the asymptotic expansions (1.12) imply that

$$A_{N,x} U - A_{N,\xi} A_{N,x} U = O\left(\sum_{k=0}^{N} \varepsilon^k r^{N-k+1} \right) = O(\varepsilon^{N+1}) \tag{1.17}$$

on $\partial \omega_\varepsilon$ (where $r < M\varepsilon$). Hence $T_N(x, \varepsilon) = O(\varepsilon^{N+1})$ on $\partial \omega_\varepsilon$. On $\partial \Omega$, on the other hand, the difference $A_{N,x} U - u(x, \varepsilon)$ is zero by construction and

$$A_{N,\xi} V - A_{N,x} A_{N,\xi} V = O\left(\sum_{i=0}^{N} \varepsilon^i \rho^{-N+i-1} \right) = O(\varepsilon^{N+1}). \tag{1.18}$$

Thus, the absolute value of the harmonic function $T_N(x, \varepsilon)$ on the boundary of the domain Ω_ε does not exceed $M\varepsilon^{N+1}$. The statement of the theorem now follows from the maximum principle. ■

Corollary. *The series* (1.4) *is a uniform asymptotic expansion of the solution* $u(x, \varepsilon)$ *of the problem* (1.1)–(1.3) *for* $x \in \overline{\Omega}$, $r \geq M\varepsilon^\gamma$, *and the series* (1.6) *is a uniform asymptotic expansion of the same solution for* $r \leq M\varepsilon^\gamma$, *where* γ *is any number such that* $0 < \gamma < 1$.

Proof. It follows from relation (1.17) that $A_{N,x} U - A_{N,\xi} A_{N,x} U = O(\varepsilon^{\gamma N})$ for $r \leq M\varepsilon^\gamma$, while (1.18) implies that $A_{N,\xi} V - A_{N,x} A_{N,\xi} V = O(\varepsilon^{(1-\gamma)N})$ for $r \geq M\varepsilon^\gamma$. These estimates together with (1.16) yield the statement of the corollary. ■

2. An elliptic equation with variable coefficients. This subsection actually repeats the above analysis applying it to the case of an equation with variable coefficients. The construction procedure is the same although some new technical details are introduced. The domains Ω, ω, ω_ε and $\Omega_\varepsilon = \Omega \backslash \omega_\varepsilon$ are the same as in the preceding subsection. The problem under consideration is of the following form

$$\mathscr{L} u \equiv \sum_{i,j=1}^{3} a_{i,j}(x) \frac{\partial^2 u}{\partial x_i \partial x_j} + \sum_{i=1}^{3} a_i(x) \frac{\partial u}{\partial x_i} + a(x) u = f(x) \quad \text{for } x \in \Omega_\varepsilon, \tag{1.19}$$

$$u(x, \varepsilon) = 0 \quad \text{for } x \in \partial \Omega, \tag{1.20}$$

$$u(x, \varepsilon) = 0 \quad \text{for } x \in \partial \omega_\varepsilon. \tag{1.21}$$

Here $a_{i,j}(x)$, $a_i(x)$, $a(x)$, $f(x) \in C^\infty(\overline{\Omega})$, and $a_{i,j}(x)$ is a positive definite matrix. The coefficients of the operator $\mathscr{L}$ are assumed to satisfy the following condition: for any $u(x) \in C^\infty(\overline{\Omega})$ and for all $\varepsilon \geq 0$ the inequality

$$\max_{\overline{\Omega}_\varepsilon} |u(x)| \leq M \left(\max_{\overline{\Omega}_\varepsilon} |\mathscr{L} u| + \max_{\partial \Omega_\varepsilon} |u| \right) \tag{1.22}$$

holds, where the constant M depends neither on $u(x)$ nor on ε and Ω_0 coincides with Ω. A sufficient condition for the estimate (1.22) is, for example, given by the inequality $a(x) \le 0$. We shall also assume that $a_{i,j}(0) = \delta_i^j$, where δ_i^j is the Kronecker delta. The last condition involves no loss of generality because the equation can, by a linear change of independent variables, be reduced to the canonical form at the point O.

The outer and inner expansions are sought in the same form as for the Laplace operator, i.e.,

$$U = \sum_{k=0}^{\infty} \varepsilon^k u_k(x), \qquad (1.23)$$

$$V = \sum_{i=0}^{\infty} \varepsilon^i v_i(\xi). \qquad (1.24)$$

The function $u_0(x)$ is the solution of the problem

$$\begin{aligned}
\mathscr{L} u_0 &= f(x) \quad \text{for } x \in \Omega, \\
u_0(x) &= 0 \quad \text{for } x \in \partial\Omega,
\end{aligned} \qquad (1.25)$$

which exists by virtue of condition (1.22) and belongs to $C^\infty(\overline{\Omega})$ (see [61, Chapter 3]).

For $k > 0$, the functions $u_k(x)$ evidently satisfy the following equations and boundary conditions:

$$\mathscr{L} u_k = 0 \quad \text{for } x \in \Omega \backslash 0, \qquad (1.26)$$

$$u_k(x) = 0 \quad \text{for } x \in \partial\Omega. \qquad (1.27)$$

For the sequel, the following notation is convenient. Denote by $P_{i,k}(x, D)$, $Q_{i,k}(x, D)$, $R_{i,k}(x, D)$ (possibly, with more subscripts) polynomials that are homogeneous of degree i with respect to x and homogeneous of degree k with respect to the differentiation symbol $D = (D_1, D_2, \dots, D_m)$, $D_j = \partial/\partial x_j$, assuming, to be definite, that the differentiation operator acts first; $P_i(x) \equiv P_{i,0}(x, D)$. It is convenient to consider $P_i(x)$, $P_{i,k}(x, D)$, etc. as defined for all integer values of the subscripts assuming that if at least one of the subscripts is negative, the polynomial is identically zero. We remind the reader that $X_i(x)$, $Y_i(x)$, $Z_i(x)$, $W_i(x)$ (possibly, with more subscripts) denote homogeneous harmonic polynomials.

Expanding the coefficients of the operator $\mathscr{L}$ in Taylor series, one can write it in the following way:

$$\mathscr{L} = \Delta - \sum_{i=1}^{\infty} Q_{i,2}(x, D) - \sum_{i=0}^{\infty} Q_{i,1}(x, D) - \sum_{i=0}^{\infty} Q_i(x), \qquad x \to 0. \quad (1.28)$$

The system of recurrence relations for $v_i(\xi)$ is obtained in the usual way from equation (1.19) after the change of variable $x = \varepsilon\xi$ with the use of (1.28):

$$\Delta v_0 = 0, \qquad \Delta v_1 = Q_{1,2}(\xi, D_\xi)v_0 + Q_{0,1}(\xi, D_\xi)v_0,$$

$$\Delta v_i = \sum_{j=1}^{i}[Q_{j,2}(\xi, D_\xi) + Q_{j-1,1}(\xi, D_\xi) + Q_{j-2}(\xi)]v_{i-j}(\xi) + R_{i-2}(\xi),$$

$$\tag{1.29}$$

where $f(x) = \sum_{j=0}^{\infty} R_j(x)$, $x \to 0$. The boundary conditions for $v_i(\xi)$ easily follow from (1.21):

$$v_i(\xi) = 0 \quad \text{for } \xi \in \partial\omega. \tag{1.30}$$

We begin with the analysis of the problems (1.26), (1.27). One has to construct, similarly to the case of the Laplace operator considered above, solutions of the homogeneous equation (1.26) having power singularities as $x \to 0$. It is, in fact, the most substantial difference of the analysis of the problem (1.19)–(1.21) with that of the preceding subsection. For the principal term of the asymptotics of the function $u_k(x)$ as $x \to 0$ we take the same expression as for the Laplace operator, i.e., $X_{k-1}(x)r^{-2k+1}$. But if in the case of the Laplace operator this term differed from $u_k(x)$ by a function harmonic everywhere in $\overline{\Omega}$, now the function $u_k(x)$ as $x \to 0$ is expanded in a more complicated asymptotic series containing, in general, other singular terms in addition to the principal one.

THEOREM 1.2. *For any positive integer k and any harmonic polynomial $Y_{k-1}(x)$ there is a function $\mathscr{E}(x) \in C^{\infty}(\overline{\Omega}\backslash O)$ such that*

$$\mathscr{L}\mathscr{E}(x) = 0 \quad \text{for } x \in \Omega\backslash O, \qquad \mathscr{E}(x) = 0 \quad \text{for } x \in \partial\Omega, \tag{1.31}$$

and the asymptotic expansion

$$\mathscr{E}(x) = \sum_{j=0}^{\infty} r^{-2k+1-2j} P_{k-1+3j}(x), \qquad x \to 0, \tag{1.32}$$

holds, where $P_{k-1}(x) = Y_{k-1}(x)$.

Before proving Theorem 1.2 we establish a simple auxiliary statement. First, the value of the Laplace operator applied to $r^\alpha Y_k(x)$, where α is a real number, is computed, taking into account the fact that the radial part of the Laplace operator in $\mathbf{R}^3$ is $r^{-2}(\partial/\partial r)(r^2(\partial/\partial r))$:

$$\Delta(r^\alpha Y_k(x)) = Y_k(x)\Delta r^\alpha + 2\sum_{j=1}^{3} \frac{\partial}{\partial x_j}(r^\alpha)\frac{\partial}{\partial x_j}Y_k(x))$$

$$= \alpha(\alpha+1)r^{\alpha-2}Y_k(x) + 2\alpha r^{\alpha-2}\sum_{j=1}^{3} x_j\frac{\partial Y_k}{\partial x_j}(x).$$

Since $Y_k(x)$ is homogeneous, one has $\sum_{j=1}^{3} x_j(\partial Y_k/\partial x_j)(x) = kY_k(x)$. Thus

$$\Delta(r^\alpha Y_k(x)) = \alpha(\alpha+1+2k)r^{\alpha-2}Y_k(x). \tag{1.33}$$

LEMMA 1.4. *For any* $Z_k(x)$, $P_j(x)$, *the equality*

$$Z_k(x)P_j(x) = \sum_{s=0}^{j} r^{2s} Y_{k+j-2s}(x) \tag{1.34}$$

holds.

PROOF. For $j = 1$, relation (1.34) can be written in the form

$$Z_k(x)P_1(x) = Y_{k+1}(x) + r^2 Y_{k-1}(x). \tag{1.35}$$

This representation is verified by applying the Laplace operator to both sides of the equality and finding the harmonic polynomial $Y_{k-1}(x)$ from (1.33). Then one finds $Y_{k+1}(x)$ from (1.35). If $P_j(x)$ is a monomial then equality (1.34) is proved by induction in j using (1.35). This implies the statement of the lemma for any j. ∎

PROOF OF THEOREM 1.2. Introduce the following notation for the terms of the series (1.32):

$$\psi_j(x) = r^{-2k+1-2j} P_{k-1+3j}(x), \quad j \leq 0, \qquad \psi_0(x) = r^{-2k+1} Y_{k-1}(x)$$

and insert the series $\mathscr{E}(x) = \sum_{j=0}^{\infty} \psi_j(x)$ into equation (1.26), where the operator $\mathscr{L}$ is written in the form (1.28). Equating the terms having the same degree of homogeneity, and taking into account the fact that the degree of homogeneity equals $j-k$ for the function $\psi_j(x)$ and $i+j-k-s$ for $Q_{i,s}\psi_j(x)$, one obtains the system which, quite understandably, looks similar to the system (1.29) for $v_i(\xi)$:

$$\Delta\psi_0 = 0,$$

$$\Delta\psi_j = \sum_{i=1}^{j}(Q_{l,2}(x, D) + Q_{l-1,1}(x, D) + Q_{l-2}(x))\psi_{j-l}, \qquad j \geq 1. \tag{1.36}$$

The first of these equations is satisfied by the choice of the function $\psi_0(x)$. Denote $\Phi(\alpha, k, x) = r^{\alpha} Y_k(x)$, and take into account that, according to (1.33), the equation $\Delta\psi = \Phi(\alpha - 2, q, x)$ has the solution $\psi = \Phi(\alpha, k, x)$ if

$$\alpha(\alpha + 1 + 2q) \neq 0. \tag{1.37}$$

Now, along with the solvability of the system (1.36), we prove by induction that

$$\psi_j(x) = \sum_{s=0}^{2j-1} \Phi(-2k + 1 - 2j + 2s, \; k - 1 + 3j - 2s, \; x), \qquad j \geq 1. \tag{1.38}$$

Taking into account that

$$\begin{aligned}
Q_{01}(x, D)\psi_0 &= \Phi(-2k - 1, k, x), \\
Q_{02}(x, D)\psi_0 &= \Phi(-2k - 3, k + 1, x),
\end{aligned} \tag{1.39}$$

one obtains, according to Lemma 1.4, the equation for $\psi_1(x)$ in the form $\Delta\psi_1 = \Phi(-2k-3, k+2, x)+\Phi(-2k-1, k, x)$. It follows from formula (1.33), that $\psi_1(x)$ of the form (1.38) exists because inequality (1.37) is satisfied.

Differentiation formulas and Lemma 1.4 imply that

$$Q_{0,1}(x, D)\Phi(\alpha, q, x) = \Phi(\alpha - 2, q + 1, x) + \Phi(\alpha, q - 1, x),$$

and

$$Q_{0,2}(x, D)\Phi(\alpha, q, x) = \Phi(\alpha - 4, q + 2, x) + \Phi(\alpha - 2, q, x) + \Phi(\alpha, q - 2, x).$$

However, if $\Phi(\alpha, q, x)$ is harmonic (i.e., $\alpha = 0$, or $\alpha + 1 + 2q = 0$), then, as in (1.39), the number of terms constituting the sum becomes smaller. This affects the computation of $Q_{l,2}(x, D)\psi_{n-l}$, because the term in $Q_{0,1}(x, D)\psi_{n-l}$ containing the harmonic polynomial of least degree is of the form $\Phi(2n-2l-2k-1, k-n+l, x)$. Thus, the induction hypothesis (1.38) for $j < n$, formula (1.36), and Lemma 1.4 yield the following equation for $\psi_n(x)$:

$$\Delta\psi_n = \sum_{s=0}^{2n-1} \Phi(-2k - 1 - n + 2s, k - 1 + 3n - 2s, x).$$

Since each term in the right-hand side satisfies inequality (1.37), formula (1.33) implies the existence of the solution $\psi_n(x)$ of the form (1.38). ∎

Thus, the series (1.32), which we denote by $\mathscr{E}$, is formally constructed. It is an f.a.s. of the equation $\mathscr{L}\mathscr{E} = 0$ as $x \to 0$. The proof of the existence of the solution for the problem (1.31) with the asymptotic expansion (1.32) is carried out in the standard way (as in Lemmas 1.1, 2.1, and Theorem 2.1 of Chapter II). Take a partial sum $B_N\mathscr{E}$, where N is sufficiently large. The construction of the series (1.32) implies that $\mathscr{L}B_N\mathscr{E} = f_N(x)$, where $f_N(x) = O(r^{N_1})$, $x \to 0$, $f_N(x) \in C^{N_1}(\overline{\Omega})$ and $N_1 \to \infty$ as $N \to \infty$. Let $z_N(x)$ be the solution of the problem $\mathscr{L}z_N = -f_N(x)$ for $x \in \overline{\Omega}$, $z_N(x) = -B_N\mathscr{E}$ for $x \in \partial\Omega$. By condition (1.22) such a solution $z_N(x)$ exists, and $z_N(x) \in C^{N_1}(\overline{\Omega})$ (see [61, Chapter 3]). The function $\mathscr{E}(x) = B_N\mathscr{E} + z_N(x)$ is the desired solution of the problem (1.31). Indeed, both the equation and boundary condition (1.31) are satisfied by virtue of the construction procedure for $z_N(x)$. It remains to verify that, for sufficiently large N, the sum $B_N\mathscr{E} + z_N(x)$ does not depend on N. This follows from (1.22) and the fact that the difference of two such sums for distinct N belongs to $C^2(\overline{\Omega})$, and satisfies both the homogenous equation and the homogeneous boundary condition. ∎

REMARK. The function $\mathscr{E}(x)$, constructed in Theorem 1.2, is not defined uniquely. If $k > 1$, then a similar function can be constructed for any polynomial $Y_{k-2}(x)$ and then added to the function $\mathscr{E}(x)$ constructed above. Similarly, one can add singular solutions with principal terms $r^{-2k+5}Y_{k-3}(x)$, $r^{-2k+7}Y_{k-4}(x), \ldots, r^1Y_0$. This actually exhausts the possibilities for constructing the solution $\mathscr{E}(x)$. This fact, however, will not be used in what follows, while the freedom of choosing the asymptotic expansion (1.32) for the function $\mathscr{E}(x)$ for $k > 1$ will be used in the construction of the functions $u_k(x)$, i.e., the coefficients of the series (1.23).

One can see without difficulty that, for $k = 1$, the function $\mathscr{E}(x)$ coincides, up to a scalar multiple, with the Green function $G(x, \overline{x})|_{\overline{x}=0}$ for the first boundary value problem for the equation $\mathscr{L}\mathscr{E} = 0$, while the other functions $\mathscr{E}(x)$ are linear combinations of the function $G(x, \overline{x})$ and its derivatives with respect to $\overline{x}_j$ for $\overline{x} = 0$.

We can now proceed with the simultaneous construction of the functions $u_k(x)$ and $v_i(\xi)$ solving the system (1.29), (1.30) such that the series (1.23) and (1.24) are matched. We will illustrate the construction process in Table 4. Its structure repeats that of Table 3 with the only difference that each square contains a more complicated expression. Therefore, we do not repeat the description of the table.

TABLE 4

V \\ U	$v_0(\xi)$	$\varepsilon v_1(\xi)$	$\varepsilon^2 v_2(\xi)$	$\varepsilon^3 v_3(\xi)$	$\ldots$
$u_0(x)$	$P_{0,0}$ ------- $P_{0,0}$	$\varepsilon P_{1,0}(\xi)$ ------- $P_{1,0}(x)$	$\varepsilon^2 P_{2,0}(\xi)$ ------- $P_{2,0}(x)$	$\varepsilon^3 P_{3,0}(\xi)$ ------- $P_{3,0}(x)$	$\ldots$
$\varepsilon u_1(x)$	$Y_{0,1}\rho^{-1}$ ------- $\varepsilon Y_{0,1} r^{-1}$	$\varepsilon \rho^{-3} P_{3,1}(\xi)$ ------- $\varepsilon r^{-3} P_{3,1}(x)$	$\varepsilon^2 \rho^{-5} P_{6,1}(\xi)$ ------- $\varepsilon r^{-5} P_{6,1}(x)$	$\varepsilon^3 \rho^{-7} P_{9,1}(\xi)$ ------- $\varepsilon r^{-7} P_{9,1}(x)$	$\ldots$
$\varepsilon^2 u_2(x)$	$\rho^{-3} Y_{1,2}(\xi)$ ------- $\varepsilon^2 r^{-3} Y_{1,2}(x)$	$\varepsilon \rho^{-5} P_{4,2}(\xi)$ ------- $\varepsilon^2 r^{-5} P_{4,2}(x)$	$\varepsilon^2 \rho^{-7} P_{7,2}(\xi)$ ------- $\varepsilon^2 r^{-7} P_{7,2}(x)$	$\varepsilon^3 \rho^{-9} P_{10,2}(\xi)$ ------- $\varepsilon^2 r^{-9} P_{10,2}(x)$	$\ldots$
$\varepsilon^3 u_3(x)$	$\rho^{-5} Y_{2,3}(\xi)$ ------- $\varepsilon^3 r^{-5} Y_{2,3}(x)$	$\varepsilon \rho^{-7} P_{5,3}(\xi)$ ------- $\varepsilon^3 r^{-7} P_{5,3}(x)$	$\varepsilon^2 \rho^{-9} P_{8,3}(\xi)$ ------- $\varepsilon^3 r^{-9} P_{8,3}(x)$	$\varepsilon^3 \rho^{-11} P_{11,3}(\xi)$ ------- $\varepsilon^3 r^{-11} P_{11,3}(x)$	$\ldots$
$\ldots$	$\ldots$	$\ldots$	$\ldots$	$\ldots$	$\ldots$

Construct the functions $u_k(x)$ for $k > 1$ according to Theorem 1.2 in such a way that

$$u_k(x) = \sum_{j=0}^{\infty} r^{-2k+1-2j} P_{k-1+3j,k}(x), \qquad x \to 0,$$

$$u_k(x) \in C^\infty(\overline{\Omega}\backslash O), \qquad P_{k-1,k}(x) = Y_{k-1,k}(x).$$

(1.40)

The function $u_0(x)$ is defined as the solution of the problem (1.25). Thus, the lower halves of all squares in Table 4 are filled. However, it is important to note that only the function $u_0(x)$, and consequently, the first row of the table is defined conclusively and uniquely. The other $u_k(x)$ are not defined uniquely. For example, the constant $Y_{0,1}$ in the second row is chosen arbitrarily, and so is the harmonic polynomial $Y_{1,2}(x)$ in the third row. In addition, according to the Remark to Theorem 1.2 a singular solution with principal term const $\cdot r^{-1}$ can be added to the function $u_2(x)$. There is, therefore, one more source of arbitrariness in the third row: one can add to the polynomial $P_{4,2}(x)$ the term cr^4, where c is any constant. Similarly, the harmonic polynomial $Y_{2,3}(x)$ in the fourth row is not unique; the polynomial $P_{5,3}(x)$ is determined up to the term $Y_1(x)r^4$, and $P_{8,3}(x)$ up to the term $Y_0 r^8$. Using Remark to Theorem 1.2 one can easily trace the degree of arbitrariness for all remaining functions $u_k(x)$.

Nonetheless, make some fixed initial choice of the functions $u_k(x)$ for $k > 0$, and then fill the upper halves of each square in Table 4. Each of them contains the same function as the corresponding lower half, but expressed in variables $\xi = \varepsilon^{-1}x$. As a result, the series $\varepsilon^i V_i(\xi)$ appear in each column of the table.

LEMMA 1.5. *Suppose the functions $u_k(x)$ of the above form solving the problems (1.26), (1.27) are chosen for all $k > 0$. Then the resulting series $V_i(\xi)$ are the f.a.s. of the system (1.29) as $\xi \to \infty$.*

PROOF. By construction, the equality

$$\mathscr{L} A_{N,x} U = f(x) \tag{1.41}$$

holds, where U is the series (1.23), $u_0(x)$ is the solution of the problem (1.25), and $u_k(x)$ are the chosen solutions for the problems (1.26), (1.27). Apply the operator $A_{N,\xi}$ to both sides of equality (1.41). The right-hand side yields $A_{N,\xi}f(x) = \sum_{i=0}^{N} \varepsilon^i R_i(\xi)$. In the left-hand side one first has to replace $A_{N,x}U$ by $\sum_{i=0}^{\infty} \varepsilon^i B_{N-i} V_i(\xi)$, and the operator $\mathscr{L}$ by

$$\varepsilon^{-2}\Delta_\xi - \varepsilon^{-2}\sum_{i=1}^{\infty}\varepsilon^i Q_{i,2}(\xi, D_\xi) - \varepsilon^{-1}\sum_{i=0}^{\infty}\varepsilon^i Q_{i,1}(\xi, D_\xi) - \sum_{i=0}^{\infty}\varepsilon^i Q_i(\xi).$$

As in the preceding examples, the equality $A_{N,\xi}\mathscr{L}A_{N,x}U = A_{N,\xi}f(x)$ contains only finitely many terms of the form $\varepsilon^j P(\xi)\rho^{-\beta}$. Equating the coefficients of the same powers of ε, one obtains a system for $B_{N-i}V_i(\xi)$ for $i < N - 2$ similar to the system (1.29). ∎

LEMMA 1.6. *Let $F(\xi) \in C^\infty(\mathbf{R}^3\backslash\omega)$, and suppose that the series*

$$V = \sum_{i=0}^{\infty} h_i(\xi) \tag{1.42}$$

is an f.a.s. of the equation $\Delta V = F$ as $\rho \to \infty$. Suppose that the series (1.42) can be repeatedly differentiated term-by-term, and the functions $h_i(\xi) \in C^\infty(\mathbf{R}^3\backslash\omega)$ are such that $\forall N \exists i_0: h_i(\xi) = O(\rho^{-N})$ for $i \geq i_0$. Then there is a function $v(\xi) \in C^\infty(\mathbf{R}^3\backslash\omega)$ such that

$$\begin{aligned} \Delta v(\xi) &= F(\xi) \quad \text{for } \xi \in \mathbf{R}^3\backslash\omega, \\ v(\xi) &= 0 \qquad \text{for } \xi \in \partial\omega \end{aligned} \tag{1.43}$$

and

$$v(\xi) = \sum_{i=0}^{\infty} h_i(\xi) + \sum_{j=0}^{\infty} X_j(\xi) \rho^{-2j-1}, \qquad \rho \to \infty. \qquad (1.44)$$

PROOF. Consider the function $F_N(\xi) = F(\xi) - \Delta(B_N V)$ for N sufficiently large. Assume that the functions $F_N(\xi)$ are smoothly extended to the entire $\mathbf{R}^3$, and consider the volume potential

$$z_N(\xi) = -\frac{1}{2\pi} \int_{\mathbf{R}^3} \frac{F_N(\bar{\xi})}{|\xi - \bar{\xi}|} d\bar{\xi}, \qquad (1.45)$$

which, as is known, (see [118, Chapter 4, §5]) is a solution of the equation $\Delta z_N(\xi) = F_N(\xi)$, $z_N(\xi) \in C^\infty(\mathbf{R}^3)$. Define $w_N(\xi)$ as a harmonic function in $\mathbf{R}^3 \backslash \omega$ vanishing at infinity and equal to $-B_N V - z_N(\xi)$ on $\partial \omega$. Let $v(\xi) = B_N V + z_N(\xi) + w_N(\xi)$. By construction, the function $v(\xi)$ is the solution of the problem (1.43). It does not depend on N because the difference of the two such functions constructed for N and $N + 1$ is a harmonic function vanishing on $\partial \omega$ and tending to zero at infinity. By Lemma 1.2, the function $w_N(\xi)$ can be expanded in the series (1.11). Expanding the kernel $|\xi - \bar{\xi}|^{-1}$ in a series as $\xi \to \infty$ and taking into account that $F_N(\xi) = O(\rho^{-N_1})$ for N sufficiently large, one can represent the integrand in (1.45) in the form of the corresponding partial sum and a small remainder. After integration a partial sum is obtained the form of which coincides with that of a partial sum of the series (1.11). This implies the asymptotic expansion (1.44). ∎

THEOREM 1.3. *There exist functions* $u_k(x)$, *solutions of the problems* (1.26), (1.27) *satisfying condition* (1.40), *and functions* $v_k(\xi)$ *solving the problems* (1.29), (1.30) *such that the series* (1.23), (1.24) *satisfy the matching condition*

$$A_{N_1, \xi} A_{N_2, x} U = A_{N_2, x} A_{N_1, \xi} V \quad \forall N_1, N_2. \qquad (1.46)$$

PROOF. The function $u_0(x)$ is the solution of the problem (1.25) which is defined uniquely. This defines all polynomials $P_{i,0}(x)$ in the first row of Table 4. Starting with the principal term at infinity, i.e., the constant $P_{0,0}$, we construct, by Lemma 1.3, the harmonic function $v_0(\xi)$ and determine from its asymptotic expansion (1.11) all the functions in the first column of Table 4. Then, starting with the principal terms $r^{-2k+1} Y_{k-1,k}(x)$, construct, according to Theorem 1.2, the functions $u_k(x)$. The function $u_1(x)$ is now defined uniquely while for the rest of $u_k(x)$ the arbitrariness referred to in the remark to Theorem 1.2 remains.

The asymptotic series arising in the upper halves of the squares of the second column is, according to Lemma 1.5, an f.a.s of equation (1.29) for the function $\varepsilon v_1(\xi)$. From this series, we construct, according to Lemma 1.6, the function $v_1(\xi)$—the solution of problem (1.29), (1.30) for $i = 1$. Under this procedure to the terms already present in the second column terms generated by the second sum in (1.44) are added. One can check without difficulty that the harmonic function $r^{2k+3}Y_{k-2}(x)$ is added to the function $r^{-2k-1}P_{k+2,k}(x)$ for $k \geq 2$. Now, according to Theorem 1.2 and the remark to it, one corrects the functions $u_k(x)$ for $k \geq 2$ by adding to them harmonic functions having the above obtained principal terms of the asymptotics. This defines the function $u_2(x)$ unambiguously.

The construction then proceeds along the same lines: next the function $v_2(\xi)$ is determined, then the corresponding modification is introduced into the asymptotics of $u_3(x)$, the function $u_3(x)$ is defined uniquely, etc. Evidently, the construction procedure ensures that the matching condition (1.46) is satisfied automatically. ∎

THEOREM 1.4. *For all positive integers N the estimate*

$$|A_{N,x}U + A_{N,\xi}V - A_{N,x}A_{N,\xi}V - u(x,\varepsilon)| < M\varepsilon^{(N-1)/2} \tag{1.47}$$

holds everywhere in $\overline{\Omega}$, where $u(x,\varepsilon)$ is the solution of the problem (1.19)–(1.21), and U and V are the series (1.23), (1.24) constructed above.

The proof almost repeats that of Theorem 1.1. Let

$$T_N(x,\varepsilon) = A_{N,x}U + A_{N,\xi}V - A_{N,x}A_{N,\xi}V - u(x,\varepsilon).$$

The estimates (1.17), (1.18) hold for the same reasons as in Theorem 1.1, and, therefore,

$$T_N(x,\varepsilon) = O(\varepsilon^{N+1}) \quad \text{on } \partial\Omega_\varepsilon. \tag{1.48}$$

Now, however, we also need to estimate the value of the operator $\mathscr{L}T_N(x,\varepsilon)$. By construction, $\mathscr{L}(A_{N,x}U - u(x,\varepsilon)) = 0$. On the other hand, the estimate $A_{N,\xi}V - A_{N,x}A_{N,\xi}V = O(\varepsilon^{(N+1)/2})$ holds for $r > \sqrt{\varepsilon}$ (see (1.18)). The derivatives of the first and second order of this difference are $O(\varepsilon^{(N-1)/2})$ for $r > \sqrt{\varepsilon}$. Hence $\mathscr{L}T_N(x,\varepsilon) = O(\varepsilon^{(N+1)/2})$ for $r > \sqrt{\varepsilon}$. For $r \leq \sqrt{\varepsilon}$, we first estimate the value of the operator $\mathscr{L}$ on $A_{N,\xi}V - u(x,\varepsilon)$. By (1.28) and (1.29), one has

$$\mathscr{L}(A_{N,\xi}V - u(x,\varepsilon))$$

$$= \left\{ \varepsilon^{-2}\Delta_\xi - \varepsilon^{-2}\sum_{i=1}^{N}Q_{i,2}(\xi,D_\xi)\varepsilon^i - \varepsilon^{-1}\sum_{i=0}^{N-1}Q_{i,1}(\xi,D_\xi)\varepsilon^i - \sum_{i=0}^{N-2}\varepsilon^iQ_i(\xi) \right\}$$

$$\times (A_{N,\xi}V) + O(r^{N-1}\|A_{N,\xi}V\|_{C^2}) - f(x)$$

$$= \varepsilon^{-2}\Delta_\xi v_0 + \varepsilon^{-1}(\Delta_\xi v_1 - Q_{1,2}(\xi,D_\xi)v_0 - Q_{0,1}(\xi,D_\xi)v_0) + \dots$$

$$+ \varepsilon^{N-2}\left(\Delta_\xi v_N - \sum_{j=1}^{N}[Q_{j,2}(\xi,D_\xi) + Q_{j-1,1}(\xi,D_\xi) + Q_{j-2,0}(\xi)]v_{N-j} \right)$$

$$- \varepsilon^{N-1} \sum_{j=1}^{N} [Q_{j,2}(\xi, D_\xi) + Q_{j-1,1}(\xi, D_\xi) + Q_{j-2,0}(\xi)] v_{N-j+1} - \ldots$$

$$- \varepsilon^{2N-2} [Q_{N,2}(\xi, D_\xi) + Q_{N-1,1}(\xi, D_\xi) + Q_{N-2,0}(\xi)] v_N$$

$$- \sum_{j=0}^{N-2} \varepsilon^j R_j(\xi) + O(r^{N-1}) + O(r^{N-1} \|A_{N,\xi} V\|_{C^2})$$

$$= O(\varepsilon^{N-1} \rho^{N-1} + \varepsilon^N \rho^N + \ldots + \varepsilon^{2N-2} \rho^{2N-2} + \varepsilon^{N-1} + r^{N-1})$$

$$+ O(r^{N-1}(1 + \varepsilon\rho + \ldots \varepsilon^N \rho^N))$$

$$= O(\varepsilon^{(N-1)/2}).$$

The difference $A_{N,x} U - A_{N,\xi} A_{N,x} U$ and its derivatives of the first and second order also do not exceed $M\varepsilon^{(N-1)/2}$ for $r \leq \varepsilon^{1/2}$ (see (1.17)). Therefore, the estimate $\mathscr{L} T_N(x, \varepsilon) = O(\varepsilon^{(N-1)/2})$ holds everywhere in $\overline{\Omega}_\varepsilon$. This and the estimates (1.22), (1.48) imply (1.47). ∎

COROLLARY. *The series* (1.23) *is a uniform asymptotic expansion of the solution* $u(x, \varepsilon)$ *of the problem* (1.19)–(1.21) *for* $x \in \overline{\Omega}$, $r \geq \sqrt{\varepsilon}$, *and the series* (1.24) *is a uniform asymptotic expansion of the same solution for* $r \leq \sqrt{\varepsilon}$.

The proof repeats that of the Corollary to Theorem 1.1 almost word by word . ∎

§2. Flow past a thin body

In this section we consider the exterior boundary value problem for the two-dimensional Laplace equation outside a small neighborhood of a closed interval. The precise formulation of the problem is as follows. Let σ be the interval $\{x_1, x_2 : 0 < x_1 < 1, x_2 = 0\}$ on the plane $\mathbf{R}^2$, $\overline{\sigma}$ its closure, and σ_ε a neighborhood of the interval σ (see Figure 10, next page). Here $\varepsilon > 0$ is a small parameter characterizing the width of the neighborhood σ_ε so that $\bigcap_{\varepsilon > 0} \sigma_\varepsilon = \sigma$. The precise form of σ_ε will be given below. Everywhere in this section the notation $x = (x_1, x_2)$, $r = \sqrt{x_1^2 + x_2^2}$ will be used. Let $u(x, \varepsilon) = u(x_1, x_2, \varepsilon)$ be a function satisfying the following conditions: $u(x, \varepsilon) \in C^\infty(\mathbf{R}^2 \backslash \sigma_\varepsilon)$,

$$\Delta u = 0 \quad \text{for } x \in \mathbf{R}^2 \backslash \sigma_\varepsilon, \tag{2.1}$$

$$u(x, \varepsilon) = 0 \quad \text{for } x \in \partial \sigma_\varepsilon, \tag{2.2}$$

$$u(x, \varepsilon) = x_2 + O(1) \quad \text{as } r \to \infty. \tag{2.3}$$

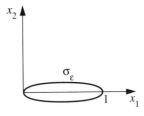

FIGURE 10

The hydrodynamic interpretation of the problem (2.1)–(2.3) is as follows. Consider a plane irrotational flow past the solid body σ_ε of an ideal incompressible fluid. Let $u(x_1, x_2, \varepsilon)$ be the flow function so that the components of the fluid velocity are given by $\partial u/\partial x_2$ and $-\partial u/\partial x_1$. Then the function $u(x_1, x_2, \varepsilon)$ satisfies equation (2.1) and is constant on the boundary of the body. One can assume, without any loss of generality, that condition (2.2) is satisfied. In order to determine the flow uniquely one has to specify the velocity of the flow at infinity. The flow with constant unit velocity parallel to the x_1-axis corresponds to the solution $u(x) = x_2$ so that a physically correct condition at infinity is $u \sim x_2$ as $x_1 \to \infty$. However, we replace it with the stronger condition (2.3) which (because there is no $\ln r$ term in the asymptotics) means, in addition, that the body causes no vorticity.

Although this interpretation is helpful, it is not necessary for the analysis that follows. In the sequel we always base our investigation on the purely mathematical formulation of the problem (2.1)–(2.3).

We are looking for the asymptotics as $\varepsilon \to 0$ of the function $u(x, \varepsilon)$ solving the problem (2.1)–(2.3). First we make the form of the neighborhood σ_ε more precise. Let $\sigma_\varepsilon = \{x : 0 < x_1 < 1,\ \varepsilon g_-(x_1) < x_2 < \varepsilon g_+(x_1)\}$, where $g_\pm(x_1) \in C^\infty(0, 1)$. In other words, σ_ε is obtained from σ_1 by contraction towards the x_1 axis with coefficient ε^{-1}. The boundary $\partial\sigma_\varepsilon$ in the vicinity of the endpoints of the closed interval $\bar{\sigma}$ is assumed to be smooth. This means that, for example, near the point $(0, 0)$ the boundary $\partial\sigma_1$ is of the form $x_1 = \psi(x_2)$, where $\psi(x_2) \in C^\infty$. Clearly,

$$\psi(0) = \psi'(0) = 0, \qquad \psi''(0) \geq 0,$$

and we also assume that $\psi''(0) > 0$, i.e., the curvature of the curve $\partial\sigma_1$ at the point $(0, 0)$ is not zero. We will assume, without any loss of generality, that $\psi''(0) = 2$. One can show, without difficulty, that these assumptions are equivalent to the following conditions on $g_\pm(x_1)$:

$$g_\pm(x_1) = \left(\sum_{j=1}^\infty g_j z^j \right)_{z = \pm\sqrt{x_1}}, \qquad x_1 \to 0,\ g_1 = 1. \qquad (2.4)$$

A similar asymptotics for the functions $g_\pm(x_1)$ will be assumed to be valid near the other end of σ, i.e., for $x_1 \to 1-0$.

1. We will look for the outer expansion of the solution of the problem (2.1)–(2.3) in the form

$$U = \sum_{k=0}^{\infty} \varepsilon^k u_k(x), \tag{2.5}$$

where

$$\Delta u_k = 0 \quad \text{for } x \in \mathbf{R}^2 \setminus \bar\sigma. \tag{2.6}$$

Condition (2.3) turns into the conditions

$$u_0(x) = x_2 + O(1) \quad \text{as } r \to \infty, \tag{2.7}$$

$$u_k(x) = O(1) \quad \text{as } r \to \infty, k > 0. \tag{2.8}$$

The derivation of the boundary conditions for the functions $u_k(x)$ on σ is slightly more complicated. Condition (2.2) can be rewritten in the following form:

$$u(x_1, \varepsilon g_\pm(x_1), \varepsilon) = 0.$$

Substituting the series (2.5) for u and expanding the functions in powers of ε, one obtains

$$u_0(x_1, \pm 0) = 0, \qquad u_1(x_1, \pm 0) + g_\pm(x_1)\frac{\partial u_0}{\partial x_2}(x_1, \pm 0) = 0,$$

$$u_k(x_1, \pm 0) + \sum_{j=1}^{k} \frac{1}{j!}[g_\pm(x_1)]^j \frac{\partial^j u_{k-j}}{\partial x_2^j}(x_1, \pm 0) = 0. \tag{2.9}$$

It follows from (2.6) and conditions (2.7), (2.9) that $u_0(x) = x_2$. The remaining functions $u_k(x)$ are harmonic in $\mathbf{R}^2 \setminus \bar\sigma$, satisfy conditions (2.9) on σ, and, by virtue of (2.8), are bounded at infinity. The solution of such problems would present no serious problems if the boundary values on $\bar\sigma$ for the functions $u_k(x)$ were continuous. However, it turns out that the functions $u_k(x)$ have singularities at the endpoints of the closed interval $\bar\sigma$. The order of these singularities increases with the growth of k.

We start with the investigation of the function $u_1(x)$. The boundary condition (2.9) for it has the form

$$u_1(x_1, \pm 0) = -g_\pm(x_1), \qquad 0 < x_1 < 1. \tag{2.10}$$

Thus, $u_1(x)$ is a bounded function which is harmonic in $\mathbf{R}^2 \setminus \bar\sigma$ and satisfies the boundary condition (2.10). The solution for such a problem exists (see [100, Chapter 3]) and is continuous everywhere, provided opposite edges of the cut along the closed interval σ are considered different. Denote the plane

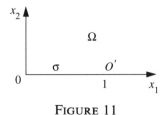

FIGURE 11

$\mathbf{R}^2$ with the cut along σ by Ω, the endpoint $(0, 0)$ of the interval σ will be denoted by O, and the other endpoint $(1, 0)$ by O' (see Figure 11).

It is also convenient to introduce the following natural notation for some classes of functions. The classes $C(\overline{\Omega})$ $(C^N(\overline{\Omega}))$ are the sets of functions which are defined in $\overline{\Omega}$ and continuous (respectively, can be differentiated N times) everywhere including the boundary of Ω. The opposite points on the different edges of the cut σ are assumed to be different (i.e., $(x_1, +0) \neq (x_1, -0)$ for $0 < x_1 < 1$). If the functions are defined just for $r \leq \delta$, the corresponding sets will be denoted $C(\overline{\Omega}_\delta)$ $(C^N(\overline{\Omega}_\delta))$. By $C^\infty(\overline{\Omega}\backslash S)$ we denote the classes of functions infinitely differentiable everywhere in $\overline{\Omega}$ with the exception of the endpoints of $\overline{\sigma}$, and by $C^\infty(\overline{\Omega}_\delta\backslash O)$ such functions defined for $r \leq \delta$. The opposite edges of the cut are again considered different.

Thus, $u_1(x) \in C(\overline{\Omega}) \cap C^\infty(\overline{\Omega}\backslash S)$, but near the endpoints of σ the function $u_1(x)$ is not smooth. It follows from (2.4) and (2.10) that

$$u_1(x_1, \pm 0) = -\left(\sum_{j=1}^\infty g_j z^j\right)_{z=\pm\sqrt{x_1}}, \qquad x_1 \to 0. \qquad (2.11)$$

Starting with this boundary condition, one can find the asymptotic expansion of the function $u_1(x)$ as $r \to 0$. However, since more general boundary conditions will be considered below, we will now find the asymptotics as $r \to 0$ for harmonic functions satisfying a wider class of boundary conditions.

LEMMA 2.1. *Let k be an integer, $h_+(x_1)$ and $h_-(x_1)$ two functions defined for $0 < x_1 < \delta < 1$ such that $h_\pm(x_1) \in C^\infty(0, \delta]$,*

$$h_\pm(x_1) = \left(\sum_{j=-k}^\infty d_j z^j\right)_{z=\pm\sqrt{x_1}}, \qquad x_1 \to 0. \qquad (2.12)$$

This equality can be repeatedly differentiated term-by-term. Then there exists a function $u(x)$ harmonic in $\Omega_\delta\backslash O$ such that $u(x_1, \pm 0) = h_\pm(x_1)$ for $x_1 \in$

$(0, \delta]$, $u(x) \in C^{\infty}(\overline{\Omega}_{\delta} \backslash O)$,

$$u(x) = \sum_{j=-k}^{\infty} d_j r^{j/2} \cos \frac{j\theta}{2} + \sum_{j=1}^{\infty} c_j r^{j/2} \sin \frac{j\theta}{2}, \qquad r \to 0, \qquad (2.13)$$

where θ is the polar angle: $0 \le \theta \le 2\pi$.

PROOF. We extend the functions $h_{\pm}(x_1)$ smoothly to the closed interval $[\delta, \delta_1]$ so that they vanish in a neighborhood of the point δ_1 identically, and let

$$w_N(x) = \sum_{j=-k}^{N} d_j r^{j/2} \cos \frac{j\theta}{2}$$

where N is sufficiently large. Evidently, $w_N(x_1, \pm 0) = h_{\pm}(x_1) + O(x_1^{(N+1)/2})$, $x_1 \to 0$, and similar estimates hold for the derivatives of this difference. Now construct a function which is harmonic in Ω_{δ_1} and satisfies the conditions

$$v_N(x_1, \pm 0) = h_{\pm}(x_1) - w_N(x_1, \pm 0), \qquad 0 \le x_1 \le \delta,$$
$$v_N|_{r=\delta_1} = -w_N|_{r=\delta_1}. \qquad (2.14)$$

As is known (see [61, Chapter 3]), a solution of this problem exists in the class $C(\overline{\Omega}_{\delta_1}) \cap C^{\infty}(\Omega)$. It is smooth on the edges of the cut for $0 < x < \delta$, and one has only to investigate the behavior of the function $v_N(x_1, x_2)$ at the point O. The simplest way to do that is by mapping the domain Ω_{δ} onto a half-disk. It is known that the mapping $\theta = 2\overline{\theta}$, $r = \overline{r}^2$ takes harmonic functions into harmonic ones. Let $\overline{v}_X(\overline{x}_1, \overline{x}_2) \equiv v_N(x_1, x_2)$, where $\overline{x}_1$, $\overline{x}_2$ denote the Cartesian coordinates of the point after the mapping $(\overline{x}_1 = x_2/\sqrt{2(r-x_1)}, \overline{x}_2 = \sqrt{r-x_1}/\sqrt{2})$. Thus, the function $\overline{v}_N(\overline{x}_1, \overline{x}_2)$ is harmonic in the upper half-disk $\overline{x}_2 \ge 0$, $\overline{r} \le \sqrt{\delta}$ and $\overline{v}_N(\overline{x}_1, 0) \in C^{N_1}(-\delta^{1/2}, \delta^{1/2})$, where $N_1 \to \infty$ for $N \to \infty$. It follows from the explicit formula for the solution of a boundary value problem for the Laplace equation in a half-plane (see [17, Chapter 4, §2]) that the function $\overline{v}_N(\overline{x}_1, \overline{x}_2)$ is sufficiently smooth in a neighborhood of the origin, $\overline{v}_N(0, 0) = 0$. Therefore, it can be represented in the form of a partial sum of the Taylor series with remainder. Since $\overline{v}_N(\overline{x}_1, \overline{x}_2)$ is a harmonic function, the terms of its Taylor series are of the form $\overline{r}^j \cos j\overline{\theta}$ and $\overline{r}^j \sin j\overline{\theta}$ for $j > 0$. Going back to the old variables one obtains the equality

$$v_N(x_1, x_2) = \sum_{j=1}^{N_2} r^{j/2} \left(\tilde{d}_j \cos \frac{j\theta}{2} + \tilde{c}_j \sin \frac{j\theta}{2} \right) + O\left(r^{N_2/2} \right), \qquad (2.15)$$

where N_2 is sufficiently large, and the equality allows repeated differentiation.

Let $u_N(x) = v_N(x) + w_N(x)$. By construction and because of (2.14), this function is harmonic in Ω_{δ_1}, equals $h_{\pm}(x_1)$ for $0 < x \le \delta_1$, and vanishes for $r = \delta_1$. Since, for N sufficiently large, $u_N(x) - u_{N+1}(x) \in C(\overline{\Omega}_{\delta_1})$ is harmonic and vanishes on the boundary of $\overline{\Omega}_{\delta_1}$, the function $u_N(x)$ does not depend on N. Formula (2.13) now follows from the explicit expression for $w_N(x)$ and the asymptotic expansion (2.15). ∎

COROLLARY. *Suppose that all the conditions of Lemma 2.1 are satisfied, and* $c_{-1}, c_{-2}, \ldots c_{-k}$, $k > 0$, *are some constants. Then there exists a*

*function $u(x)$ satisfying all conditions of Lemma 2.1 with the asymptotic
expansion (2.13) replaced by the asymptotic expansion*

$$u(x) = \sum_{j=-k}^{\infty} d_j r^{j/2} \cos \frac{j\theta}{2} + \sum_{j=-k}^{\infty} c_j r^{j/2} \sin \frac{j\theta}{2}, \qquad r \to 0, \qquad (2.16)$$

where $c_{-1}, \dots, c_{-k}$ are the given constants.

The proof differs from that of the lemma only in that one has to take for $w_N(x)$
the function

$$\sum_{j=-k}^{N} d_j r^{j/2} \cos \frac{j\theta}{2} + \sum_{j=-k}^{-1} c_j r^{j/2} \sin \frac{j\theta}{2}. \quad \blacksquare$$

THEOREM 2.1. *Let k be an integer, and suppose that the functions $h_{\pm}(x_1) \in
C^{\infty}(0, 1)$ have the asymptotic expansions (2.12) as $x_1 \to +0$ and similar
asymptotic expansions as $x_1 \to 1 - 0$. In the case $k > 0$, let constants
$c_{-1}, c_{-2}, \dots, c_{-k}, c_{-1}', c_{-2}', \dots, c_{-k}'$ be given. Then there exists a function $u(x) \in C^{\infty}(\overline{\Omega} \backslash S)$ which is harmonic in Ω, bounded at infinity, and
satisfies the boundary conditions*

$$u(x_1, \pm 0) = h_{\pm}(x_1) \quad \text{for } 0 < x_1 < 1. \qquad (2.17)$$

*For $r \to 0$, the asymptotic expansion (2.13) is valid for the function $u(x)$ for
$k \le 0$, and the asymptotic expansion (2.16) is valid for $u(x)$ for $k > 0$, while
for $x \to O'$ a similar asymptotic expansion holds with c_{-j} replaced by c_{-j}'.*

PROOF. According to Lemma 2.1 and the corollary thereto, one can construct a
function $\tilde{u}(x) \in C^{\infty}(\overline{\Omega} \backslash O)$ that is harmonic in $\Omega_\delta \backslash O$, satisfies condition (2.17),
and has either the asymptotic expansion (2.13) or (2.16) depending on the sign of k.
A similar function $\tilde{\tilde{u}}(x)$ can be constructed in a neighborhood of the point O'. Now
construct a compactly supported function $\overline{u}(x) \in C^{\infty}(\overline{\Omega} \backslash S)$ coinciding with $\tilde{u}(x)$ in
a neighborhood of the point O, and with $\tilde{\tilde{u}}(x)$ in a neighborhood of the point O'.

To conclude the proof one has to construct the function $v(x)$ such that

$$\begin{aligned}
\Delta v(x) &= -\Delta \overline{u}(x) \quad \text{in } \Omega, \\
v(x_1, \pm 0) &= h_{\pm}(x_1) - \overline{u}(x_1, \pm 0) \quad \text{for } 0 < x_1 < 1,
\end{aligned} \qquad (2.18)$$

$v(x) \in C(\overline{\Omega})$, and $v(x)$ is bounded at infinity. As is known, this problem can
be solved, because the boundary value function in the right-hand side of equality
(2.18) is continuous on the closed interval $\overline{\sigma}$, and, moreover, identically vanishes
in a neighborhood of its endpoints. The proof can, for example, be achieved by
conformally mapping the plane minus the straight-line segment $\overline{\sigma}$ onto the region
outside the circle, and then taking the standard solution of the exterior boundary
value problem (see [100, §32]). Evidently, the sum $v(x) + \overline{u}(x)$ is the desired function
$u(x)$. In a neighborhood of the endpoints of the straight-line segment σ, the function
$v(x)$ can be expanded in the series (2.13), where $d_j = 0$ for all j, so that in the
asymptotic expansion (2.16) of the sum $v(x) + \overline{u}(x)$, as compared with the same
asymptotic expansion of the function $\overline{u}(x)$ (or, which is the same, of the function
$\tilde{u}(x)$), only the coefficients c_j for $j > 0$ are modified. $\blacksquare$

Thus, according to Theorem 2.1 and condition (2.11), one has

$$u_1(x_1, x_2) = -\sum_{j=1}^{\infty} g_j r^{j/2} \cos \frac{j\theta}{2} + \sum_{j=1}^{\infty} c_{j,1} r^{j/2} \sin \frac{j\theta}{2}, \qquad r \to 0. \quad (2.19)$$

A similar asymptotic expansion is valid near the point O'.

The boundary condition for the function $u_2(x)$ is of the following form:

$$u_2(x_1, \pm 0) = -g_{\pm}(x_1) \frac{\partial u_1}{\partial x_2} - \frac{[g_{\pm}(x_1)]^2}{2} \frac{\partial^2 u_0}{\partial x_2^2}(x_1, \pm 0). \quad (2.20)$$

Taking into account the asymptotic expansion (2.19) and the fact that $u_0(x) = x_2$, one has

$$u_2(x_1, \pm 0) = \left(\sum_{j=1}^{\infty} g_j z^j \right)_{z=\pm\sqrt{x_1}} \cdot \left(\sum_{j=1}^{\infty} c_{j,1} \frac{j}{2} z^{j-2} \right)_{z=\pm\sqrt{x_1}}$$

$$= \left(\sum_{j=0}^{\infty} d_{j,2} z^j \right)_{z=\pm\sqrt{x_1}}, \qquad x_1 \to 0.$$

According to Theorem 2.1, we can construct a function $u_2(x)$ which is harmonic and bounded in Ω and satisfies condition (2.20). One has $u_2(x) \in C^{\infty}(\overline{\Omega}\backslash S)$ and

$$u_2(x) = \sum_{j=0}^{\infty} d_{j,2} r^{j/2} \cos \frac{j\theta}{2} + \sum_{j=1}^{\infty} c_{j,2} r^{j/2} \sin \frac{j\theta}{2}, \qquad r \to 0.$$

A similar asymptotic expansion is valid in a neighborhood of the point O'.

The boundary function for the solution $u_3(x)$ already has singularities at the endpoints of $\overline{\sigma}$. Indeed, by virtue of condition (2.9), one has $u_3(x_1, \pm 0)$

$$= -g_{\pm}(x_1) \frac{\partial u_2}{\partial x_2}(x_1, \pm 0) - \frac{g_{\pm}^2(x_1)}{2} \frac{\partial^2 u_1}{\partial x_2^2}(x_1, \pm 0) - \frac{g_{\pm}^3(x_1)}{3!} \frac{\partial^3 u_0}{\partial x_2^3}(x_1, \pm 0)$$

$$= \left(\sum_{j=-1}^{\infty} d_{j,3} z^j \right)_{z=\pm\sqrt{x_1}}, \qquad x_1 \to 0.$$

Therefore, the function $u_3(x)$ is not bounded near the points O and O'. In the vicinity of the point O it grows at least as $r^{-1/2}$. In accordance with Theorem 2.1, in the class of such functions there is a solution $u_3(x)$ defined to within two arbitrary constants c_{-1} and c'_{-1}.

THEOREM 2.2. *There exist functions* $u_k(x)$ *satisfying conditions* (2.6), (2.8), (2.9) *and having asymptotic expansions*

$$u_k(x) = \sum_{j=-k+2}^{\infty} d_{j,k} r^{j/2} \cos \frac{j\theta}{2} + \sum_{j=-k+2}^{\infty} c_{j,k} r^{j/2} \sin \frac{j\theta}{2}, \qquad r \to 0, \quad (2.21)$$

for $k \geq 1$, *and similar asymptotic expansions near the point* O'. *Each function* $u_k(x)$ *for* $k \geq 3$ *is defined to within arbitrary constants* $c_{-1,k}$, $c_{-2,k}$, $\dots$, $c_{-k+2,k}$, $c'_{-1,k}$, $c'_{-2,k}$, $\dots$, $c'_{-k+2,k}$.

The proof is easily achieved by induction using Theorem 2.1. The statement is already proved for $k = 3$. If it holds for $p < k$, then in the construction of $u_k(x)$ it only remains to verify that the asymptotic expansion of the boundary function (2.9) starts with the term $x_1^{(-k+2)/2}$. Indeed, the principal term of the asymptotics in the factor $[g_\pm(x_1)]^j$ equals $x_1^{j/2}$ while, by the induction hypothesis and the asymptotic expansion (2.21), the principal term in the factor $(\partial^j u_{k-j}/\partial x_2^j)(x_1, \pm 0)$ equals $c x_1^{-k/2+1-j/2}$. This and Theorem 2.1 imply the existence of the desired function $u_k(x)$. ∎

2. Thus, the problem (2.1)–(2.3), like the preceding problems, is bisingular: the coefficients $u_k(x)$ of the outer expansion (2.5) have increasing singularities near the endpoints of the straight-line segment $\bar{\sigma}$. Clearly, a different asymptotic expansion has to be used in the vicinity of these endpoints. We make a detailed investigation in a neighborhood of the point O only. The equation of the boundary $\partial\sigma_\varepsilon$ in the vicinity of this point is of the form $x_2 = \varepsilon g_\pm(x_1) = \varepsilon(\pm\sqrt{x_1} + O(x_1))$. The inner, "stretched" coordinates have to be chosen in such a way that the Laplace equation is preserved and the equation of the boundary, in its principal term, does not depend on the parameter ε. Hence, the inner variables are $\xi = \varepsilon^{-2}x_1$, $\eta = \varepsilon^{-2}x_2$. The Laplace equation preserves its form in the variables ξ, η, while the equation of the boundary $\partial\sigma_\varepsilon$ is of the form

$$\eta = \pm\sqrt{\xi} + \varepsilon\Phi_\pm(\xi, \varepsilon), \qquad (2.22)$$

where

$$\Phi_\pm(\xi, \varepsilon) = \left(\sum_{j=0}^{\infty} g_{j+2}\varepsilon^j z^{j+2}\right)_{z=\pm\sqrt{\xi}}, \qquad \varepsilon\sqrt{\xi} \to 0. \qquad (2.23)$$

The boundary condition (2.2) for the function $v(\xi, \eta, \varepsilon) \equiv u(x_1, x_2, \varepsilon)$ turns into the equality

$$v(\xi, \pm\sqrt{\xi} + \varepsilon\Phi_\pm(\xi, \varepsilon), \varepsilon) = 0. \qquad (2.24)$$

We look for the inner expansion in the form

$$V = \sum_{i=2}^{\infty} \varepsilon^i v_i(\xi, \eta). \qquad (2.25)$$

(The series starts with the term $i = 2$ because $u_0(0, 0) = 0$, $u_0(x) \in C^\infty$.) Equation (2.1) implies that

$$\Delta_{\xi,\eta} v_i = 0. \qquad (2.26)$$

On inserting the series (2.25) into the boundary condition (2.24), one formally obtains the boundary conditions for $v_i(\xi, \eta)$:

$$v_2(\xi, \pm\sqrt{\xi}) = 0, \qquad (2.27)$$

$$v_i(\xi, \pm\sqrt{\xi}) - \left(\sum_{l=2}^{i-1}\sum_{q=1}^{i-l} c_{q,l,i} \frac{\partial^q v_l(\xi, z)}{\partial \eta^q} z^{i-l+q}\right)_{z=\pm\sqrt{\xi}} = 0, \qquad i \geq 3. \qquad (2.28)$$

Here $c_{q,l,i}$ are constants expressed through g_j. The explicit form of $c_{q,l,i}$ is of no importance.

Approximately replacing the boundary (2.22) with the parabola $\eta = \pm\sqrt{\xi}$, find the functions $v_i(\xi, \eta)$ for $\xi < \eta^2$. As in other bisingular problems, the functions $v_i(\xi, \eta)$ grow at infinity. Moreover, in this case, the solutions of the problems (2.26)–(2.28) are not unique. One could approach these problems by finding the general form of their solutions in the class of slowly growing functions and then establishing the extent of indeterminacy of the solutions. It is, however, more convenient to find $v_i(\xi, \eta)$ using the already constructed functions $u_k(x)$, and the matching condition for the series U and V. It is, therefore, sufficient to find the asymptotics at infinity for the solution of the Poisson equation with rapidly decaying right-hand side. Denote by D the domain $\{\xi, \eta : \xi < \eta^2, \eta \in \mathbf{R}^1\}$.

LEMMA 2.2. *Let* $F(\xi, \eta) \in C^\infty(\overline{D})$,

$$F(\xi, \eta) = O\left((\xi^2 + \eta^2)^{-N}\right) \quad \text{for } \xi^2 + \eta^2 \to \infty, \ N > 0, \qquad (2.29)$$

and let $v(\xi, \eta)$ *be a bounded solution of the following boundary value problem*

$$v(\xi, \eta) \in C^\infty(\overline{D}), \qquad v(\eta^2, \eta) = 0, \qquad (2.30)$$

$$\Delta_{\xi, \eta} v = F(\xi, \eta) \quad \text{for } (\xi, \eta) \in D. \qquad (2.31)$$

Then

$$v(\xi, \eta) = \sum_{j=1}^{2N_1} \rho^{-j/2}\left(c_i \sin\frac{j\theta}{2} + d_j \cos\frac{j\theta}{2}\right) + O\left(\rho^{-N_1}\right), \qquad \rho \to \infty, \qquad (2.32)$$

uniformly in the domain $\overline{D}$, *where* ρ, θ *are the polar coordinates:* $\xi = \rho\cos\theta$, $\eta = \rho\sin\theta$, $N_1 \to \infty$ *as* $N \to \infty$, *and equality* (2.32) *can be differentiated sufficiently many times.*

PROOF. Map the domain D onto the half-plane $t > 0$ by the change of independent variables

$$\xi = s^2 - t^2 - t, \qquad \eta = s + 2ts. \qquad (2.33)$$

Under this change, equation (2.31) goes into

$$\Delta_{s, t} v = [4s^2 + (2t + 1)^2]F(s^2 - t^2 - t, \ s + 2ts).$$

Denoting the right-hand side of this equation by $f(s, t)$, and $v(s^2 - t^2 - t, s + 2ts)$ by $w(s, t)$, we conclude that the problem (2.30), (2.31) is equivalent to the problem

$$\Delta_{s,t} w = f(s, t) \quad \text{for } t \geq 0, w(s, 0) = 0. \tag{2.34}$$

By virtue of (2.29), one has $f(s, t) = O\left((s^2 + t^2)^{-N_2}\right)$. The solution of the problem (2.34) can be written out explicitly:

$$w(s, t) = \frac{1}{4\pi} \int_0^\infty \int_{-\infty}^\infty f(\sigma, \tau) \ln \frac{(s - \sigma)^2 + (t - \tau)^2}{(s - \sigma)^2 + (t + \tau)^2} \, d\sigma \, d\tau. \tag{2.35}$$

Introduce the polar coordinates on the plane s, t: $s = \lambda \cos \omega$, $t = \lambda \sin \omega$ and represent the kernel in the integral (2.35) in the form of a partial sum of the Taylor series for $\lambda^{-1} \to 0$ plus the remainder:

$$\ln \frac{(s - \sigma)^2 + (t - \tau)^2}{(s - \sigma)^2 + (t + \tau)^2} = \ln \left(1 - 2\frac{\sigma \cos \omega}{\lambda} - 2\frac{\tau \sin \omega}{\lambda} + \frac{\tau^2 + \sigma^2}{\lambda^2}\right)$$

$$- \ln \left(1 - 2\frac{\sigma \cos \omega}{\lambda} + 2\frac{\tau \sin \omega}{\lambda} + \frac{\tau^2 + \sigma^2}{\lambda^2}\right)$$

$$= \sum_{k=1}^{2N_3} \lambda^{-k} P_k(\sigma, \tau, \sin \omega, \cos \omega) + O\left(\left(\frac{\sigma^2 + \tau^2}{\lambda^2}\right)^{N_3}\right).$$

Inserting this expression into (2.35), one has

$$w(s, t) = v(\xi, \eta) = \sum_{j=1}^{N_4} \lambda^{-j} \Phi_j(\omega) + O\left(\lambda^{-N_4}\right). \tag{2.36}$$

Relations (2.33) in polar coordinates are of the form

$$\rho \cos \theta = \lambda^2 \cos 2\omega - \lambda \sin \omega, \qquad \rho \sin \theta = \lambda \cos \omega + \lambda^2 \sin 2\omega,$$

whereby

$$\rho^2 - \frac{1}{2}\rho \cos \theta + \frac{1}{16} = \left(\lambda^2 + \lambda \sin \omega + \frac{1}{4}\right)^2.$$

One easily concludes from this that

$$\lambda = \rho^{1/2} + \sum_{j=0}^\infty A_j(\theta) \rho^{-j/2}, \qquad \omega = \frac{\theta}{2} + \sum_{j=1}^\infty B_j(\theta) \rho^{-j/2}.$$

Inserting these asymptotics into (2.36), one obtains the equality

$$v(\xi, \eta) = \sum_{j=1}^{2N_1} \rho^{-j/2} \Phi_j(\theta).$$

This equality, like all the others obtained above, can be repeatedly differentiated term by term. The form of $\Phi_j(\theta)$ appearing in (2.32) follows from the fact that the function $v(\xi, \eta)$ is "almost" harmonic. ∎

LEMMA 2.3. *Suppose that the series* $\tilde{v} = \sum_{j=-k}^{\infty} \rho^{-j/2} \psi_j(\theta)$ *is an f.a.s. of the boundary value problem*

$$\Delta \tilde{v} = F(\xi, \eta) \quad \text{in the domain } D, \tag{2.36}$$

$$\tilde{v}(\eta^2, \eta) = \varphi(\eta), \qquad \eta \in \mathbf{R}^1, \tag{2.37}$$

as $\rho \to \infty$, *where* $F(\xi, \eta) \in C^\infty(\overline{D})$, $\varphi(\eta) \in C^\infty(\mathbf{R}^1)$, $\psi_j(\theta) \in C^\infty[0, 2\pi]$. *Suppose also that equalities* (2.36), (2.37) *admit differentiation of any order in the sense that their right-hand sides can be expanded into asymptotic series obtained by termwise differentiation of the series* $\tilde{v}$. *Then there exists a function* $v(\xi, \eta) \in C^\infty(\overline{D})$ *satisfying relations* (2.36), (2.37) *such that*

$$v(\xi, \eta) = \tilde{v} + \sum_{j=1}^{\infty} \rho^{-j/2} \left(C_j \sin \frac{j\theta}{2} + D_j \cos \frac{j\theta}{2} \right), \qquad \rho \to \infty.$$

PROOF. Let $\chi(\rho) \in C^\infty[0, \infty)$ be a truncating function vanishing in a neighborhood of zero and equal to 1 for $\rho > 1$. Denote by $v_N(\xi, \eta)$ the product $\chi(\rho) B_N \tilde{v}$, where $B_N \tilde{v}$, as usual, denotes the partial sum of the series $\tilde{v}$. By hypothesis, $\Delta v_N = F(\xi, \eta) + f_N(\xi, \eta)$, $v_N(\eta^2, \eta) = \varphi(\eta) + h_N(\eta)$, where $f_N(\xi, \eta) = O(\rho^{-N_1})$, $h_N(\eta) = O(|\eta|^{-N_1})$, $N_1 \to \infty$ as $N \to \infty$. Corresponding estimates are also valid for the derivatives of the functions f_N and h_N.

Let $w_N(\xi, \eta)$ be a bounded solution of the boundary value problem $\Delta w_N = f_N$ for $(\xi, \eta) \in D$, $w_N(\eta^2, \eta) = h_N(\eta)$ for $\eta \in \mathbf{R}^1$. Such a solution $w_N(\xi, \eta) \in C^\infty(\overline{D})$ exists and, by Lemma 2.2, has the asymptotic expansion (2.32). The difference $v_N(\xi, \eta) - w_N(\xi, \eta)$ satisfies equation (2.36) and the boundary condition (2.37). Since the bounded solution of the boundary value problem (2.36),(2.37) is unique, the function $v(\xi, \eta) = v_N(\xi, \eta) - w_N(\xi, \eta)$ does not depend on N for N large enough and is the desired function. ∎

3. The coefficients of the series (2.5) and (2.25), i.e., the functions $u_k(x)$ and $v_i(\xi, \eta)$, can be constructed using Table 5 (next page) which presents the matching procedure in convenient form. Note that along with the series (2.25) one has to construct a similar series V' with coefficients $v'_i(\xi', \eta')$ in the vicinity of the endpoint O' making sure that the series U and V' match.

The structure of Table 5 is the same as that of the preceding tables and and needs no additional explanation. The functions $u_k(x)$ for $k \geq 1$ are constructed according to Theorem 2.2. The functions $u_1(x)$ and $u_2(x)$ are defined uniquely, while for $k \geq 3$ each of the functions depends on $(2k - 2)$ arbitrary constants $c_{1,k}, \ldots, c_{-k+2,k}, c'_{-1,k}, \ldots, c'_{-k+2,k}$. Make some temporary choice of these constants and write the asymptotic expansion of the function $\varepsilon^k u_k(x)$ as $r \to 0$ in the lower halves of the squares in the corresponding row of Table 5. After passing to the inner variables ξ, η (so that $r = \varepsilon^2 \rho$, while the polar angle θ retains its value) the series $\varepsilon^i V_i$ appear in the columns of Table 5.

TABLE 5

V \\ U	$\varepsilon^2 v_2(\xi, \eta)$	$\varepsilon^3 v_3(\xi, \eta)$	$\varepsilon^4 v_4(\xi, \eta)$	$\cdots$
$u_0(x)$	$\dfrac{\varepsilon^2 \eta}{x_2}$	0	0	$\cdots$
$\varepsilon u_1(x)$	$\dfrac{\varepsilon^2 \rho^{1/2}\Phi_{-1,2}(\theta)}{\varepsilon r^{1/2}\Phi_{-1,2}(\theta)}$	$\dfrac{\varepsilon^3 \rho\Phi_{-2,3}(\theta)}{\varepsilon r\Phi_{-2,3}(\theta)}$	$\dfrac{\varepsilon^4 \rho^{3/2}\Phi_{-3,4}(\theta)}{\varepsilon r^{3/2}\Phi_{-3,4}(\theta)}$	$\cdots$
$\varepsilon^2 u_2(x)$	$\dfrac{\varepsilon^2 \Phi_{0,2}(\theta)}{\varepsilon^2 \Phi_{0,2}(\theta)}$	$\dfrac{\varepsilon^3 \rho^{1/2}\Phi_{-1,3}(\theta)}{\varepsilon^2 r^{1/2}\Phi_{-1,3}(\theta)}$	$\dfrac{\varepsilon^4 \rho\Phi_{-2,4}(\theta)}{\varepsilon^2 r\Phi_{-2,4}(\theta)}$	$\cdots$
$\varepsilon^3 u_3(x)$	$\dfrac{\varepsilon^2 \rho^{-1/2}\Phi_{1,2}(\theta)}{\varepsilon^3 r^{-1/2}\Phi_{1,2}(\theta)}$	$\dfrac{\varepsilon^3 \Phi_{0,3}(\theta)}{\varepsilon^3 \Phi_{0,3}(\theta)}$	$\dfrac{\varepsilon^4 \rho^{1/2}\Phi_{-1,4}(\theta)}{\varepsilon^3 r^{1/2}\Phi_{-1,4}(\theta)}$	$\cdots$
$\varepsilon^4 u_4(x)$	$\dfrac{\varepsilon^2 \rho^{-1}\Phi_{2,2}(\theta)}{\varepsilon^4 r^{-1}\Phi_{2,2}(\theta)}$	$\dfrac{\varepsilon^3 \rho^{-1/2}\Phi_{1,3}(\theta)}{\varepsilon^4 r^{-1/2}\Phi_{1,3}(\theta)}$	$\dfrac{\varepsilon^4 \Phi_{0,4}(\theta)}{\varepsilon^4 \Phi_{0,4}(\theta)}$	$\cdots$
$\cdots$	$\cdots$	$\cdots$	$\cdots$	$\cdots$

LEMMA 2.4. *The series V_i are f.a.s. of the boundary value problem* (2.26)–(2.28).

PROOF. The verification of the fact that the differential equation is satisfied is, in this case, achieved very easily because the equation is of a very simple form. By construction, each square of the table contains a harmonic function so that each partial sum of the series V_i satisfies the Laplace equation exactly. To check the boundary conditions (2.27), (2.28), note that, by construction,

$$A_{N,\xi,\eta} A_{N,x} U = \sum_{j=2}^{N} \varepsilon^j (B_{(N-j)/2} V_j), \tag{2.38}$$

and

$$A_{N,x} U(x_1, \varepsilon g_\pm(x_1), \varepsilon) = \sum_{k=0}^{N} \varepsilon^k \sum_{j=0}^{N-k} \frac{\varepsilon^j [g_\pm(x_1)]^j}{j!} \frac{\partial^j u_k}{\partial x_2^j}(x_1, \pm 0)$$

$$+ \sum_{k=0}^{N} \varepsilon^{N+1} \frac{[g_\pm(x_1)]^{N+1-k}}{(N-k)!}$$

$$\times \int_0^1 (1-\mu)^{N-k} \frac{\partial^{N+1-k} u_k}{\partial x_2^{N+1-k}}(x_1, \mu \varepsilon g_\pm(x_1)) \, d\mu. \tag{2.39}$$

The double sum in the right-hand side vanishes by virtue of conditions (2.9). Hence and from (2.38) and (2.22) one obtains, after applying the operator $A_{N,\xi}$ to both sides of equality (2.39):

$$A_{N,\xi} \left(\sum_{j=2}^{N} \varepsilon^j \sum_{q=0}^{N-j} \frac{1}{q!} [\varepsilon \Phi_\pm(\xi, \varepsilon)]^q \frac{\partial^q}{\partial \eta^q}(B_{(N-j)/2} V_j)(\xi, \pm\sqrt{\xi}) \right) = A_{N,\xi} \Lambda, \tag{2.40}$$

where Λ is the last sum in relation (2.39). According to the definition of the operator $A_{N,\xi}$, one has to substitute $\varepsilon^2 \xi$ for x_1 in the expression for Λ, and then expand Λ in an asymptotic series as $\varepsilon \to 0$. Taking into account the asymptotics (2.4) and (2.21), one has $\Lambda = \sum_{j=2}^{\infty} \varepsilon^j \Lambda_j(\xi)$, where $\Lambda_j(\xi) = O(\xi^{-(N+1-j)/2})$, $\xi \to \infty$. The left-hand side of equality (2.40) contains the polynomial in ε with the coefficients of ε^j equal to the left-hand sides of equalities (2.27), (2.28). This implies the statement of the lemma. ∎

Now, by Lemma 2.3, one can construct from the asymptotic series V_i the functions $v_i(\xi, \eta)$ solving the problems (2.26)–(2.28). If the V_i were asymptotic series for the functions $v_i(\xi, \eta)$ for $\rho \to \infty$ then the matching condition for the series (2.5) and (2.25) would have been satisfied by virtue of the construction procedure for the series V_i. However, as mentioned in Lemma 2.3, the asymptotic series for $v_i(\xi, \eta)$, in general, differs from V_i by the series $\sum_{j=1}^{\infty} \rho^{-j/2}(c_{j,i} \sin \frac{i\theta}{2} + d_{j,i} \cos \frac{i\theta}{2})$. (This series satisfies the homogeneous boundary condition $V_i(\xi, \pm\sqrt{\xi}) = 0$ whence $d_{1,i} = 0$.) The functions $v_i(\xi, \eta)$ have, therefore, to be constructed step by step.

First, we construct the function $v_2(\xi, \eta)$ from the series V_2. As a result, the first column in Table 5, starting with the term $\varepsilon^2 \rho^{-1/2} \Phi_{1,2}(\theta)$, is modified: the summand $c_{1,2} \sin \frac{\theta}{2}$ is added to the function $\Phi_{1,2}(\theta)$. The same change in the asymptotics of the function $u_3(x)$ takes place in the vicinity of the point O'. This, by virtue of Theorem 2.2, completes the construction of the function $u_3(x)$.

The modification of the function $u_3(x)$ results in the modification of the boundary conditions for the succeeding functions; in general, all of them are altered. Nevertheless, the principal terms of the asymptotics appearing in the first column of the table are defined irrevocably. The series V_i for $i \geq 3$ are also modified, but, by virtue of Lemma 2.4, they remain f.a.s. of the problem (2.26)–(2.28). Then, according to Lemma 2.3, the function $v_3(\xi, \eta)$ is constructed from the series V_3 appearing in the second column. This column is modified starting with the term $\varepsilon^3 \rho^{-1/2} \Phi_{1,3}(\theta)$. It is now possible to determine the term $u_4(x)$ in its final form etc.

Thus, the functions $u_k(x)$ solving the problems (2.6)–(2.9), and $v_i(\xi, \eta)$ solving the problems (2.26)–(2.28), are constructed in such a way that the matching conditions

$$A_{N_1, \xi, \eta} A_{N_2, x} U = A_{N_2, x} A_{N_1, \xi, \eta} V \quad \forall N_1, N_2 \qquad (2.41)$$

are satisfied for the series (2.5) and (2.25). A similar series V' is constructed near the point O'.

The construction of the asymptotic expansions for the solution of the problem (2.1)–(2.3) is essentially completed. It remains to note that the functions $u_k(x)$ are defined in $\overline{\Omega} \setminus S$, i.e., in a wider domain than $\Omega \setminus \sigma_\varepsilon$, while the functions $v_i(\xi, \eta)$ are defined just in the domain D, i.e., for $\xi < \eta^2$. However, even in a neighborhood of the point O, one has to approximate the solution $u(x, \varepsilon)$ for $-\sqrt{\xi} + \varepsilon \Phi_-(\xi, \varepsilon) \leq \eta \leq \sqrt{\xi} + \varepsilon \Phi_+(\xi, \eta)$, i.e., in a domain that may be wider than D. We will assume, for simplicity sake, that $g_+(x_1) \leq \sqrt{x_1}$ and $g_-(x_1) \geq -\sqrt{x_1}$ in a fixed neighborhood of zero so that the functions $v_i(\xi, \eta)$ are defined everywhere in the intersection of this neighborhood with Ω_ε. Similar conditions will be assumed to be satisfied near the point O'. (If these conditions are not satisfied, then, in order to construct the asymptotics, one has to extend the functions $v_i(\xi, \eta)$ beyond the domain D. No essential difficulties are involved, but some new technical details arise, which we do not consider here.)

Denote by $S(\delta)$ the intersection of the domain Ω_ε with the disk of radius δ and center at the point O, and by $S'(\delta)$ the intersection of the domain Ω with the disk of the same radius with center at the point O'. (Of course, the sets $S(\delta)$, $S'(\delta)$ also depend on ε, but we do not include this dependence in the notation.)

Let $\chi(x) \in C^\infty(\mathbf{R}^2)$, $\chi(x) \equiv 0$ outside $S(2\delta)$, and $\chi(x) \equiv 1$ inside $S(\delta)$, where δ is a fixed small positive number, and let $\tilde{\chi}(x)$ be a similar cut-off function in a neighborhood of the point O'. Let

$$T_N(x, \varepsilon) = A_{N,x} U + (A_{N, \xi, \eta} V - A_{N, \xi, \eta} A_{N, x} U) \chi(x)$$
$$+ (A_{N, \xi', \eta'} V' - A_{N, \xi', \eta'} A_{N, x} U) \tilde{\chi}(x) - u(x, \varepsilon),$$

where U is the series (2.5), V is the series (2.25), V' is a similar series constructed near the point O', and ξ', η' are inner coordinates in the vicinity of this point.

THEOREM 2.3. *For all sufficiently large N everywhere in $\overline{\Omega}_\varepsilon$ the estimate $|T_N(x, \varepsilon)| \leq M\varepsilon^{N/4}$ holds, where the constant M depends neither on x, nor on ε.*

PROOF. Relations (2.3), (2.7), and (2.8) imply that

$$T_N(x, \varepsilon) = O(1) \quad \text{as } r \to \infty. \tag{2.42}$$

We now prove that the boundary values of $T_N(x, \varepsilon)$ on $\partial \sigma_\varepsilon$ are small. For $x \in \partial\sigma\backslash\{S(\delta)\cup S'(\delta)\}$, one has $A_{N,x}U = \sum_{k=0}^{N}\varepsilon^k u_k(x_1, \varepsilon g_\pm(x_1)) = O(\varepsilon^{N+1})$ in view of conditions (2.9) and because the function $g_\pm(x_1)$ is uniformly smooth on this section of the boundary. Hence, for $x \in \partial\sigma_\varepsilon\backslash\{S(2\delta)\cup S'(2\delta)\}$, we have

$$T_N(x, \varepsilon) = A_{N,x}U = O\left(\varepsilon^{N+1}\right). \tag{2.43}$$

For $x \in \partial\sigma_\varepsilon \cap S(\varepsilon^{3/2})$, by virtue of (2.21), (2.23), (2.37), and (2.28), one has

$$
\begin{aligned}
T_N(x, \varepsilon) &= A_{N,x}U - A_{N,\xi,\eta}A_{N,x}U + A_{N,\xi,\eta}V \\
&= \sum_{i=0}^{N}\varepsilon^i\left[\sum_{j=0}^{N-i}\frac{1}{j!}\frac{\partial^j v_i}{\partial\eta^j}(\xi, \pm\sqrt{\xi})[\varepsilon\Phi_\pm(\xi, \varepsilon)]^j + O\left([\varepsilon(1+\xi)]^{N-i+1}\right)\right] \\
&\quad + O\left(\varepsilon^{3N/4}\right) \\
&= O\left(\varepsilon^{N/2}\right). \tag{2.44}
\end{aligned}
$$

For $x \in \partial\sigma_\varepsilon \cap \{S(2\delta)\backslash S(\varepsilon^{3/2})\}$, by (2.4), (2.9), (2.21), and (2.41), we have

$$
\begin{aligned}
T_N(x, \varepsilon) &= A_{N,x}U + (A_{N,\xi,\eta}V - A_{N,x}A_{N,\xi,\eta}V)\chi(x) \\
&= \sum_{k=0}^{N}\varepsilon^k u_k(x_1, \varepsilon g_\pm(x_1)) + \chi(x)\sum_{i=0}^{N}\varepsilon^i(v_i(\xi, \eta) - B_{(N-i)/2}v_i(\xi, \eta)) \\
&= \sum_{k=0}^{N}\varepsilon^k\left\{\sum_{j=0}^{N-k}\frac{1}{j!}[\varepsilon g_\pm(x_1)]^j\frac{\partial^j u_k}{\partial x_2^j}(x_1, \pm 0)\right. \\
&\qquad\qquad \left. + O(x_1^{(k-2)/2-N+k-1})\left(\varepsilon x_1^{1/2}\right)^{N-k+1}\right\} \\
&\quad + O\left(\sum_{i=0}^{N}\varepsilon^i\rho^{-(N-1-i)/2}\right) \\
&= O\left(\sum_{k=0}^{N}\varepsilon^{N+1}x_1^{-N/2+k-3/2}\right) + O\left(\rho^{-(N-1)/2}\right) \\
&= O\left(\sum_{k=0}^{N}\varepsilon^{N+1+(3/2)(-N/2+k-3/2)} + \varepsilon^{(N-1)/4}\right) = O\left(\varepsilon^{N/4-2}\right). \tag{2.45}
\end{aligned}
$$

Thus, it follows from (2.43)–(2.45) and similar estimates in a neighborhood of the point O' that

$$T_N(x, \varepsilon)|_{\partial\sigma_\varepsilon} = O\left(\varepsilon^{N/4-2}\right). \tag{2.46}$$

The function $T_N(x, \varepsilon)$, is, by construction, harmonic everywhere except the set $\Omega_\varepsilon \cap \{[S(2\delta)\backslash S(\delta)] \cup [S'(2\delta)\backslash S'(\delta)]\}$, and it only remains to estimate the Laplace operator on this set. Let $x \in \Omega_\varepsilon \cap \{S(2\delta)\backslash S(\delta)\}$. Then

$$\Delta T_N(x, \varepsilon) = (A_{N,\xi,\eta}V - A_{N,x}A_{N,\xi,\eta}V)\Delta\chi$$

$$+ 2\sum_{j=1}^{2} \frac{\partial\chi}{\partial x_j}\frac{\partial}{\partial x_j}(A_{N,\xi,\eta}V - A_{N,x}A_{N,\xi,\eta}V).$$

Since the derivatives of χ vanish outside the annulus $\delta < r < 2\delta$, one has

$$\Delta T_N(x, \varepsilon) = O\left(\sum_{i=0}^{N}\varepsilon^i\|v_i(\xi, \eta) - B_{(N-i)/2}v_i(\xi, \eta)\|_{C^1}\right) = O\left(\varepsilon^{N-1}\right).$$

A similar estimate holds in the annulus $S'(2\delta)\backslash S(\delta)$. The statement of the theorem follows from this estimate, estimate (2.46), and condition (2.42). ∎

COROLLARY. *The series* (2.5) *is a uniform asymptotic expansion of the solution of the problem* (2.1)–(2.3) *in the domain* $\Omega_\varepsilon\backslash\{S(\varepsilon^\gamma) \cup S'(\varepsilon^\gamma)\}$. *The series* (2.25) (*respectively, the series* V') *is a uniform asymptotic expansion for the same problem in the domain* $\Omega_\varepsilon \cap S(\varepsilon^\gamma)$ (*respectively,* $\Omega_\varepsilon \cap S'(\varepsilon^\gamma)$), *where* γ *is any number such that* $0 < \gamma < 2$.

PROOF. It is sufficient to verify that the difference $A_{N,\xi,\eta}V - A_{N,\xi,\eta}A_{N,x}U$ is uniformly small in the domain $\Omega_\varepsilon\cap\{S(2\delta)\backslash S(\varepsilon^\gamma)\}$, and that the difference $A_{N,x}U - A_{N,\xi,\eta}A_{N,x}U$ is uniformly small in the domain $\Omega_\varepsilon \cap S(\varepsilon^\gamma)$. Indeed, in the first domain:

$$|A_{N,\xi,\eta}V - A_{N,\xi,\eta}A_{N,x}U| \le M\sum_{i=2}^{N}\varepsilon^i\left|v_i(\xi, \eta) - B_{(N-i)/2}v_i(\xi, \eta)\right|$$

$$\le M\sum_{i=0}^{N}\varepsilon^i\rho^{-(N-1-i)/2} \le M\sum_{i=0}^{N}\varepsilon^{i+(N-1-i)(2-\gamma)/2} \le M\varepsilon^{(N-1)(1-\gamma/2)}.$$

In the second domain:

$$|A_{N,x}U - A_{N,\xi,\eta}A_{N,x}U| \le M\sum_{k=1}^{N}\varepsilon^k r^{(N-k+1)/2}$$

$$\le M\sum_{k=1}^{N}\varepsilon^{k+\gamma(N-k+1)/2} \le M\varepsilon^{\gamma(N+1)/2}. \quad ∎$$

§3. Two-dimensional boundary value problem in a domain with a small hole

In this section we consider a boundary value problem which is stated in exactly the same terms as that studied in §1.1. The only difference is that

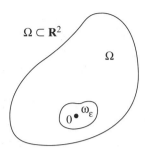

FIGURE 12

instead of three-dimensional domains two-dimensional domains are considered. This change seems insignificant at the first glance but actually leads to considerable and interesting complications.

We, therefore, repeat the notation (see Figure 12): $\Omega \subset \mathbf{R}^2$ is a bounded domain, $\partial \Omega \in C^\infty$, $O \in \omega \subset \Omega$, $x \in \omega_\varepsilon \Leftrightarrow \varepsilon^{-1} x \in \omega$, $\Omega_\varepsilon = \Omega \backslash \overline{\omega}_\varepsilon$,

$$\Delta u = 0 \quad \text{for } x \in \Omega_\varepsilon, \tag{3.1}$$

$$u(x, \varepsilon) = \psi(x) \quad \text{for } x \in \partial \Omega, \tag{3.2}$$

$$u(x, \varepsilon) = 0 \quad \text{for } x \in \partial \omega_\varepsilon, \tag{3.3}$$

$$x = (x_1, x_2), \quad \psi(x) \in C^\infty(\partial \Omega), \quad r = |x|.$$

The limit of the solution $u(x, \varepsilon)$ as $\varepsilon \to 0$ equals, as before, $u_0(x) \in C^\infty(\overline{\Omega})$ everywhere in $\overline{\Omega} \backslash O$, where $u_0(x)$ is the solution of the problem

$$\begin{aligned} \Delta u_0 &= 0 \quad \text{for } x \in \overline{\Omega}, \\ u_0(x) &= \psi(x) \quad \text{for } x \in \partial \Omega. \end{aligned} \tag{3.4}$$

However, this is where the analogy with the three-dimensional case ends. An attempt to find an asymptotic expansion of the solution $u(x, \varepsilon)$ as $\varepsilon \to 0$ in the form of a power series in ε, i.e., in the form (1.4), fails. Indeed, consider a particular case where Ω and ω are unit disks with the center at the origin, $\psi(x) \equiv 1$. Then $u(x, \varepsilon) = 1 + \left(\ln \frac{1}{\varepsilon}\right)^{-1} \ln r$. It is now clear that one has to look for the asymptotic expansion in a more general form than (1.4).

Let the series

$$U = \sum_{k=0}^{\infty} \nu_k(\varepsilon) u_k(x) \tag{3.5}$$

be an f.a.s. of the problem (3.1)–(3.2), where $\nu_k(\varepsilon)$ is a gauge sequence and, naturally, $\nu_0(\varepsilon) = 1$. Then

$$\Delta u_k = 0 \quad \text{for } x \in \overline{\Omega} \backslash O, \tag{3.6}$$

$$u_k(x) = 0 \quad \text{for } x \in \partial \Omega, k \geq 1. \tag{3.7}$$

Evidently, if such functions do not vanish identically, they must have singularities for $x \to 0$. As in the three-dimensional case, the singular terms of the asymptotics of such solutions having at most power growth for $x \to 0$

are given by linear combinations of the fundamental solution of the Laplace equation and its derivatives. However, in the two-dimensional case the fundamental solution of the Laplace equation is of the form $c \ln r$ which makes the two-dimensional case substantially different from all other dimensions.

We again denote by $X_l(x)$, $Y_l(x)$, $Z_l(x)$, $W_l(x)$, $X_{l,j}(x)$, etc. homogeneous harmonic polynomials of degree l. As in the three-dimensional case, one can easily check that the lth derivative of $\ln r$ is of the form $X_l(x)r^{-2l}$, and for each $X_l(x)$ the function $X_l(x)r^{-2l}$ is a harmonic one.

LEMMA 3.1. *Let* $z(x)$ *be a linear combination of functions of the form* $\ln r$ *and* $X_l(x)r^{-2l}$ *for* $l \geq 1$. *Then there exists a function* $u(x) = \tilde{u}(x) + z(x)$ *such that* $\tilde{u}(x) \in C^\infty(\overline{\Omega})$,

$$\begin{aligned} \Delta u &= 0 \quad \textit{for } x \in \Omega\backslash O, \\ u(x) &= 0 \quad \textit{for } x \in \partial\Omega. \end{aligned} \tag{3.8}$$

The proof repeats that of Lemma 1.1. ∎

We look for the inner expansion in the form

$$V = \sum_{i=0}^{\infty} \mu_i(\varepsilon)v_i(\xi), \tag{3.9}$$

where $\mu_i(\varepsilon)$ is a gauge sequence, and $\xi = \varepsilon^{-1}x$. As in §1.1, the functions $v_i(\xi)$ are solutions of the boundary value problems

$$\Delta v_i = 0 \quad \text{for } \xi \in \mathbf{R}^2\backslash\omega, \tag{3.10}$$
$$v_i(\xi) = 0 \quad \text{for } \xi \in \partial\omega. \tag{3.11}$$

The statements and proofs of the following two lemmas are completely analogous to those of Lemmas 1.2 and 1.3. What one has to take into account is that in two dimensions, unlike the three-dimensional case, the solution of the Dirichlet problem is unique in the class of bounded functions (see [100, §32]). In what follows, the notation $\rho = |\xi|$ is used.

LEMMA 3.2. *Suppose that the function* $v(\xi)$ *is bounded and harmonic in a neighborhood of infinity. Then the asymptotic expansion*

$$v(\xi) = \sum_{j=0}^{\infty} X_j(\xi)\rho^{-2j}, \qquad \rho \to \infty,$$

holds. ∎

LEMMA 3.3. *Let* δ *be a constant, and* $H(\xi)$ *a harmonic polynomial. Then there is a function* $v(\xi) = \tilde{v}(\xi) + H(\xi) + \delta \ln r$ *such that* $v(\xi) \in C^\infty(\mathbf{R}^2\backslash\omega)$,

$$\Delta v = 0 \quad \textit{for } \xi \in \mathbf{R}^2\backslash\omega, \tag{3.12}$$
$$v(\xi) = 0 \quad \textit{for } \xi \in \partial\omega, \tag{3.13}$$
$$\tilde{v}(\xi) = \sum_{j=0}^{\infty} X_j(\xi)\rho^{-2j}, \qquad \rho \to \infty. \quad \blacksquare \tag{3.14}$$

The difference of this lemma from Lemma 1.3 is not limited to the presence of the logarithmic term in the asymptotics. The main distinction is that the zero order term (i.e., the constant) in the asymptotics of $\tilde{v}(\xi)$ differs from the same term in the asymptotics of $v(\xi)$, while in the three-dimensional case both $H(\xi)$ and $v(\xi)$ have the same zero order term in the asymptotics.

Among the functions constructed in Lemmas 3.1 and 3.3, a special place is occupied by those functions which have singularities of the least order at the origin and infinity, respectively. The corresponding singularities are of the types $\ln r$ and $\ln \rho$. Denote such functions by $\Lambda(x)$ and $\Gamma(x)$.

Thus, $\Lambda(x) = \tilde{u}(x) + \ln r$, where $\tilde{u}(x) \in C^{\infty}(\overline{\Omega})$ and $\Lambda(x)$ satisfies equalities (3.8). The function $\Lambda(x)$ exists by virtue of Lemma 3.1, and, expanding the function $\tilde{u}(x)$ in a Taylor series, one obtains the asymptotic expansion

$$\Lambda(x) = \ln r + P_0 + \sum_{j=1}^{\infty} X_j(x), \qquad x \to 0. \tag{3.15}$$

Here all harmonic polynomials, including the constant P_0, depend only on the domain Ω.

We construct the function $\Gamma(\xi)$ in accordance with Lemma 3.3, so that it satisfies relations (3.12), (3.13), and, by (3.14),

$$\Gamma(\xi) = \ln \rho + Q_0 + \sum_{j=1}^{\infty} Y_j(\xi)\rho^{-2j}, \qquad \rho \to \infty. \tag{3.16}$$

Here also the constant Q_0 depends only on the domain ω.

It is natural to take $a\Lambda(x)$ for the principal singular term of the series (3.5), i.e., the function $u_1(x)$, and the function $b\Gamma(\xi)$ for the principal term of the series (3.9), i.e., the function $v_1(\xi)$. Suppose that $\nu_1(\varepsilon)$ and $\mu_0(\varepsilon)$ tend to zero slower than ε^{α}, and $\nu_2(\varepsilon)$ and $\mu_1(\varepsilon)$ faster than ε^{α_1}, where $0 < \alpha < \alpha_1 < 1$, i.e., $\varepsilon^{\alpha} = o(\nu_1(\varepsilon))$, $\varepsilon^{\alpha} = o(\mu_0(\varepsilon))$ and $\nu_2(\varepsilon) = o(\varepsilon^{\alpha_1})$, $\mu_1(\varepsilon) = o(\varepsilon^{\alpha_1})$. Apply the matching condition

$$A_{\alpha,\xi}A_{\alpha,x}U = A_{\alpha,x}A_{\alpha,\xi}V \tag{3.17}$$

to the series (3.5) and (3.9). The asymptotic expansions (3.5), (3.8), (3.15), and (3.16) imply that

$$A_{\alpha,\xi}A_{\alpha,x}U = A_{\alpha,\xi}[u_0(x) + \nu_1(\varepsilon)a\Lambda(x)] = u_0(0) + \nu_1(\varepsilon)a[\ln \rho + \ln \varepsilon + P_0],$$
$$A_{\alpha,x}A_{\alpha,\xi}V = A_{\alpha,x}\mu_0(\varepsilon)v_0(\xi) = \mu_0(\varepsilon)b[\ln \rho + Q_0].$$

It follows hence and from (3.17) that

$$\left.\begin{array}{l} a\nu_1(\varepsilon) = b\mu_0(\varepsilon), \\ u_0(0) + a\nu_1(\varepsilon)[\ln \varepsilon + P_0] = b\mu_0(\varepsilon)Q_0 \end{array}\right\} \Rightarrow a\nu_1(\varepsilon) = u_0(0)\left[\ln \frac{1}{\varepsilon} - P_0 + Q_0\right]^{-1}.$$

The constants a and b can be chosen arbitrarily. It is natural to set $a = b = u_0(0)$ which ensures that $\nu_1(\varepsilon)$ and $\mu_0(\varepsilon)$ depend just on the domains

Ω and ω and not on the boundary function $\psi(x)$: $\nu_1(\varepsilon) = \mu_0(\varepsilon) = \lambda(\varepsilon)$, where

$$\lambda(\varepsilon) = \left[\ln\frac{1}{\varepsilon} - P_0 + Q_0\right]^{-1}. \qquad (3.18)$$

Thus, the first two terms of the series (3.5) and the first term of the series (3.9) are found. Similar considerations can be used to find the following terms of the gauge sequences $\nu_k(\varepsilon)$ and $\mu_i(\varepsilon)$. Although the reasoning is cumbersome, it is not complicated and can serve as a good exercise for the reader. Skipping the detailed argument, we write out the series (3.5) and (3.9):

$$U = u_0(x) + \lambda(\varepsilon)u_1(x) + \sum_{k=1}^{\infty}\varepsilon^k\sum_{j=0}^{k+1}[\lambda(\varepsilon)]^j u_{k,j}(x), \qquad (3.19)$$

$$V = \sum_{i=0}^{\infty}\varepsilon^i\sum_{j=0}^{i+1}[\lambda(\varepsilon)]^j v_{i,j}(\xi). \qquad (3.20)$$

The functions $u_{k,j}(x)$ and $v_{i,j}(\xi)$ will be constructed according to Lemmas 3.1 and 3.3 in such a way that they satisfy relations (3.6),(3.7) and (3.10),(3.11), respectively, and have the following asymptotic expansions:

$$u_{k,j}(x) = \sum_{l=1}^{k}X_{l,k,j}(x)r^{-2l} + \sum_{l=0}^{\infty}Z_{l,k,j}(x) + a_{k,j}\ln r, \qquad x\to 0, \quad (3.21)$$

$$v_{i,j}(\xi) = \sum_{l=1}^{i}Y_{l,i,j}(\xi) + \sum_{l=0}^{\infty}W_{l,i,j}(\xi)\rho^{-2l} + a_{i,j}\ln\rho, \qquad \rho\to\infty,\ i\geq 1. \qquad (3.22)$$

The first terms in the series (3.19), (3.20) are already constructed in such a way that (3.17) is satisfied. The function $u_0(x)$ is the solution of the problem (3.4). The functions $u_{1,0}(x)$ and $v_{0,0}(\xi)$ must be set identically equal to zero. The functions $u_1(x)$ and $v_{0,1}(\xi)$ were constructed above. One only has to rename the function $v_0(\xi)$ because the structure of the series (3.20) now requires that it have the indices 0, 1. Thus, $u_1(x) = u_0(0)\Lambda(x)$, and $v_{0,1} = u_0(0)\Gamma(\xi)$, where $\Lambda(x)$ and $\Gamma(\xi)$ were defined above and can be expanded in the asymptotic series (3.15) and (3.16).

THEOREM 3.1. *There exist functions* $u_{k,j}(x)$ *and* $v_{i,j}(\xi)$ *possessing the above properties such that the matching condition*

$$A_{N_1,\xi}A_{N_2,x}U = A_{N_2,x}A_{N_1,\xi}V \qquad (3.23)$$

is satisfied for the series (3.19), (3.20).

PROOF. The functions $u_{k,j}(x)$ and $v_{i,j}(\xi)$ will be defined by induction. As before, it is convenient to use Table 6 (pp. 100–101). Its structure needs no explanation. We only recall that the presence of the logarithmic terms often causes the values of the functions in the lower and upper parts of the squares to differ: the change of variable moves some summands, arising because of the appearance of the factor $[\lambda(\varepsilon)]^{-1}$, to the neighboring squares. However, the coincidence rule holds for the larger boxes (bounded by solid lines), which contain all terms of the asymptotic expansions with a fixed power of ε and all powers of $\lambda(\varepsilon)$. The matching condition (3.23) ensures the equality of the sum of all functions appearing in the lower halves of smaller squares and the corresponding sum in the upper halves of the same squares.

Before we begin with the construction of the functions $u_{k,j}(x)$ and $v_{i,j}(\xi)$, a few comments are in order. For each polynomial in the table, the first subscript, as usual, indicates the degree of the polynomial, while other subscripts repeat the subscripts of the corresponding function u_k, $u_{k,j}$, or $v_{i,j}$. However, in most of the squares, the latter subscripts are omitted for technical reasons.

Denote by β the constant $Q_0 - P_0$ in formula (3.18) which depends on the domains Ω and ω only. It follows from (3.18) that

$$\ln \varepsilon = \beta - [\lambda(\varepsilon)]^{-1}, \qquad \ln r = \ln \rho + \beta - [\lambda(\varepsilon)]^{-1}. \qquad (3.24)$$

We also recall that the construction procedure for $\lambda(\varepsilon)$ and the functions $\Lambda(x)$, $\Gamma(\xi)$ (see (3.15), (3.16)) matched these functions in the terms of zero order to within λ^{-1}:

$$\ln r + P_0 = \ln \rho + Q_0 - [\lambda(\varepsilon)]^{-1}. \qquad (3.25)$$

Finally, we rewrite the series (3.21), (3.22) in an equivalent but more convenient form:

$$u_{k,j}(x) = \sum_{l=1}^{k} X_{l,k,j}(x) r^{-2l} + \sum_{l=0}^{\infty} \tilde{Z}_{l,k,j}(x) + a_{k,j}\Lambda(x), \qquad x \to 0, \quad (3.26)$$

$$v_{i,j}(\xi) = \sum_{l=1}^{i} Y_{l,i,j}(\xi) + \sum_{l=0}^{\infty} \tilde{W}_{l,i,j}(\xi) \rho^{-2l} + a_{i,j}\Gamma(\xi), \qquad i \geq 1, \; \rho \to \infty, \quad (3.27)$$

where $\Lambda(x)$ and $\Gamma(\xi)$ are the functions defined above (see (3.15), (3.16)).

We begin the construction with the function $v_{1,0}(\xi)$. Setting in formula (3.27) $a_{1,0} = 0$, and $Y_{1,1,0} = X_{1,0}$ (to match $u_0(x)$ to $\varepsilon v_{1,0}(\xi)$ (see Table 6)), one defines, by Lemma 3.3, the function $v_{1,0}(\xi)$. The constant $W_{0,1,0}$ is hereby defined uniquely. Taking into account condition (3.24), one can, therefore, set $a_{1,1} = -W_{0,1,0}$, and then define the functions $u_{1,1}(x)$ and $v_{1,1}(\xi)$ uniquely by setting $Y_{1,1,1} = X_{1,1}$ and $X_{1,1,1} = W_{1,0,1}$. The constants $W_{0,1,1}$ and $Z_{0,1,1}$ are again defined uniquely so that the expression appearing in the upper half of the intersection of the 1,1 row with the 1,0 and 1,1 columns is not matched to the expression in the lower halves of the same squares. The upper sum differs from the lower one by $\varepsilon\lambda(\varepsilon)[W_{0,1,1} - Z_{0,1,1} - a_{1,1}\beta]$. We set, therefore, $a_{1,2} = Z_{0,1,1} + a_{1,1}\beta - W_{0,1,1}$, $v_{1,2}(\xi) = a_{1,2}\Gamma(\xi)$, $u_{1,2}(x) = a_{1,2}\Lambda(x)$. The matching condition

TABLE 6a

$\diagdown$ V $\diagup$ U	$v_{0,0} \equiv 0$	$\lambda v_{0,1}$	$\varepsilon v_{1,0}$	$\varepsilon\lambda v_{1,1}$	$\varepsilon\lambda^2 v_{1,2}$
$u_{0,0}$	$X_{0,0}$ – – – – $X_{0,0}$		$\varepsilon Y_{1,1,0}$ – – – – $X_{1,0}$		
λu_1	$-a$ – – – –	$\lambda(Y_0 + a\ln\rho)$ – – – – $\lambda(X_0 + a\ln r)$		$\varepsilon\lambda Y_1$ – – – – $\lambda X_{1,1}$	
$\varepsilon\lambda u_{1,1}$		$\lambda W_1\rho^{-2}$ – – – – $\varepsilon\lambda X_1 r^{-2}$	$\varepsilon W_{0,1,0}$	$\varepsilon\lambda(W_0 + a\ln\rho)$ – – – – $\varepsilon\lambda(Z_0 + a\ln r)$	
$\varepsilon\lambda^2 u_{1,2}$					$\varepsilon\lambda^2(W_0 + a\ln\rho)$ – – – – $\varepsilon\lambda^2(Z_0 + a\ln r)$
$\varepsilon^2 u_{2,0}$			$\varepsilon W_1\rho^{-2}$ – – – – $\varepsilon^2 X_1 r^{-2}$		– – – –
$\varepsilon^2\lambda u_{2,1}$		$\lambda W_2\rho^{-4}$ – – – – $\varepsilon^2\lambda X_2 r^{-4}$		$\varepsilon\lambda W_1\rho^{-2}$ – – – – $\varepsilon^2\lambda X_1 r^{-2}$	
$\varepsilon^2\lambda^2 u_{2,2}$					$\varepsilon\lambda^2 W_1\rho^{-2}$ – – – – $\varepsilon^2\lambda^2 X_1 r^{-2}$
$\varepsilon^2\lambda^3 u_{2,3}$					
$\varepsilon^3 u_{3,0}$			$\varepsilon W_2\rho^{-4}$ – – – – $\varepsilon^3 X_2 r^{-4}$		
$\varepsilon^3\lambda u_{3,1}$		$\lambda W_3\rho^{-6}$ – – – – $\varepsilon^3\lambda X_3 r^{-6}$		$\varepsilon\lambda W_2\rho^{-4}$ – – – – $\varepsilon^3\lambda X_2 r^{-4}$	
$\varepsilon^3\lambda^2 u_{3,2}$					$\varepsilon\lambda^2 W_2\rho^{-4}$ – – – – $\varepsilon^3\lambda^2 X_2 r^{-4}$
$\varepsilon^3\lambda^3 u_{3,3}$					
$\varepsilon^3\lambda^4 u_{3,4}$					

Table 6b

$\epsilon^2 v_{2,0}$	$\epsilon^2\lambda v_{2,1}$	$\epsilon^2\lambda^2 v_{2,2}$	$\epsilon^2\lambda^3 v_{2,3}$	$\epsilon^3 v_{3,0}$	$\epsilon^3\lambda v_{3,1}$	...
$\epsilon^2 Y_2$ ‒ ‒ ‒ $X_{2,0}$				$\epsilon^3 Y_3$ ‒ ‒ ‒ $X_{3,0}$		...
	$\epsilon^2\lambda Y_2$ ‒ ‒ ‒ $\lambda X_{2,1}$				$\epsilon^3\lambda Y_3$ ‒ ‒ ‒ $\lambda X_{3,1}$	...
	$\epsilon^2\lambda Y_1$ ‒ ‒ ‒ $\epsilon\lambda Z_1$				$\epsilon^3\lambda Y_2$ ‒ ‒ ‒ $\epsilon\lambda Z_2$	...
		$\epsilon^2\lambda^2 Y_1$ ‒ ‒ ‒ $\epsilon\lambda^2 Z_1$				...
$\epsilon^2 W_0$ ‒ ‒ ‒ $\epsilon^2 Z_0$				$\epsilon^2 W_1$ ‒ ‒ ‒ $\epsilon^2 Z_1$		...
	$\epsilon^2\lambda(W_0+a\ln\rho)$ ‒ ‒ ‒ $\epsilon^2\lambda(Z_0+a\ln r)$				$\epsilon^3\lambda Y_1$ ‒ ‒ ‒ $\epsilon^2\lambda Z_1$	...
		$\epsilon^2\lambda^2(W_0+a\ln\rho)$ ‒ ‒ ‒ $\epsilon^2\lambda^2(Z_0+a\ln r)$				...
			$\epsilon^2\lambda^3(W_0+a\ln\rho)$ ‒ ‒ ‒ $\epsilon^2\lambda^3(Z_0+a\ln r)$			...
$\epsilon^2 W_1\rho^{-2}$ ‒ ‒ ‒ $\epsilon^3 X_1 r^{-2}$				$\epsilon^3 W_0$ ‒ ‒ ‒ $\epsilon^3 Z_0$		...
	$\epsilon^2\lambda W_1\rho^{-2}$ ‒ ‒ ‒ $\epsilon^3\lambda X_1 r^{-2}$				$\epsilon^3\lambda(W_0+a\ln\rho)$ ‒ ‒ ‒ $\epsilon^3\lambda(Z_0+a\ln r)$	...
		$\epsilon^2\lambda^2 X_1\rho^{-2}$ ‒ ‒ ‒ $\epsilon^3\lambda^2 X_1 r^{-2}$				...

(3.23) is now satisfied, by virtue of (3.25), for $N_1 \leq 1$, $N_2 \leq 1$. The construction of the functions $u_{k,j}(x)$ and $v_{i,j}(\xi)$ proceeds by induction. Suppose that the functions are constructed for $k \leq n$, $i \leq n$ so that the matching condition (3.23) is satisfied for $N_1 \leq n$, $N_2 \leq n$. This means that all the terms in the asymptotics of the functions $u_{n+1,j}(x)$ for $j \leq n + 1$ appearing in the first sum in (3.26) are determined, as well as the corresponding terms in the asymptotics of $v_{n+1,j}(\xi)$ in (3.27). According to Lemmas 3.1 and 3.3, in order to complete the determination of the functions $u_{n+1,j}(x)$ and $v_{n+1,j}(\xi)$, it remains to determine the constants $a_{n+1,j}$.

Setting $a_{n+1,0} = 0$ we construct $u_{n+1,0}(x)$ and $v_{n+1,0}(\xi)$. This matching gives rise to the residual $\varepsilon^{n+1}(W_{0,n+1,0} - Z_{0,n+1,0})$ in the square at the intersection of the corresponding row and column. It can be eliminated by setting $a_{n+1,1} = Z_{0,n+1,0} - W_{0,n+1,0}$ whereby by Lemmas 3.1 and 3.3, the functions $u_{n+1,1}(x)$ and $v_{n+1,1}(\xi)$ are determined uniquely. The residual $\varepsilon^{n+1}\lambda(\varepsilon)[W_{0,n+1,1} - Z_{0,n+1,1} - a_{n+1,1}\beta]$ arising in the zero order terms of their asymptotics is eliminated by setting $a_{n+1,2} = Z_{0,n+1,1} + a_{n+1,1}\beta - W_{0,n+1,1}$, and the procedure goes on until the functions $u_{n+1,n+1}(x)$ and $v_{n+1,n+1}(\xi)$ are determined. The residual in the zero order terms in the asymptotics of these functions is eliminated with the aid of the functions $u_{n+1,n+2}(x) = a_{n+1,n+2}\Lambda(x)$ and $v_{n+1,n+2}(\xi) = a_{n+1,n+2}\Gamma(\xi)$, while the zero order terms in the asymptotics of these functions are matched by virtue of (3.25).

This completes the construction of all the functions $u_{k,j}(x)$, $v_{i,j}(\xi)$ for $k \leq n + 1$, $i \leq n + 1$, whereby condition (3.23) is satisfied for $N_1 \leq n + 1$, $N_2 \leq n + 1$. ∎

THEOREM 3.2. *For all positive integers N, the estimate*

$$|A_{N,x}U + A_{N,\xi}V - A_{N,x}A_{N,\xi}V - u(x,\varepsilon)| < M\varepsilon^{N+1}$$

holds everywhere in $\overline{\Omega}_\varepsilon$, *where $u(x,\varepsilon)$ is the solution of the problem* (3.1)–(3.3), *and U and V are the series* (3.19), (3.20).

COROLLARY. *The series* (3.19) *is a uniform asymptotic expansion of the solution $u(x,\varepsilon)$ of the problem* (3.1)–(3.3) *for $x \in \overline{\Omega}$, $r > M\varepsilon^\gamma$, and the series* (3.20) *is the uniform asymptotic expansion of the same solution for $r \leq M\varepsilon^\gamma$, where γ is any number such that $0 < \gamma < 1$.*

The proof of Theorem 3.2 and the corollary repeats that of Theorem 1.1 and the corollary thereto almost word by word. ∎

§4. Analysis of the asymptotics in the case where the limit problem has no solution

In this section, we proceed with the analysis of the problem (1.19)–(1.21) of §1, subsection 2, in a peculiar and, in a sense, exceptional situation. We will now assume that the limit problem (1.25) has no solution. In particular, this means that zero is an eigenvalue of the operator $\mathscr{L}$ with the boundary condition (1.20). In order not to overload the exposition with unnecessary details, we limit our analysis to the three-dimensional case, and consider the simplest case of an equation with constant coefficients.

Thus, as before, ω and Ω are bounded domains, $0 \in \omega \subset \Omega \subset \mathbf{R}^3$, $\partial\omega \in C^\infty$, $\partial\Omega \in C^\infty$, the complement of ω is connected, $x = (x_1, x_2, x_3)$, $r = |x|$, $x \in \omega_\varepsilon$, $x \in \omega_\varepsilon \Leftrightarrow \varepsilon^{-1} x \in \omega$, $\Omega_\varepsilon = \Omega \backslash \overline{\omega}_\varepsilon$, $u(x, \varepsilon)$ is the solution of the boundary value problem

$$\mathscr{L}u \equiv \Delta u + a^2 u = f(x) \quad \text{for } x \in \Omega_\varepsilon, \tag{4.1}$$

$$u(x, \varepsilon) = 0 \quad \text{for } x \in \partial\Omega, \tag{4.2}$$

$$u(x, \varepsilon) = 0 \quad \text{for } x \in \partial\omega_\varepsilon, \tag{4.3}$$

$$f(x) \in C^\infty(\overline{\Omega}), \qquad a = \text{const}, 0 < \varepsilon < \varepsilon_0.$$

Suppose that a^2 is a simple eigenvalue of the operator $-\Delta$ with the boundary condition (4.2), i.e., such that there exists a unique (up to a scalar multiple) function $h(x) \not\equiv 0$ such that

$$\mathscr{L}h = 0 \quad \text{for } x \in \overline{\Omega}, \qquad h(x) = 0 \quad \text{for } x \in \partial\Omega. \tag{4.4}$$

Then, as is known, the boundary value problem

$$\mathscr{L}u_0 = \tilde{f}(x) \quad \text{for } x \in \overline{\Omega}, \qquad u_0(x) = 0 \quad \text{for } x \in \partial\Omega, \tag{4.5}$$

has no solution, in general, and the construction of the asymptotics for $u(x, \varepsilon)$ encounters new difficulties as compared with those of §1. A sufficient and necessary condition for the solvability of the problem (4.5) is given by the following equality (see [17, Chapter 4, §10]):

$$\int_\Omega h(x)\tilde{f}(x)\,dx = 0. \tag{4.6}$$

Suppose, to be definite, that this equality does not hold for $f(x) = \tilde{f}(x)$. Then the problem (4.5), which can naturally be considered as the limit problem for the problem (4.1)–(4.3), has no solution. On the other hand, for each positive $\varepsilon < \varepsilon_0$, the problem (4.1)–(4.3) has a unique solution (this will be proved below in the corollary to Lemma 4.6 under the extra condition (4.7)). Since the limit problem (4.5) has no solution, there is no reason to expect that the limit of $u(x, \varepsilon)$ as $\varepsilon \to 0$ exists. Nevertheless, this does not void the question about the asymptotics of $u(x, \varepsilon)$ as $\varepsilon \to 0$. This asymptotics essentially depends on whether the eigenfunction $h(x)$ vanishes at the origin or not. We restrict our attention to the second, simpler case $h(0) \neq 0$ only. Then one can assume, without any loss of generality, that

$$h(0) = 1. \tag{4.7}$$

Everywhere in this section we denote by $h(x)$ the eigenfunction defined above and satisfying conditions (4.4), (4.7). As before, $P_l(x)$, $Q_l(x)$, $R_l(x)$, $P_{l,j}(x)$, etc. denote homogeneous polynomials of degree l, and $X_l(x)$, $Y_l(x)$, $Z_l(x)$, $W_l(x)$, $X_{i,j}(x)$, etc. homogeneous harmonic polynomials of degree l.

We now begin with the construction of the asymptotics of $u(x, \varepsilon)$. One is led to conclude from the plausible considerations given above that this function is unbounded for $\varepsilon \to 0$. Therefore, we will look for its inner and outer expansions in the form

$$U = \sum_{k=-1}^{\infty} \varepsilon^k u_k(x) \tag{4.8}$$

and

$$V = \sum_{i=-1}^{\infty} \varepsilon^i v_i(\xi), \tag{4.9}$$

where $\xi = \varepsilon^{-1}x$. If the series U and V are f.a.s. of equation (4.1), then necessarily

$$\mathscr{L} u_0 = f(x), \tag{4.10}$$

$$\mathscr{L} u_k = 0, \text{ for } k \neq 0, x \in \overline{\Omega}\backslash O, \tag{4.11}$$

and

$$\Delta v_{-1} = 0, \quad \Delta v_0 = 0, \quad \Delta v_1 + a^2 v_{-1} = 0,$$
$$\Delta v_i = -a^2 v_{i-2} + R_{i-2}(\xi), \qquad i \geq 2, \text{ for } \xi \in \mathbf{R}^3\backslash\omega, \tag{4.12}$$

where $f(x) = \sum_{j=0}^{\infty} R_j(x)$ for $x \to 0$.

The boundary conditions are induced by conditions (4.2), (4.3) in a natural way:

$$u_k(x) = 0 \quad \text{for } x \in \partial\Omega, \tag{4.13}$$

$$v_i(\xi) = 0 \quad \text{for } \xi \in \partial\omega. \tag{4.14}$$

It is natural to set the principal term of the series U, i.e., the function $u_{-1}(x)$, equal to $c_{-1}h(x)$, but the value of c_{-1} is not yet clear. The problem (4.10), (4.13) has no solution that is smooth everywhere in $\overline{\Omega}$. Accordingly, a solution for this problem will be sought in the class of functions having singularities at the point O. The situation is, therefore, very similar to that considered in §1, but the singularities of the functions now begin with $k = 0$ instead of $k = 1$. As to the problems (4.12), (4.14), they do not differ, at least externally, from similar problems of §1. Thus, our first task is to find the structure of the solutions of equations (4.10), (4.11) having singularities at the origin.

LEMMA 4.1. *For any function* $f(x) \in C^\infty(\overline{\Omega})$, *there exists a function* $u(x) \in C^\infty(\overline{\Omega}\backslash O)$ *such that*

$$u(x) = f(x) \quad \text{in } \overline{\Omega}\backslash O, \qquad u(x) = 0 \quad \text{on } \partial\Omega$$

and

$$u(x) = \sum_{j=0}^{\infty} X_{0,j} r^{-1+2j} + \sum_{j=0}^{\infty} P_j(x), \qquad r \to 0, \tag{4.15}$$

where

$$X_{0,0} = (4\pi)^{-1} \int_\Omega f(x)h(x)\,dx. \tag{4.16}$$

PROOF. We begin with the construction of an f.a.s. of the equation $\Delta u + a^2 u = 0$ as $r \to 0$ such that the asymptotic series starts with the term $X_{0,0}r^{-1}$, where the constant $X_{0,0}$ is defined by formula (4.16). Denote this f.a.s. by

$$\mathscr{E} = \sum_{j=0}^{\infty} X_{0,j} r^{-1+2j}. \tag{4.17}$$

Inserting this series into the equation, one obtains the recurrence system

$$\Delta(X_{0,j}r^{-1+2j}) = -a^2 X_{0,j-1}r^{-3+2j}, \qquad j \geq 1.$$

All $X_{0,j}$ are now found uniquely, which is easily verified directly, although one can also use equality (1.33). Denoting $\mathscr{E}_N(x) = \chi(x)B_N\mathscr{E}$, where $\mathscr{E}$ is the series (4.17), N is a sufficiently large number, and $\chi(x) \in C^\infty(\Omega)$ is a truncating function equal to 1 in a neighborhood of the origin and vanishing in a neighborhood of $\partial\Omega$, we conclude that

$$\mathscr{L}\mathscr{E}_N(x) = g_N(x) \in C^N(\overline{\Omega}), \qquad \mathscr{E}_N(x) = 0 \quad \text{on } \partial\Omega. \tag{4.18}$$

Construct a function $w_N(x) \in C^N(\overline{\Omega})$ vanishing on $\partial\Omega$ such that

$$\mathscr{L}w_N = f(x) - g_N(x) + b_N h(x), \tag{4.19}$$

where the constant b_N is chosen in such a way that the right-hand side of this equation $\tilde{f}(x) = f(x) - g_N(x) + b_N h(x)$ satisfies condition (4.6). This is sufficient for a solution of the problem (4.19) to exist.

Let x_0 be a point in $\Omega\backslash O$, where $h(x_0) \neq 0$. Denote $u_N(x) = w_N(x) + \mathscr{E}_N(x) + d_N h(x)$, where the constant d_N is chosen in such a way that $u_N(x_0) = 0$. By construction, $u_N(x) \in C^N(\Omega\backslash O)$, $u_N(x) = 0$ on $\partial\Omega$, and

$$\mathscr{L}u_N = f(x) + b_N h(x). \tag{4.20}$$

In addition, for $x \to 0$, the function $u_N(x)$ has an asymptotic representation in the form of a partial sum of the series (4.15). The constants $X_{0,j}$ are constructed above, and the polynomials $P_j(x)$, in general, depend on N and owe their origin to the Taylor expansions of the functions $w_N(x)$ and $h(x)$.

Denote by S_δ the ball of radius δ with center at the origin, and by G_δ the domain $\Omega\backslash S_\delta$. Multiply both sides of equation (4.20) by $h(x)$ and integrate the resulting equality over G_δ for δ small. Integrating by parts, one has

$$\int_{G_\delta} [f(x)h(x) + b_N h^2(x)]\,dx = \int_{G_\delta} h(x)(\Delta u_N + a^2 u_N)\,dx$$

$$= \int_{r=\delta} \sum_{j=1}^{3} \left[u_N(x)\frac{\partial h}{\partial x_j}(x) - h(x)\frac{\partial u_N}{\partial x_j}(x)\right] \frac{x_j}{\delta}\,ds. \tag{4.21}$$

Replacing the function $u_N(x)$ in the right-hand side of this equality by its asymptotic representation as $x \to 0$, one obtains $4\pi X_{0,0} + O(\delta)$ for the value of the integral. Passing to the limit as $\delta \to 0$ in equality (4.21), one has

$$\int_\Omega f(x)h(x)\,dx + b_N \int_\Omega h^2(x)\,dx = 4\pi X_{0,0}.$$

It follows from this and from condition (4.16) that $b_N = 0$.

Thus, all functions $u_N(x)$ satisfy the equation $\mathscr{L} u_N = f(x)$ in $\Omega \backslash O$. The difference $u_N(x) - u_{N+1}(x)$ satisfies the homogeneous equation and is bounded. The theorem on a removable singularity (see [82, Chapter 4]) implies that $u_N(x) - u_{N+1}(x) \in C^\infty(\overline{\Omega})$. Since this difference vanishes on $\partial\Omega$, one has $u_N(x) - u_{N+1}(x) = \beta_N h(x)$. Since, by construction, $u_N(x_0) = u_{N+1}(x_0) = 0$, we have $u_N(x) \equiv u_{N+1}(x)$. Thus, $u_N(x)$ does not depend on N, and, therefore, is the desired function $u(x)$. ∎

Remark. Suppose that the condition $f(x) \in C^\infty(\Omega)$ in Lemma 4.1 is replaced by the condition $f(x) \in C^N(\overline{\Omega})$, where N is a sufficiently large integer. The conclusion of the lemma is then valid in a slightly modified form: the relation $u(x) \in C^\infty(\overline{\Omega}\backslash O)$ is replaced by $u(x) \in C^{N_1}(\overline{\Omega}\backslash O)$, and the series (4.15) by the relation

$$u(x) = \sum_{j=0}^{N_1} X_{0,j} r^{-1+2j} + \sum_{j=0}^{N_1} P_j(x) + O(r^{N_1}),$$

where N_1 grows unboundedly together with N. The proof virtually repeats that of Lemma 4.1.

LEMMA 4.2. *Let m be a positive integer, and $Y_m(x)$ a homogeneous harmonic polynomial. Then there exists a function $\mathscr{E}(x) \in C^\infty(\overline{\Omega}\backslash O)$ which vanishes on $\partial\Omega$, satisfies the equation $\mathscr{L}\mathscr{E} = 0$ in $\overline{\Omega}\backslash O$, and expands, as $x \to 0$, in the asymptotic series*

$$\mathscr{E}(x) = \sum_{j=0}^{\infty} X_{m,j}(x) r^{-1+2j-2m} + \sum_{j=0}^{\infty} X_{0,j} r^{-1+2j} + \sum_{j=0}^{\infty} P_j(x), \qquad r \to 0,$$
(4.22)

where $X_{m,0}(x)$ equals $Y_m(x)$ given above.

PROOF. Construct an f.a.s. of the equation $\mathscr{L}\mathscr{E} = 0$ in the form

$$\mathscr{E} = \sum_{j=0}^{\infty} X_{m,j}(x) r^{-1+2j-2m}, \quad \text{where } X_{m,0}(x) = Y_m(x).$$

Inserting this series into the equation, one obtains the recurrence system

$$\Delta(X_{m,j}(x) r^{-1+2j-2m}) = -a^2 X_{m,j-1}(x) r^{-3+2j-2m}, \qquad j \geq 1,$$

which, according to formula (1.33), defines all $X_{m,j}(x)$. As in the proof of Lemma 4.1, one now constructs the function $\mathscr{E}_N(x) = \chi(x) B_N \mathscr{E}$ satisfying relation (4.18).

According to the remark to Lemma 4.1, we construct the function $w_N(x) \in C^{N_1}(\overline{\Omega}\backslash O)$, which vanishes on $\partial\Omega$, satisfies equation $\mathscr{L} w_N = -g_N(x)$ in $\overline{\Omega}\backslash O$, and, as $x \to 0$, admits the asymptotic representation

$$w_N(x) = \sum_{j=0}^{N_1} X_{0,j,N} r^{-1+2j} + \sum_{j=0}^{N_1} P_{j,N}(x) + O(r^{N_1}),$$

where $X_{0,0,N} = -\frac{1}{4\pi} \int_\Omega g_N(x) h(x)\, dx$. Suppose that $h(x_0) \neq 0$. Let $u_N(x) = w_N(x) + \mathscr{E}_N(x) + d_N h(x)$, where the constant d_N is chosen in such a way that

$u_N(x_0) = 0$. By construction, the function $u_N(x)$ vanishes on $\partial\Omega$, and satisfies the equation $\mathscr{L}u_N = 0$ in $\overline{\Omega}\backslash O$. The function $z_N(x) = u_N(x) - u_{N+1}(x)$ also satisfies the equation $\mathscr{L}z_N = 0$ in $\overline{\Omega}\backslash O$, and has the asymptotic representation

$$z_N(x) = \sum_{j=0}^{N_1}(X_{0,j,N} - X_{0,j,N+1})r^{-1+2j} + \sum_{j=0}^{N_1}(P_{j,N}(x) - P_{j,N+1}(x)) + O(r^{N_1}).$$

Multiply both sides of the equation for $z_N(x)$ by $h(x)$, and integrate the resulting equality over the domain G_δ. Proceeding now as in the proof of Lemma 4.1 (see (4.21)), and passing to the limit as $\delta \to 0$, one obtains the equality $X_{0,0,N} = X_{0,0,N+1}$. Therefore, $z_N(x)$ is a bounded solution of the homogeneous equation, $z_N(x) = 0$ on $\partial\Omega$, whence $z_N(x) = c_1 h(x)$. The equality $z_N(x_0) = 0$ implies that $z_N(x) \equiv 0$. Thus, $u_N(x)$ does not depend on N, and is the desired function $\mathscr{E}(x)$ (see (4.22)). ∎

Having concluded the discussion of preliminary results necessary for the construction of the coefficients of the outer expansion, we now begin with the analysis of the problems (4.12), (4.14). In the following constructions, an important role is played by the harmonic function defined in $\mathbf{R}^3\backslash\omega$, equal to 1 on $\partial\omega$, and vanishing at infinity. Denote this function by $\tilde{\Gamma}(\xi)$. Such a function exists (see [118, Chapter 4, §5]), and, according to Lemma 1.2, has the following asymptotic expansion at infinity:

$$\tilde{\Gamma}(\xi) = \sum_{j=0}^{\infty} Z_j(\xi)\rho^{-1-2j}, \qquad \rho = |\xi| \to \infty. \tag{4.23}$$

The coefficient Z_0 is called the capacity of the surface $\partial\omega$ (see [17, p. 305]). Let us prove that $Z_0 > 0$. Indeed, in the domain $S_\delta\backslash\omega$, where δ is sufficiently large, one obtains, by virtue of (4.23), the equality

$$0 = \int_{\partial(S_\delta\backslash\omega)} \frac{\partial\tilde{\Gamma}}{\partial n} ds = \int_{\partial\omega} \frac{\partial\tilde{\Gamma}}{\partial n} ds + \int_{\partial S_\delta} \frac{\partial\tilde{\Gamma}}{\partial n} ds = \int_{\partial\omega} \frac{\partial\tilde{\Gamma}}{\partial n} ds - 4\pi Z_0 + O\left(\delta^{-1}\right),$$

where $\partial/\partial n$ denotes the derivative along the outer normal to $S_\delta\backslash\omega$. Since the function $\tilde{\Gamma}(\xi)$ assumes its maximum value on $\partial\omega$, we have $\partial\tilde{\Gamma}/\partial n|_{\partial\omega} > 0$ (see [100, §28]), and, therefore, $Z_0 > 0$. Let $\Gamma(\xi) = Z_0^{-1}[\tilde{\Gamma}(\xi) - 1]$. Thus,

$$\Delta\Gamma(\xi) = 0 \quad \text{for } \mathbf{R}^3\backslash\omega, \qquad \Gamma(\xi) = 0 \quad \text{for } \partial\omega \text{ and}$$
$$\Gamma(\xi) = \rho^{-1} - Z_0^{-1} + \sum_{j=1}^{\infty} W_j(\xi)\rho^{-1-2j}, \qquad \rho \to \infty. \tag{4.24}$$

Recall the meaning of Lemma 1.3 proved in §1: given the principal terms of the asymptotics of a harmonic function at infinity, one can construct that function $v(\xi)$ in such a way that it vanishes on $\partial\omega$. By the principal terms there we meant all terms that do not tend to zero, i.e., simply said, the harmonic polynomial. In the present section, another way of reconstructing a harmonic function from its asymptotics at infinity will be needed. Indeed, the outer expansion will be constructed according to Lemmas 4.1 and 4.2, which show that, in contrast to §1, the coefficient of r^{-1} can no longer be chosen arbitrarily (see formula (4.16)). But the zero term of the asymptotics at the point C (i.e., the constant) can be varied by adding a function of the form $c_1 h(x)$ to the solution $u(x)$. Lemma 1.3 has to be modified accordingly: one has to be able to define the term $c\rho^{-1}$ arbitrarily. However, this has to be paid for by the loss of control over the zero term of the asymptotics at infinity, i.e, the constant: it is now determined by the other terms of the asymptotics uniquely. The precise statement is given by the following lemma.

LEMMA 4.3. *Given a harmonic polynomial* $Y(\xi)$ *and a constant* Y_0, *there exists a function* $w(\xi) = \tilde{v}(\xi) + Y(\xi)$ *such that* $\Delta w(\xi) = 0$ *for* $\xi \in \mathbf{R}^3 \backslash \omega$, $w(\xi) = 0$ *for* $\xi \in \partial \omega$, *and*

$$\tilde{v}(\xi) = X_0 + Y_0 \rho^{-1} + \sum_{j=1}^{\infty} Y_j(\xi) \rho^{-2j-1}, \qquad \rho \to \infty.$$

PROOF. Construct the function $\tilde{v}(\xi)$ according to Lemma 1.3 in such a way that the following relations are valid for the function $v(\xi) = \tilde{v}(\xi) + Y(\xi)$:

$$\Delta v(\xi) = 0 \quad \text{for } \xi \in \mathbf{R}^3 \backslash \omega, \qquad v(\xi) = 0 \quad \text{for } \xi \in \partial \omega,$$

$$v(\xi) = Y(\xi) + \sum_{j=0}^{\infty} X_j(\xi) \rho^{-2j-1}, \qquad \rho \to \infty.$$

It remains to set $w(\xi) = v(\xi) + (Y_0 - X_0)\Gamma(\xi)$, where $\Gamma(\xi)$ is the function constructed above, and satisfying relations (4.24). ■

Lemma 1.6 has also to be modified in exactly the same way.

LEMMA 4.4. *Suppose that the conditions of Lemma* 1.6 *are satisfied. Then there exists a function* $w(\xi) \in C^{\infty}(\mathbf{R}^3 \backslash \omega)$ *such that*

$$\Delta w(\xi) = F(\xi) \quad \text{for } \xi \in \mathbf{R}^3 \backslash \omega, \qquad w(\xi) = 0 \quad \text{for } \xi \in \partial \omega$$

and

$$w(\xi) = \sum_{j=0}^{\infty} h_i(\xi) + Y_0 + \sum_{j=1}^{\infty} Y_j(\xi) \rho^{-2j-1}, \qquad \rho \to \infty.$$

PROOF. One should add the function $-X_0 \Gamma(\xi)$ to the function $v(\xi)$ constructed in Lemma 1.6 and having the asymptotic expansion (1.44). Here $\Gamma(\xi)$ is the same function as above. ■

We can now begin with the construction of the series U and V. As before, we will use Table 7 (next page), whose structure repeats that of the preceding tables.

We will begin the construction of the series U with the term $u_0(x)$, and not with the principal term $u_{-1}(x)$ as in §1. By Lemma 4.1, construct a solution of the problem (4.10), (4.13) which, as $r \to 0$, has the asymptotic expansion (4.15). This function is not defined uniquely but up to a summand

TABLE 7

U \ V	$\varepsilon^{-1}v_{-1}(\xi)$	$v_0(\xi)$	$\varepsilon v_1(\xi)$	$\varepsilon^2 v_2(\xi)$	$\varepsilon^3 v_3(\xi)$
	$\vdots$	$\vdots$	$\vdots$	$\vdots$	$\vdots$
$\varepsilon^{-1}u_{-1}(x)$	$\varepsilon^{-1}R_0$ $\varepsilon^{-1}P_{0,-1}$	$P_{1,-1}(\xi)$ $\varepsilon^{-1}P_{1,-1}(x)$	$\varepsilon P_{2,-1}(\xi)$ $\varepsilon^{-1}P_{2,-1}(x)$	$\varepsilon^2 P_{3,-1}(\xi)$ $\varepsilon^{-1}P_{3,-1}$	$\varepsilon^3 P_{4,-1}(\xi)$ $\varepsilon^{-1}P_{4,-1}(x)$
$u_0(x)$	$\varepsilon^{-1}X_{0,0}\rho^{-1}$ $X_{0,0}r^{-1}$	$Q_{0,0}$ $Q_{0,0}$	$\varepsilon(X_{0,1}\rho + Q_{1,0}(\xi))$ $X_{0,1}r + Q_{1,0}(x)$	$\varepsilon^2 Q_{2,0}(\xi)$ $Q_{2,0}(x)$	$\varepsilon^3(X_{0,2}\rho^3 + Q_{3,0}(\xi))$ $X_{0,2}r^3 + Q_{3,0}(x)$
$\varepsilon u_1(x)$	$\varepsilon^{-1}X_{1,0}(\xi)\rho^{-3}$ $\varepsilon X_{1,0}(x)r^{-3}$	$X_{0,0,1}\rho^{-1}$ $\varepsilon X_{0,0,1}r^{-1}$	$\varepsilon(X_{1,1}(\xi)\rho^{-1} + Q_{0,1})$ $\varepsilon(X_{1,1}(x)r^{-1} + Q_{0,1})$	$\varepsilon^2(Q_{1,1}(\xi) + X_{0,1,1}\rho)$ $\varepsilon(Q_{1,1}(x) + X_{0,1,1}r)$	$\varepsilon^3(X_{1,2}(\xi)\rho + Q_{2,1}(\xi))$ $\varepsilon(X_{1,2}(x)r + Q_{2,1}(x))$
$\varepsilon^2 u_2(x)$	$\varepsilon^{-1}X_{2,0}(\xi)\rho^{-5}$ $\varepsilon^2 X_{2,0}(x)r^{-5}$	$X_{1,0}(\xi)\rho^{-3}$ $\varepsilon^2 X_{1,0}(x)r^{-3}$	$\varepsilon(X_{2,1}(\xi)\rho^{-3} + X_{0,0,2}\rho^{-1})$ $\varepsilon^2(X_{2,1}(x)r^{-3} + X_{0,0,2}r^{-1})$	$\varepsilon^2 Q_{0,2}$ $\varepsilon^2 Q_{0,2}$	$\varepsilon^3(X_{2,2}(\xi)\rho^{-1} + X_{0,1,2}\rho + Q_{1,2}(\xi))$ $\varepsilon^2(X_{2,2}(x)r^{-1} + X_{0,1,2}r + Q_{1,2}(x))$
$\varepsilon^3 u_3(x)$	$\varepsilon^{-1}X_{3,0}(\xi)\rho^{-7}$ $\varepsilon^3 X_{3,0}(x)r^{-7}$	$X_{2,0}(\xi)\rho^{-5}$ $\varepsilon^3 X_{2,0}(x)r^{-5}$	$\varepsilon(X_{3,1}(\xi)\rho^{-5} + X_{1,1}(\xi)\rho^{-3})$ $\varepsilon^3(X_{3,1}(x)r^{-5} + X_{1,1}(x)r^{-3})$	$\varepsilon^2(X_{0,0,3}\rho^{-1} + X_{2,1}\rho^{-3})$ $\varepsilon^3(X_{0,0,3}r^{-1} + X_{2,1}r^{-3})$	$\varepsilon^3(X_{3,2}(\xi)\rho^{-3} + Q_{0,3})$ $\varepsilon^3(X_{3,2}(x)r^{-3} + Q_{0,3})$

$$u_{-1}(x) = -\frac{h(x)}{4\pi Z_0}\int_\Omega f(x)h(x)\,dx$$

of the form $c_1 h(x)$. We will, for the time being, choose some solution $\tilde{u}_0(x)$. This function and other similar functions will be called "pre-solutions" to the corresponding solutions $u_k(x)$. Expanding the pre-solution $\tilde{u}_0(x)$ into the asymptotic series as $x \to 0$, one determines the functions appearing in the lower halves of the squares in the $u_0(x)$ row. Rewriting these functions in terms of the inner variables, one obtains the functions in the upper halves of these squa.es, e.g., $\varepsilon^{-1} X_{0,0} \rho^{-1}$. Proceeding now from the matching condition, we set $v_{-1}(\xi) = X_{0,0} \Gamma(\xi)$, where $\Gamma(\xi)$ is the function satisfying conditions (4.24). This determines all functions in the first column of the table, and it follows from (4.16) and (4.24) that

$$R_0 = -\frac{1}{4\pi Z_0} \int_\Omega f(x) h(x) \, dx. \tag{4.25}$$

One has to set, therefore, $u_{-1}(x) = R_0 h(x)$. The functions $v_{-1}(\xi)$ and $u_{-1}(x)$ are finally determined.

Next, according to Lemma 4.2, one constructs the functions $\tilde{u}_k(x)$ (pre-solutions for $u_k(x)$) from the principal terms of the asymptotic expansions, i.e., the functions $X_{k,0}(x) r^{-1-2k}$ for $k \geq 1$. This fills, although not conclusively, all the lower halves of the squares in Table 7. Rewrite them in the inner variables in the upper half-squares of the table. In what follows, this operation will be performed without special mention: we will always assume that both upper and lower parts of the same square contain one and the same function but expressed in different variables.

Thus, all squares of the Table 7 are provisionally filled. Since the functions $\tilde{u}_k(x)$ satisfy the system (4.10), (4.11), then, according to Lemma 1.5, the series $V_i(\xi)$ appearing in the columns are f.a.s. of the system (4.12). According to Lemma 4.4, we can now construct the function $v_0(\xi)$, i.e., the solution of the problem (4.12) and (4.14), from the series $V_0(\xi)$. This does not change the term $X_{0,0} \rho^{-1}$ in the second column, but, in general, alters the constant $Q_{0,0}$, and, beginning with the fourth row, gives rise to the terms $X_{j,0}(\xi) \rho^{-2j-1}$, $j \geq 1$. Now the function $v_0(\xi)$ is determined conclusively. For the final determination of the function $u_0(x)$, one adds the term $c_1 h(x)$ to the function $\tilde{u}_0(x)$ in order to match the constants $Q_{0,0}$.

We now alter the functions $\tilde{u}_k(x)$ for $k \geq 2$ by $X_{j,0}(\xi) \rho^{-1-2j}$ according to Lemma 4.1, adding the terms compensating for the alterations in these functions. Next, according to Lemmas 1.5 and 4.4 one finally constructs the function $v_1(\xi)$, which changes the constant $Q_{0,1}$ in the table, and adds the terms $X_{j,1}(\xi) \rho^{-2j-1}$, $j \geq 1$, to the corresponding squares. Finally, the function $u_1(x)$ is determined by adding to the function $\tilde{u}_1(x)$ the term $c_1 h(x)$, and modifying the function $u_k(x)$ for $k \geq 3$. The process can evidently be continued indefinitely.

Thus, we have constructed the functions $u_k(x)$ and $v_i(\xi)$—the solutions of the problems (4.10), (4.11), (4.13) and (4.12), (4.14), respectively—so that the matching condition for the series (4.8) and (4.9)

$$A_{N_1,\xi} A_{N_2,x} U = A_{N_2,x} A_{N_1,\xi} V \quad \forall N_1, N_2$$

is satisfied.

THEOREM 4.1. *For all positive integers* N, *the estimate*

$$|A_{N,x}U + A_{N,\xi}V - A_{N,x}A_{N,\xi}V - u(x,\varepsilon)| \leq M\varepsilon^{N_1} \qquad (4.26)$$

holds everywhere in Ω_ε, *where* $u(x,\varepsilon)$ *is the solution of the problem* (4.1)–(4.3), U *and* V *are the series* (4.8) *and* (4.9) *constructed above, and* $N_1 \to \infty$ *as* $N \to \infty$.

PROOF. Both $\mathscr{L} T_N(x,\varepsilon)$ and the values of $T_N(x,\varepsilon)$ on $\partial\Omega_\varepsilon$, where $T_N(x,\varepsilon)$ denotes the expression the absolute value of which appears on left-hand side of (4.26), are estimated as in Theorem 1.4. Thus,

$$\max_{x\in\overline{\Omega}_\varepsilon} |\mathscr{L} T_N(x,\varepsilon)| + \max_{x\in\partial\Omega} |T_N(x,\varepsilon)| < M\varepsilon^{(N-1)/2}. \qquad (4.27)$$

However, the rest of the proof is more complicated, because in this case there is no bound of the form (1.22) for the norm of the inverse operator. Accordingly, the estimate (4.27) does not imply a similar bound for $T_N(x,\varepsilon)$. However, no uniform bound for the inverse operator is necessary for the justification of the asymptotics. It is enough to prove that the norm of the inverse operator does not exceed $M\varepsilon^{-q}$, where q is a fixed positive number. Below we shall prove the estimate $\|u(x,\varepsilon)\|_{C(\overline{\Omega}_\varepsilon)} \leq M\varepsilon^{-1}\|f(x)\|_{C^n(\overline{\Omega})}$, where n is a fixed number, and $u(x,\varepsilon)$ is any solution of the problem (4.1)–(4.3) (Lemma 4.6). If one follows the proof of Theorem 1.4 step-by-step, it is not difficult to show that the estimates of $\mathscr{L} T_N(x,\varepsilon)$ and $T_N(x,\varepsilon)|_{\partial\Omega}$ can be obtained in a stronger norm. The expression $\|\mathscr{L} T_N(x,\varepsilon)\|_{C(\overline{\Omega})} + \|T_N(x,\varepsilon)\|_{C(\partial\Omega)}$ in inequality (4.27) can be replaced with $\|\mathscr{L} T_N(x,\varepsilon)\|_{C^m(\overline{\Omega}_\varepsilon)} + \|T_N(x,\varepsilon)\|_{C^m(\partial\Omega)}$, where m is any given number, and the estimate may suffer only a slight deterioration: the expression $M\varepsilon^{(N-1)/2}$ on the right-hand side is replaced by $M\varepsilon^{N_2}$, where $N_2 \to \infty$ as $N \to \infty$. The function $T_N(x,\varepsilon)|_{\partial\Omega_\varepsilon}$ can now be continued by the function $\overline{T}_N(x,\varepsilon)$ smoothly everywhere in $\overline{\Omega}$ so that $\|\overline{T}_N(x,\varepsilon)\|_{C^m(\overline{\Omega}_\varepsilon)} \leq M\varepsilon^{N_3}$ where $N_3 \to \infty$ as $N \to \infty$. Applying Lemma 4.5 to the difference $T_N(x,\varepsilon) - \overline{T}_N(x,\varepsilon)$, one obtains the estimate (4.26). ∎

LEMMA 4.5. *There exists a constant* $d > 0$ *depending only on* a^2 *such that for any function* $u(x,\varepsilon) \in C^2\left(\overline{S_d\backslash\omega_\varepsilon}\right)$ *satisfying condition* (4.3), *and for all sufficiently small* ε, *the estimate*

$$|u(x,\varepsilon)| \leq M\left(\max_{x\in S_d\backslash\omega_\varepsilon} |\mathscr{L}u(x,\varepsilon)| + \max_{|x|=d} |u(x,\varepsilon)|\right) \qquad (4.28)$$

holds, where the constant M is independent of ε.

PROOF. Make the change $u(x, \varepsilon) = (4d^2 - r^2)v(x, \varepsilon)$, and note that $\mathscr{L}(4d^2 - r^2) = -6 + a^2(4d^2 - r^2) < -5$ for sufficiently small d and $r \le d$. Inequality (4.28) now follows from the maximum principle for $v(x, \varepsilon)$. ∎

The estimate (4.28) is actually the estimate (1.22) of §1, so that in the domain $S_d \backslash \omega_\varepsilon$ for the solution of equation (4.1) the asymptotic expansion considered in subsection 2 of §1 is valid. Although the zero condition on the boundary $|x| = d$ is now replaced by another function, the latter, provided it is uniformly smooth with respect to ε, can be extended inside $S_d \backslash \omega_\varepsilon$ which reduces the problem to the problem (1.19)–(1.21). Since the function $v_0(\xi)$ in the asymptotic expansion (1.24) satisfies the homogeneous equation (1.29), it only slightly depends on the boundary function for $|x| = d$. The function $v_0(\xi)$, as the proof of Theorem 1.3 shows, is a harmonic function which vanishes on $\partial \omega$ and tends to $u_0(0)$ as $\xi \to \infty$. Theorem 1.4 implies that, for $x \in S_{M\varepsilon} \backslash \omega_\varepsilon$, where M is sufficiently large fixed number, the relations

$$u(x, \varepsilon) = v_0(\xi) + O(\varepsilon), \qquad \frac{\partial u}{\partial n}\bigg|_{\partial \omega_\varepsilon} = \frac{1}{\varepsilon}\frac{\partial v_0}{\partial n}\bigg|_{\partial \omega_\varepsilon} + O(1) \qquad (4.29)$$

hold, where the estimate $O(\cdot)$ is uniform with respect to the values of $u(x, \varepsilon)|_{|x|=d}$ provided these values are uniformly bounded together with a finite number of their derivatives. We can now proceed with the proof of the estimate of $u(x, \varepsilon)$ in the domain $\overline{\Omega}_\varepsilon$.

LEMMA 4.6. *Let $u(x, \varepsilon)$ be a solution of the problem (4.1)–(4.3). Then the estimate*

$$|u(x, \varepsilon)| \le M\varepsilon^{-1}\|f(x)\|_{C^n(\overline{\Omega})} \qquad (4.30)$$

holds for all sufficiently small ε, where n is a number which does not depend on ε.

PROOF. Suppose that the estimate (4.30) does not hold. Then there exist sequences $\varepsilon_k \to 0$ and $f_k(x)$ such that

$$\varepsilon_k^{-1}\|f_k\|_{C^n(\overline{\Omega})} \xrightarrow[k\to\infty]{} 0, \qquad (4.31)$$

and the corresponding solutions $u_k(x, \varepsilon_k)$ satisfy the relations:

$$\max_{\overline{\Omega}_{\varepsilon_k}} |u_k(x, \varepsilon_k)| = 1. \qquad (4.32)$$

Here n is a sufficiently large number, the choice of which will be made clear by what follows.

Using a priori estimates for the solution of equation (4.1) and estimates of the derivatives in domains adjoining $\partial \Omega$ (see [61, Chapter 3]), choose a subsequence $u_k(x, \varepsilon_k)$ converging in $C^n\{\mathscr{K}\}$ for any compact $\mathscr{K} \subset \overline{\Omega}\backslash O$. The limit function $u(x)$ satisfies the equation $\mathscr{L}u = 0$ in $\overline{\Omega}\backslash O$, vanishes on $\partial \Omega$ and is bounded. The theorem on a removable singularity (see [82, Chapter 4]) implies that $u(x) \in C^\infty(\overline{\Omega})$, and, consequently, $u(x) = c_1 h(x)$. One can assume that the sequence $u_k(x, \varepsilon_k)$ coincides with the subsequence chosen above. Thus, $u_k(x, \varepsilon_k) \to c_1 h(x)$ as $k \to \infty$ uniformly on any compact $\mathscr{K} \subset \overline{\Omega}\backslash O$ together with their derivatives to order n. It follows from Lemma 4.5 and condition (4.32) that $c_1 \ne 0$. We can assume, without any loss of generality, that $c_1 > 0$. Applying Green's formula to the functions $u_k(x, \varepsilon_k)$ and $h(x)$, and omitting the subscript k in ε_k in the notation for domains, one has

$$\int\limits_{\Omega_\varepsilon} f_k(x)h(x)\,dx = \int\limits_{\Omega_\varepsilon} h(x)\mathscr{L}u_k(x\,,\varepsilon_k)\,dx = \int\limits_{\partial\omega_\varepsilon} h(x)\frac{\partial u_k}{\partial n}\,ds.$$

Hence, by (4.31),

$$\varepsilon_k^{-1}\int\limits_{\partial\omega_\varepsilon} h(x)\frac{\partial u_k}{\partial n}\,ds \xrightarrow[k\to\infty]{} 0. \tag{4.33}$$

On the other hand, by virtue of (4.29), this integral equals

$$\varepsilon_k^{-1}\int\limits_{\partial\omega_\varepsilon} (1+O(\varepsilon_k))\left(\frac{1}{\varepsilon_k}\frac{\partial v_{0,k}}{\partial n}+O(1)\right)\,ds = \int\limits_{\partial\omega}\frac{\partial v_{0,k}}{\partial n}\,ds + O(\varepsilon_k). \tag{4.34}$$

Since, by assumption, $u_k(x\,,\varepsilon_k) \to c_1 h(x)$ for $|x| = d$ and $k \to \infty$, we have $u_{0,k}(0\,,\varepsilon_k) = c_{1,k} \to c_1$ as $k \to \infty$. Therefore, $v_{0,k}(\xi)$ are harmonic functions vanishing on $\partial\omega$, tending to $c_{1,k}$ at infinity, and uniformly converging to $v_0(\xi)$. The last harmonic function also vanishes on $\partial\omega$, and tends to $c_1 > 0$ at infinity. It follows from (4.33) and (4.34) that $\int_{\partial\omega}(\partial v_0/\partial n)\,ds = 0$. Since for this function $v_0(\xi)$ one actually has $\partial v_0/\partial n|_{\partial\omega} < 0$ (see [100, §28]), the contradiction proves the lemma. ■

COROLLARY TO LEMMA 4.6. *The estimate* (4.30) *implies that the solution of the problem* (4.1)–(4.3) *is unique, and, consequently (see* [82, Chapter 4]*), this problem is solvable for all sufficiently small* ε.

COROLLARY TO THEOREM 4.1. *The series* (4.8) *is a uniform asymptotic expansion of the solution* $u(x\,,\varepsilon)$ *of the problem* (4.1)–(4.3) *for* $x \in \Omega$, $r > \varepsilon^\gamma$, *and the series* (4.9) *is a uniform asymptotic expansion of the same solution for* $r < \varepsilon^\gamma$, *where* γ *is any number such that* $0 < \gamma < 1$.

The proof repeats that of the corollary to Theorem 1.1 almost word by word. ■

It follows from Theorem 4.1 and the form of the series (4.8), (4.9) that the estimate

$$\left|u(x\,,\varepsilon) - \frac{R_0}{\varepsilon}(h(x) - 1 - C(\omega)\Gamma(\xi)) - u_0(x) - v_0(\xi)\right.$$
$$\left.+ R_0\left(\sum_{i=1}^{3} x_i\frac{\partial h}{\partial x_i}(0) - \frac{C(\omega)}{r}\right) + Q_{0,0}\right| < M\varepsilon \tag{4.35}$$

is valid everywhere in $\overline{\Omega}_\varepsilon$. Here the constant M depends neither on x, nor on ε, and $\Gamma(\xi)$ is a function satisfying conditions (4.24). According to (4.25),

$$R_0 = -\frac{1}{4\pi C(\omega)}\int\limits_\omega f(x)h(x)\,dx, \tag{4.36}$$

$C(\omega)$ is the capacity of the surface $\partial\omega$, and the functions $u_0(x)$, $v_0(\xi)$, and the constant $Q_{0,0}$ are those constructed above (see Table 7). Let us also write out the principal term of the asymptotic expansion of the solution $u(x, \varepsilon)$. By virtue of (4.35),

$$u(x, \varepsilon) \sim R_0 h(x)\varepsilon^{-1}, \tag{4.37}$$

at each point $x \in \overline{\Omega}\backslash O$ (R_0 is given by formula (4.36)). It is of interest to examine the behavior of the principal term (4.37) in the following situation. Suppose that a domain Ω and the right-hand side $f(x)$ are fixed; fix also the eigenfunction $h^*(x)$ and the form of the domain ω. Move the point O with a small domain ω_ε cut out around it over the domain Ω. Then $h(x) = [h^*(0)]^{-1}h^*(x)$ (assuming that $h^*(0) \neq 0$), and the principal term of the asymptotic expansion equals

$$-\frac{h^*(x)}{4\pi C(\omega_\varepsilon)} \int\limits_{\Omega} f(x)h^*(x)[h^*(0)]^{-2}\,dx,$$

where $C(\omega_\varepsilon) = \varepsilon C(\omega)$ is the capacity of the surface $\partial\omega_\varepsilon$. The first factor does not depend on the point O, while the second tends to infinity as O nears the boundary. It should be remembered that if the point O is close to the boundary $\partial\Omega$, one has to fix it first, and then to consider $\varepsilon \to 0$. The examination of the double asymptotics as $O \to \partial\Omega$ and $\varepsilon \to 0$ requires a new analysis, which can also be carried out using the method of matched asymptotic expansions, but this will not be considered here.

In conclusion we consider a specific example. Let Ω and ω be two unit balls such that the distance from the center O' of the ball Ω to the origin O equals $b\,(0 \leq b < 1)$. Suppose that the coordinates of the point O' are $(0, 0, -b)$, $f(x) \equiv -1$, and $a^2 = \pi^2$ is the first eigenvalue of the operator $-\Delta$ with the zero boundary conditions. In this case the first terms of both the outer and inner expansions can be written out explicitly.

Denote the eigenfunction $(r_1)^{-1}\sin \pi r_1$, where $r_1 = |x - O'|$, by $h^*(x)$. Then $h(x) = h^*(x)[h^*(0)]^{-1}$. The computations will be conducted as described above for the construction of the series U and V. Thus, $\tilde{u}_0(x)$ is the solution of the equation $\mathscr{L}\tilde{u}_0 = -1$, which, as $x \to 0$, has the function $r^{-1}X_{0,0}$ as the principal term of its asymptotic expansion, where

$$X_{0,0} = -\frac{1}{4\pi}\int\limits_{\Omega} h(x)\,dx = -[\pi h^*(0)]^{-1} = -b(\pi \sin \pi b)^{-1}.$$

It is convenient to represent $\tilde{u}_0(x)$ as the sum $u_0^*(x) + w(x)$, where

$$w(x) = X_{0,0}r^{-1}\cos \pi r - \pi^{-2}. \tag{4.38}$$

Since the principal term of the asymptotics of $w(x)$ at the point O is of the form $r^{-1}X_{0,0}$ and satisfies equation $\mathscr{L}w = -1$, the function $u_0^*(x)$ is a solution of the equation $\Delta u_0^* + \pi^2 u_0^* = 0$. It is continuous in Ω, and equals $-w(x)$ on the boundary $\partial\Omega$. Such a solution exists by virtue of Lemma

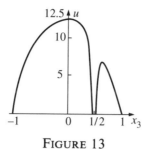

FIGURE 13

4.1, is defined up to $ch(x)$, and can be obtained by separation of variables. Without going into detailed computations, we write out the result

$$u_0^*(x) = \sum_{n=1}^{\infty} c_n \frac{1}{\sqrt{r_1}} \frac{J_{n+1/2}(\pi r_1)}{J_{n+1/2}(\pi)} P_n(\cos \theta), \tag{4.39}$$

where

$$c_n = \left(n + \frac{1}{2}\right) \frac{b}{\pi \sin \pi b} \int_{-1}^{1} \cos \pi b t P_n \left(b \frac{1-t^2}{2} - t\right) dt.$$

Here $r_1 = \sqrt{x_1^2 + x_2^2 + (x_3 + b)^2}$ is the distance from the point x to the center of the ball Ω, θ is the polar angle so that $\cos \theta = (x_3 + b) r_1^{-1}$, $P_n(y)$ are the Legendre polynomials, and $J_\nu(y)$ the Bessel functions. The series (4.39) converges, and for the interior points of the ball very rapidly.

Since ω is a unit ball, we have $\Gamma(\xi) = \rho^{-1} - 1$. Therefore, $v_{-1}(\xi) = X_{0,0}(\rho^{-1} - 1)$, $R_0 = -X_{0,0}$, and $u_{-1}(x) = R_0 h(x) = (1/\pi)(b^2/\sin^2 \pi b)$ $\cdot (\sin \pi r_1 / r_1)$. One has $\tilde{u}_1(x) \equiv 0$. The polynomial $P_{1,-1}(x)$ equals $x_3 \partial u_{-1}(0)/\partial x_3$ (see Table 7). Therefore, $P_{1,-1}(x) = \alpha x_3$, where

$$\alpha = \frac{\sin \pi b - \pi b \cos \pi b}{\pi \sin^2 \pi b}. \tag{4.40}$$

We also have $v_0(\xi) = \alpha \xi_3 (1 - \rho^{-1})$, whence $Q_{0,0} = 0$ and $u_0(x) = \tilde{u}_0(x) - (u_0^*(0) - \pi^{-2}) h(x)$. The functions $u_{-1}(x)$, $u_0(x)$, $v_{-1}(\xi)$, and $v_0(\xi)$ are thus constructed. We write out the final estimate:

$$\left| u(x, \varepsilon) - \frac{b^2 \sin \pi r_1}{\varepsilon \pi r_1 \sin^2 \pi b} - \tilde{u}_0(x) + (u_0^*(0) - \pi^{-2}) \frac{b \sin \pi r_1}{\sin \pi b} + \alpha \xi_3 \rho^{-3} \right| < M\varepsilon$$

everywhere in Ω_ε. Here the constant M depends neither on x, nor on ε, $r_1 = [x_1^2 + x_2^2 + (x_3 + b)^2]^{1/2}$, $\tilde{u}_0(x) = u_0^*(x) + w(x)$, the functions $u_0^*(x)$ and $w(x)$ are defined by formulas (4.38), (4.39), and α by formula (4.40).

Figure 13 shows the graph of the function $u(0, 0, x_3, \varepsilon)$ for $b = 0.5$; $\varepsilon = 0.02$.

§5. Example of solving a boundary value problem with a complex asymptotics

In the preceding sections we have constructed the uniform asymptotics of the solutions of some typical problems with singular perturbations of the boundary of the domain. The method of matched asymptotic expansions described above succeeds for a very wide range of such problems. One can consider elliptic equations of high order, rounded angles and conic points on the boundary, narrow bridges, etc. Nevertheless, the method is not all-embracing. There are some very simply stated problems, where it cannot be applied.

First we note that a necessary requirement for the applicability of the method is the existence of the outer expansion. This presumes the existence of a gauge sequence $\nu_k(\varepsilon)$, which may depend on the domain Ω and the coefficients of the equation, but does not depend on the right-hand side of the equation or the boundary functions, such that, in each interior point of the domain $\Omega \backslash \sigma$, the solution can be expanded in an asymptotic series of the form (3.5). It turns out that no such sequence $\nu_k(\varepsilon)$, which could asymptotically approximate the solution at least to within $O(\varepsilon)$, exists in the case of a second order elliptic equation in a three-dimensional domain Ω, and a straight-line segment $\sigma \subset \overline{\Omega}$, if the width of the neighborhood σ_ε is of order ε. This section is devoted to the analysis of this example.

Let Ω be the cylinder $\{x_1, x_2, y: x_1^2 + x_2^2 < 1, 0 < y < \pi\}$, and Ω_ε be the same cylinder with a hole inside: $\Omega_\varepsilon = \{x_1, x_2, y: \varepsilon^2 < x_1^2 + x_2^2 < 1, 0 < y < \pi\}$ (see Figure 14). Introduce the notation: $r = \sqrt{x_1^2 + x_2^2}$, σ is the straight-line segment $\{x, y: r = 0, 0 \leq y \leq \pi\}$. The boundary value problem for $u(x_1, x_2, y, \varepsilon) \in C(\overline{\Omega}_\varepsilon)$ is of the form:

$$
\begin{aligned}
\Delta u = 0 \quad & \text{for } (x_1, x_2, y) \in \Omega_\varepsilon, \\
u = 0 \quad & \text{for } y = 0, \text{ for } y = \pi \text{ and for } r = \varepsilon, \qquad (5.1) \\
u = f(y) \quad & \text{for } r = 1.
\end{aligned}
$$

Suppose, for simplicity's sake, that the function $f(y)$ consists of finitely many harmonics: $f(y) = \sum_{m=1}^{n} c_m \sin my$. Then the solution of the problem (5.1) is of the form

$$
u(x_1, x_2, y, \varepsilon) = \sum_{m=1}^{m} c_m \frac{I_0(mr)K_0(m\varepsilon) - I_0(m\varepsilon)K_0(mr)}{I_0(m)K_0(m\varepsilon) - I_0(m\varepsilon)K_0(m)} \sin my, \qquad (5.2)
$$

where $I_0(z)$ is the Bessel function of the imaginary argument, $K_0(z)$ is Macdonald's function. Since $I_0(m\varepsilon) = 1 + O(\varepsilon^2)$, $K_0(m\varepsilon) = \ln \frac{2}{m\varepsilon} - \gamma + O\left(\varepsilon^2|\ln \varepsilon|\right)$, where γ is Euler's constant (see [64, §5.7]), it follows from

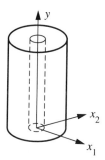

FIGURE 14

(5.2) that

$$u(x_1, x_2, y, \varepsilon)$$

$$= \sum_{m=1}^{n} c_m [I_0(m)]^{-1} [I_0(mr) + \varphi_m(r)(|\ln \varepsilon| + h_m)^{-1}] \sin my + O(\varepsilon^2).$$

Here

$$\varphi_m(r) = I_0(mr) K_0(m) [I_0(m)]^{-1} - K_0(mr),$$

$$h_m = \ln \frac{2}{m} - \gamma - K_0(m)[I_0(m)]^{-1}.$$

It is now evident that, for any fixed point $\Omega \backslash \sigma$, there is no gauge sequence $\nu_k(\varepsilon)$ yielding the approximation $O(\varepsilon)$. To be more precise, for any fixed point in $\Omega \backslash \sigma$ and any sequence $\nu_k(\varepsilon) \to 0$ such that $\nu_{k+1}(\varepsilon)[\nu_k(\varepsilon)]^{-1} \xrightarrow[\varepsilon \to 0]{} 0$ and $\nu_k(\varepsilon) = O(\varepsilon)$ for $k > N$, there is a boundary function $f(y)$ such that the solution of the problem (5.1) admits no asymptotic expansion with respect to $\nu_k(\varepsilon)$ at this point.

Indeed, assuming the contrary, one has

$$\left| u(x_1, x_2, y, \varepsilon) - \sum_{k=0}^{N} d_k \nu_k(\varepsilon) \right| < M\varepsilon, \qquad (5.3)$$

where the constants d_k depend on the solution u. Take $N + 2$ integers m such that $\sin my \neq 0$, and all h_m are distinct. Choose $N + 2$ different boundary functions such that each function $f_j(y)$ consists of a single harmonic $c_{m_j} \sin m_j y$, and the corresponding solution equals

$$I_0(m_j r) + \varphi_{m_j}(r)[|\ln \varepsilon| + h_{m_j}]^{-1}.$$

Recall that both y and r are fixed whereby $I_0(m_j r)$, $\varphi_{m_j}(r)$, and h_{m_j} are fixed numbers, which will be denoted a_j, b_j, and g_j, respectively. Here $b_j \neq 0$ for all j. Each of the solutions obtained has its own coefficients d_k so that

$$a_j + b_j(|\ln \varepsilon| + g_j)^{-1} = \sum_{k=0}^{N} d_{k,j} \nu_k(\varepsilon) + O(\varepsilon), \qquad j = 0, 1, \ldots, N+1.$$

There exist λ_j, not all equal to zero, such that $\sum_{j=0}^{N+1} \lambda_j d_{k,j} = 0$ for $k = 0, 1, \ldots, N$. Thus,

$$\sum_{j=0}^{N+1} \lambda_j (a_j + b_j(|\ln \varepsilon| + g_j)^{-1}) = O(\varepsilon), \qquad \varepsilon \to 0.$$

Hence $\sum_{j=0}^{N+1} \lambda_j (a_j + b_j(|\ln \varepsilon| + g_j)^{-1}) \equiv 0$. Since $b_j \neq 0$ for all j, and all g_j are distinct, one has $\lambda_j = 0$ for all j. The contradiction proves that (5.3) is impossible. ∎

CHAPTER IV

Elliptic Equation
with Small Parameter at Higher Derivatives

In this chapter we consider boundary value problems for the equation

$$\varepsilon \mathcal{M} u + l u = f \tag{0.1}$$

in a bounded domain Ω. Here $\mathcal{M}$ is an elliptic operator, $\varepsilon > 0$ a small parameter, and l a differential operator of the first order. We will use the method of matched asymptotic expansions to study the asymptotics of solutions of such problems as $\varepsilon \to 0$. This approach makes it possible to obtain the asymptotics up to any power of ε which is uniform in $\overline{\Omega}$ for a wide class of problems: the operator $\mathcal{M}$ may be of a high order, and there is no restriction on the dimension of the domain Ω. The form of the boundary conditions can also be quite diverse. However, for simplicity sake, in what follows we consider only plane domains Ω, second order operators $\mathcal{M}$, and the first boundary value problem. Short comments on other cases can be found at the end of the book.

Thus, let Ω be a bounded domain in $\mathbf{R}^2$ with piecewise smooth boundary, and consider the equation of the form (0.1), where

$$
\begin{aligned}
\mathcal{M} &= a_{1,1}(x,y)\frac{\partial^2}{\partial x^2} + 2a_{1,2}(x,y)\frac{\partial^2}{\partial x \partial y} + a_{2,2}(x,y)\frac{\partial^2}{\partial y^2}, \\
l &= a_1(x,y)\frac{\partial}{\partial x} + a_2(x,y)\frac{\partial}{\partial y}, \qquad a_1^2(x,y) + a_2^2(x,y) > 0.
\end{aligned} \tag{0.2}
$$

We will assume that the field of characteristics of the limit equation $l u = f$ is diffeomorphic to the field of parallel straight lines. Then an appropriate change of independent variables turns the operator l into $\partial/\partial y$. It is, therefore, sufficient to consider the boundary value problem

$$\varepsilon \mathcal{M} u - \frac{\partial u}{\partial y} = f(x,y), \qquad (x,y) \in \Omega, \tag{0.3}$$

$$u(x,y) = 0 \quad \text{on } \partial\Omega, \tag{0.4}$$

where $\mathcal{M}$ is the operator (0.2). The characteristics of the limit equation are lines parallel to the y-axis, while the limit equation itself is, in fact, an

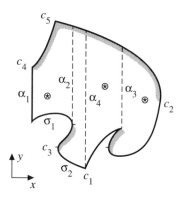

FIGURE 15

ordinary differential equation of the form $\partial u/\partial y = f(x, y)$ along each of the straight-line segments parallel to the y-axis and lying in $\overline{\Omega}$ (see Figure 15).

If the operator $\mathscr{M}$ were equal to $\partial^2/\partial y^2$, the problem (0.3), (0.4) on each such straight-line segment would coincide with the problem for an ordinary differential equation similar to those considered in Chapter I, §1 (Examples 1 and 2). Then the variable x would be simply a parameter. The boundary condition $u = 0$ would have been preserved for the limit problem at the lower endpoints of the segments, and the exponential boundary layer would arise in the vicinity of the upper endpoints (the shaded area in Figure 15). It turns out that for the elliptic operator $\mathscr{M}$ the picture is, on the whole, the same. One can easily construct the outer asymptotic expansion everywhere in Ω. Its coefficient functions $u_k(x, y)$ are smooth, for example, in the part of the domain marked by $\circledast$ in Figure 15. In this area, exponentially decaying boundary layer functions are constructed without difficulty. However, the coefficients $u_k(x, y)$ have various singularities in Ω. Figure 15 shows typical examples of how such a singularity can arise. The straight-line segment α_1 is a part of the boundary $\partial\Omega$, and one can easily see that the residual in the boundary condition on α_1 cannot be eliminated using a solution of an ordinary differential equation. On the segments α_2 and α_3, all the coefficients $u_k(x, y)$ are, in general, discontinuous. Indeed, to the left of the segment α_2, the function $u_0(x, y)$ is a solution of the equation $\partial u_0/\partial y = f(x, y)$ vanishing on σ_1, while to the right of α_2 it is a solution of the same equation vanishing on σ_2. The same phenomenon is observed for the segment α_3. If the boundary is not smooth at the point c_1, then, near the segment α_4, the outer asymptotic expansion also does not approximate the solution $u(x, y, \varepsilon)$ to all powers of ε. In addition, the functions $u_k(x, y, \varepsilon)$ turn out to have growing singularities at the point c_2, where the tangent to the boundary is parallel to the y-axis. Segments similar to $\alpha_1 - \alpha_4$ will be called *singular characteristics* of the problem (0.3),(0.4).

Thus, the problem (0.3), (0.4) is, in general, bisingular. The uniform asymptotic approximation to the solution in $\overline{\Omega}$ to any power of ε can be obtained by considering appropriate inner expansions and applying the method of matched asymptotic expansions. Typical cases of the bisingular problems (0.3), (0.4) are studied below.

§1 is devoted to the analysis of the asymptotics in a neighborhood of the boundary α_1 for one example of the problem of this kind. The behavior of the solution in the neighborhood of singular characteristics of types α_3 and α_4 is studied in §2. The asymptotics of the solution in a neighborhood of a point of type c_2 or c_3 is examined in §3, and in a neighborhood of a singular characteristic of type α_2 in §4.

Note that the corner points can lie on the upper part of the boundary $\partial\Omega$ (e.g., the points c_4 and c_5). The asymptotics in a neighborhood of such a point lying on a singular boundary (the point c_4) is studied in §1. In a neighborhood of a point of type c_5 the asymptotics of the solution is even simpler—in this case one has to add to the outer expansion the corner boundary layer which is essentially the same as in Chapter I, §2, Example 4. A brief comment on that can be found in §5. This section also includes some auxiliary lemma used to justify the f.a.s.

§1. The case where a characteristic of the limit equation coincides with a part of the boundary

1. Consider the boundary value problem

$$\mathscr{L}_\varepsilon u \equiv \varepsilon^2 \left(\frac{\partial^2 u}{\partial x^2} + \frac{\partial^2 u}{\partial y^2} \right) - a(x,y)\frac{\partial u}{\partial y} = f(x,y), \qquad (x,y) \in \Omega,$$

$$\tag{1.1}$$

$$u(x,y) = 0 \quad \text{for } (x,y) \in \partial\Omega, \tag{1.2}$$

where Ω is the square $\{x,y: 0 < x < 1,\ 0 < y < 1\}$, the functions $a(x,y)$ and $f(x,y)$ are infinitely differentiable in $\overline{\Omega}$, $a(x,y) > 0$, $0 < \varepsilon \ll 1$. This problem is a particular case of the problem (0.3), (0.4): the Laplace operator is taken for the operator $\mathscr{M}$ for simplicity, but in order to retain the steps of the investigation typical for variable coefficients, the coefficient of the first derivative is considered to be variable. For convenience of further notation the parameter ε in equation (1.1) is replaced with ε^2. The singular characteristics of the limit equation in the problem (1.1), (1.2) are the straight-line segments $\{x,y: x = 0,\ 0 \le y \le 1\}$ and $\{x,y: x = 1,\ 0 \le y \le 1\}$ coinciding with the sides of the square.

The outer expansion of the solution of the problem (1.1), (1.2) is of the form

$$U = \sum_{k=0}^{\infty} \varepsilon^{2k} u_{2k}(x,y), \tag{1.3}$$

where

$$a(x, y)\frac{\partial u_0}{\partial y} = -f(x, y),$$

$$a(x, y)\frac{\partial u_{2k}}{\partial y} = \Delta u_{2k-2}, \qquad k \geq 1, \quad u_k(x, 0) = 0. \tag{1.4}$$

The boundary condition (1.4) is chosen for the same reasons as those given above (see, e.g., Chapter I, §1, Example 2), namely, that the residuals arising in the boundary data for the outer expansion can be eliminated by boundary layer functions of exponential type at $y = 1$, but cannot be eliminated in this way at $y = 0$. Therefore, the series U has to satisfy the boundary conditions (1.2) for $y = 0$ yielding the last equalities in (1.4).

In the vicinity of the top side of the square Ω, the series

$$S = \sum_{k=0}^{\infty} \varepsilon^{2k} s_{2k}(x, \tau), \tag{1.5}$$

where $\tau = \varepsilon^{-2}(1 - y)$, should be added to the outer expansion U. On inserting this series into the homogeneous equation $\mathcal{L}_\varepsilon S = 0$, and expanding the coefficient $a(x, y) = a(x, 1 - \varepsilon^2\tau)$ in a Taylor series $(a(x, 1 - \varepsilon^2\tau) = \sum_{j=0}^{\infty} b_j(x)\varepsilon^{2j}\tau^j)$, one obtains the recurrence system of equations for $s_{2k}(x, \tau)$:

$$\frac{\partial^2 s_0}{\partial \tau^2} + b_0(x)\frac{\partial s_0}{\partial \tau} = 0,$$

$$\frac{\partial^2 s_{2k}}{\partial \tau^2} + b_0(x)\frac{\partial s_{2k}}{\partial \tau} = -\frac{\partial^2 s_{2k-2}}{\partial x^2} - \sum_{j=1}^{k} b_j(x)\tau^j\frac{\partial s_{2k-2j}}{\partial \tau}, \qquad k \geq 1. \tag{1.6}$$

In order for the sum $U + S$ of the series U and S to satisfy condition (1.2) at $y = 1$, the equalities

$$s_{2k}(x, 0) = -u_{2k}(x, 1) \tag{1.7}$$

must be satisfied. Evidently, there exist functions $s_{2k}(x, \tau) \in C^\infty([0, 1] \times [0, \infty))$ solving the system (1.6) subject to conditions (1.7), and rapidly decaying as $\tau \to \infty$.

It is easily verified that the series $U + S$ is an f.a.s. of equation (1.1):

$$|\mathcal{L}_\varepsilon(A_{2n,x,y}U + A_{2n,x,\tau}S) - f(x, y)| \leq M\varepsilon^{2n+2} \quad \text{in } \overline{\Omega} \tag{1.8}$$

and, moreover,

$$(A_{2n,x,y}U + A_{2n,x,\tau}S)_{y=1} = 0,$$

$$(A_{2n,x,y}U + A_{2n,x,\tau}S)_{y=0} = O\left(e^{-\gamma/\varepsilon}\right), \qquad \gamma > 0. \tag{1.9}$$

So far we have encountered no singularities in the coefficients of the asymptotic expansions, and one might think that the problem (1.1), (1.2) belongs to the relatively simple type of problems considered in Chapter I. However, such a conclusion would be premature. There remains the task of eliminating the residuals in the boundary condition (1.2) on the sides $x = 0$ and $x = 1$. Since both sides are on an equal footing, we can consider only the side $x = 0$.

2. In the vicinity of the side $x = 0$, one has to add to the outer expansion (1.3) another series, the coefficients of which depend on new, "stretched" variables. To be more precise, the variable y does not change, but a new variable $\zeta = x\varepsilon^{-\alpha}$ is introduced instead of x. Here the constant α has to be chosen in such a way that at least two terms in the homogeneous equation (1.1) are of the same, and the highest, order. Let $u(x, y, \varepsilon) \equiv v(\zeta, y, \varepsilon)$. Then the homogeneous equation (1.1) in the new variables is of the form

$$\varepsilon^{2-2\alpha}\frac{\partial^2 v}{\partial \zeta^2} + \varepsilon^2 \frac{\partial^2 v}{\partial y^2} - a(\varepsilon^\alpha \zeta, y)\frac{\partial v}{\partial y} = 0.$$

Since $a(x, y) \geq \text{const} > 0$, one evidently has to set $\alpha = 1$.

Thus, in the inner variables $\zeta = \varepsilon^{-1}x$, y, the equation is of the form

$$\frac{\partial^2 v}{\partial \zeta^2} - a(\varepsilon\zeta, y)\frac{\partial v}{\partial y} + \varepsilon^2 \frac{\partial^2 v}{\partial y^2} = 0, \qquad 0 \leq \zeta < \infty, \ 0 \leq y \leq 1,$$

with the inner expansion

$$V = \sum_{k=0}^{\infty} \varepsilon^k v_k(\zeta, y). \tag{1.10}$$

The sum of the series U and V must satisfy the boundary conditions (1.2) for $y = 0$ and for $x = \zeta = 0$. Taking this into account, and expanding the function $a(\varepsilon\zeta, y)$ in a series, one obtains the following recurrence system of boundary value problems for the coefficients $v_k(\zeta, y)$:

$$Lv_0 = 0, \tag{1.11}$$

$$Lv_k = -\frac{\partial^2 v_{k-2}}{\partial y^2} + \sum_{j=1}^{k} a_j(y)\zeta^j \frac{\partial v_{k-j}}{\partial y}, \qquad k \geq 1, \tag{1.12}$$

$$v_k(\zeta, 0) = 0, \tag{1.13}$$

$$v_{2l}(0, y) = -u_{2l}(0, y), \qquad v_{2l+1}(0, y) = 0. \tag{1.14}$$

Here $L = \partial^2/\partial\zeta^2 - a_0(y)\partial/\partial y$, and $a_k(y)$ are the Taylor coefficients for the function $a(x, y)$: $a(x, y) = \sum_{j=0}^{\infty} a_j(y)x^j$, $x \to 0$. By assumption, $a_0(y) = a(0, y) > \text{const} > 0$, so that (1.11)–(1.14) are usual problems for parabolic equations in the half-strip $\{\zeta, y: 0 \leq \zeta < \infty, 0 \leq y \leq 1\}$. Condition (1.13) implies that the solutions $v_k(\zeta, y)$ decay exponentially as $\zeta \to \infty$. (This can be easily verified, for example, as follows. After the change of the variable y, the operator L goes into the operator $\partial^2/\partial\zeta^2 - \partial/\partial y_1$. Therefore, the solution of the equation $Lv = f$ in any half-strip $\zeta \geq \text{const}$,

$0 \leq y_1 \leq Y$ can be expressed by an explicit formula (see, for example, [118, Chapter 3, §3], or formulas (1.15), (1.16) below) which implies that the solution v decays rapidly as $\zeta \to +\infty$ provided the right-hand side f has the same property.) Thus, the coefficients of the series (1.10) have no singularities as $\zeta \to \infty$, which is quite natural, because the coefficients of the outer expansion (1.3), in turn, have no singularities as $x \to 0$. (The series S can, for the time being, be ignored because its coefficients are exponentially small for all values of y outside a small neighborhood of the point $y = 1$.)

Nevertheless, difficulties typical for bisingular problems arise in the coefficients of the series V, although not at infinity, but at the point $(0, 0)$. Since, in general, $(\partial u_0/\partial y)(0, 0) = -[a(0, 0)]^{-1} f(0, 0) \neq 0$, one has, by (1.14), that $(\partial v_0/\partial y)(0, y)|_{y=0} \neq 0$. On the other hand, equation (1.11) and condition (1.13) imply that $(\partial v_0/\partial y)(\zeta, 0) = [a_0(0)]^{-1}(\partial^2 v_0/\partial \zeta^2)(\zeta, 0) \equiv 0$ for $\zeta > 0$. In view of this discord of the boundary conditions $((\partial v_0/\partial y)(0, +0) \neq (\partial v_0/\partial y)(+0, 0))$, the function $v_0(\zeta, y)$, although continuous everywhere, is not smooth at the point $(0, 0)$. One can show that the smoothness properties of the functions $v_k(\zeta, y)$ at the point $(0, 0)$ deteriorate with the increase of k. An imprecise, but essentially correct explanation of this phenomenon is the fact that the right-hand side of equations (1.12) contains the operator $\partial^2/\partial y^2$ applied to the preceding function, while the inversion of the operator L corresponds to a single integration with respect to y. Only if infinitely many special equalities in the boundary conditions for each of the functions $v_k(\zeta, y)$ are satisfied at the point $(0, 0)$, will all functions $v_k(\zeta, y)$ be smooth in the closed strip $\{\zeta, y : 0 \leq \zeta < \infty, 0 \leq y \leq 1\}$ (for example, it is sufficient that the function $f(x, y)$ and all its derivatives vanish at the point $(0, 0)$). Otherwise, beginning with some number k, the functions $v_k(\zeta, y)$ have singularities at the point $(0, 0)$, and the order of these singularities increases with k. Thus, the problem (1.1), (1.2) is, on the whole, bisingular.

Consider the behavior of the functions $v_k(\zeta, y)$ in the case where $(\partial u_0/\partial y)(0, 0) \neq 0$. After the change of variable $y_1 = \int_0^y [a_0(\theta)]^{-1} d\theta$, equation (1.11) goes into the equation $\partial^2 v_0/\partial \zeta^2 - \partial v_0/\partial y_1 = 0$. Denoting $u_0(0, y) = -\psi(y_1)$, $v_0(\zeta, y) = \overline{v}_0(\zeta, y_1)$, write out the explicit formula for the solution of the problem (1.11), (1.13), (1.14):

$$
\begin{aligned}
\overline{v}_0(\zeta, y_1) &= \frac{2}{\sqrt{\pi}} \int_{\zeta/2\sqrt{y_1}}^{\infty} e^{-\theta^2} \psi\left(y_1 - \frac{\zeta^2}{4\theta^2}\right) d\theta \\
&= \frac{1}{2\sqrt{\pi}} \int_0^{y_1} \frac{\zeta}{(y_1 - \theta)^{3/2}} e^{-\zeta^2/4(y_1 - \theta)} \psi(\theta) \, d\theta.
\end{aligned}
\tag{1.15}
$$

For any smooth bounded function $F(\zeta, y_1)$ one can also easily write out the explicit formula for the solution of the problem $\partial^2 v/\partial \zeta^2 - \partial v/\partial y_1 = F(\zeta, y_1)$, $v(0, y_1) = 0$, $v(\zeta, 0) = 0$:

$$v(\zeta, y_1) = \frac{1}{2\sqrt{\pi}} \int_0^{y_1} \int_0^\infty \frac{1}{\sqrt{y_1 - \theta}} \left[\exp\left(-\frac{(\zeta - \xi)^2}{4(y_1 - \theta)}\right) \right.$$
$$\left. - \exp\left(-\frac{(\zeta + \xi)^2}{4(y_1 - \theta)}\right) \right] F(\xi, \theta) \, d\xi \, d\theta.$$

$$(1.16)$$

These formulas imply that the functions $\overline{v}_0(\zeta, y_1)$ and $v(\zeta, y)$ are sufficiently smooth in the strip $\{\zeta, y_1 : 0 \le \zeta < \infty, \ 0 \le y_1 \le Y\}$, provided the functions $\psi(y_1)$ and $F(\zeta, y_1)$ vanish at the origin together with their derivatives of sufficiently high order. Otherwise, these functions have singularities at the origin.

For example, set $a(x, y) \equiv 1$, $f(x, y) \equiv 1$. Then $u_0(x, y) \equiv -y$, $u_{2k}(x, y) \equiv 0$ for $k > 0$. According to formula (1.15),

$$v_0(\zeta, y) = \frac{2}{\sqrt{\pi}} \int_{\zeta/(2\sqrt{y})}^\infty \left(y - \frac{\zeta^2}{4\theta^2}\right) e^{-\theta^2} \, d\theta, \qquad v_1(\zeta, y) \equiv 0.$$

The function $v_2(\zeta, y)$ satisfies the equation

$$\frac{\partial^2 v_2}{\partial \zeta^2} - \frac{\partial v_2}{\partial y} = -\frac{\partial^2 v_0}{\partial y^2} = -\frac{\zeta}{2\sqrt{\pi} y^{3/2}} \exp\left(-\frac{\zeta^2}{4y}\right)$$

and zero boundary conditions. The right-hand side of this equation has a singularity at the origin—on the parabolas $y = \text{const} \cdot \zeta^2$ it has a singularity of type y^{-1}. One can easily check that the only bounded solution of this problem is

$$v_2(\zeta, y) = \frac{\zeta}{2\sqrt{\pi y}} \exp\left(\frac{-\zeta^2}{4y}\right).$$

The right-hand side of the equation for $v_4(\zeta, y)$ is the function

$$\frac{\zeta}{2\sqrt{\pi y}} \left(-\frac{3}{4y^2} + \frac{3\zeta^2}{4y^3} - \frac{\zeta^4}{16y^4}\right) \exp\left(-\frac{\zeta^2}{4y}\right)$$

which on the same parabolas has a singularity of the type y^{-2}. If this function is substituted into the integral (1.16), one can see without difficulty that the resulting integral diverges. Thus, the function $v_4(\zeta, y)$ cannot be obtained directly from formula (1.16). One can, however, find a self-similar solution:

$$v_4(\zeta, y) = \frac{1}{2\sqrt{\pi y}} \left(-\frac{\zeta^3}{8y^2} + \frac{\zeta^5}{32y^3}\right) \exp\left(\frac{-\zeta^2}{4y}\right),$$

but this solution is not unique. One can, for example, add to it the function $c\zeta y^{-3/2} \exp(-\zeta^2/4y)$ with any constant c. Any solution $v_4(\zeta, y)$ has at least y^{-1} singularity on the parabolas $y = \text{const} \cdot \zeta^2$. Further, the order of singularity of the functions $v_k(\zeta, y)$ increases.

The situation in the general case is the same. This can be seen by expanding the boundary functions and the coefficients $a_i(y)$ in Taylor series, using formulas (1.15) and (1.16) for the remainders, and writing out the self-similar solutions for the resulting polynomials similar to those obtained above.

Here, as in Chapter III, we encounter two difficulties. First, how can one construct the solution of the parabolic equation (1.12) with a very strong singularity in the right-hand side? The usual convolution with Green's function of type (1.16) fails because the integral diverges. However, this is only a technical problem, which can be overcome, without much difficulty, in one way or another by the actual regularization of the diverging integrals.

The second difficulty is of a more fundamental nature. Since solutions of equations (1.12) have strong singularities for large k, the solution of problems (1.12)–(1.14) in the class of such functions is not unique. Indeed, let

$$\Gamma(\zeta, y_1) = 1\sqrt{y_1} \exp\left(-\zeta^2 4y_1\right), \qquad y_1 = \int_0^y [a_0(\theta)]^{-1} d\theta.$$

Then any function of the form $(\partial^{2j+1}/\partial\zeta^{2j+1})\Gamma(\zeta, y_1)$ is the solution of equation (1.11) vanishing for $\zeta = 0$ and for $y = 0$. Any linear combination of these functions can be added to the solution of the problems (1.12)–(1.14), so that the question which of the solutions are the correct ones remains open (if one starts only from the problems (1.12)–(1.14)). The answer to this question is provided by matching the asymptotic expansion (1.10) to a new inner expansion.

3. One has to introduce new variables $\xi = \varepsilon^{-2}x$, $\eta = \varepsilon^{-2}y$ in a neighborhood of the origin, and seek the inner expansion

$$W = \sum_{k=1}^{\infty} \varepsilon^{2k} w_{2k}(\xi, y). \tag{1.17}$$

The choice of scale is made in a natural way ensuring that all three terms in the homogeneous equation are of the same order. The series (1.17) begins with the term ε^2 because both solutions $u(x, y, \varepsilon)$ and $u_0(x, y)$ vanish at the origin. The requirement imposed on the series (1.17) is as follows: its sum with the series (1.3) must be an f.a.s. of the problem (1.1), (1.2) in a neighborhood of the origin. Since the series (1.3) is the f.a.s. of the equation (1.1), and vanishes for $y = 0$, the series (1.17) must be the f.a.s. of the homogeneous equation which, in the variables ξ, η, is of the following form:

$$\frac{\partial^2 w}{\partial \xi^2} + \frac{\partial^2 w}{\partial \eta^2} - a(\varepsilon^2\xi, \varepsilon^2\xi)\frac{\partial w}{\partial \eta} = 0, \tag{1.18}$$

the series (1.17) must vanish for $\eta = 0$, and, moreover,

$$\sum_{k=1}^{\infty} \varepsilon^{2k} w_{2k}(0, \eta) + \sum_{k=0}^{\infty} \varepsilon^{2k} u_{2k}(0, \varepsilon^2\eta) = 0.$$

Let us assume, for simplicity, that $a(0, 0) = 1$. The last two equalities (which are, of course, asymptotic) yield the recurrence system of boundary

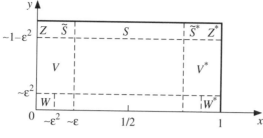

FIGURE 16

value problems

$$L_1 w_2 = 0, \tag{1.19}$$

$$L_1 w_{2k} = \sum_{j=1}^{k-1} P_j(\xi, \eta) \frac{\partial w_{2k-2j}}{\partial \eta}, \qquad k > 1, \ 0 < \xi, \ \eta < \infty, \tag{1.20}$$

$$w_{2k}(\xi, 0) = 0, \qquad w_{2k}(0, \eta) = -\sum_{j=1}^{k} \frac{\eta^j}{j!} \frac{\partial^j u_{2k-2j}}{\partial y^j}(0, 0). \tag{1.21}$$

Here $L_1 = \Delta_{\xi, \eta} - \partial/\partial\eta$, $P_j(\xi, \eta)$ are homogeneous polynomials of degree j obtained from the Taylor expansion of the coefficient $a(x, y)$:

$$a(x, y) = \sum_{j=0}^{\infty} P_j(x, y), \qquad x \to 0, \ y \to 0.$$

One should note the intermediate position occupied by the series (1.10): the expansion V is an inner one with respect to U, and an outer one with respect to the expansion W. A similar situation was observed in Chapter II, §3. There are, however, some differences. While in Chapter II, §3, both interactions of the intermediate expansion with its neighbors were bisingular, now the coefficients of the series U are smooth functions, and those of the series V decay exponentially as $\zeta \to \infty$. That is, the interaction between them is the same as in the examples of Chapter I, and the matching of the expansion V to W is of the same kind as in other bisingular problems. Besides that, in Chapter II, §3, no requirements were imposed on the intermediate expansion except that it should satisfy the equation and the matching condition with respect to its neighbors. Now the expansion V must satisfy the additional boundary condition for $\zeta = 0$. Figure 16 shows the characteristic dimensions of the domains in which the asymptotic expansion V and W are valid.

We now proceed with the investigation of the problems (1.19)–(1.21). The solutions $w_k(\xi, \eta)$, in general, grow at infinity, but unlike the corresponding problems of Chapter III, now the solutions are written out explicitly by means of converging integrals. One can easily verify that the function

$$-(2\pi)^{-1} K_0 \left(\frac{\sqrt{(\xi - \xi_1)^2 + (\eta - \eta_1)^2}}{2} \right) \exp \left(\frac{\eta - \eta_1}{2} \right),$$

where K_0 is Macdonald's function, is the solution of the equation $L_1 w = \delta(\xi - \xi_1, \eta - \eta_1)$ (see, for example, [64, p. 140]; [118, Chapter 7, §2, p. 498]). Therefore, the function

$$G(\xi, \eta, \xi_1, \eta_1) = (2\pi)^{-1} \left\{ -K_0 \left(\frac{\sqrt{(\xi - \xi_1)^2 + (\eta - \eta_1)^2}}{2} \right) \right.$$

$$+ K_0 \left(\frac{\sqrt{(\xi - \xi_1)^2 + (\eta + \eta_1)^2}}{2} \right)$$

$$+ K_0 \left(\frac{\sqrt{(\xi + \xi_1)^2 + (\eta - \eta_1)^2}}{2} \right)$$

$$\left. - K_0 \left(\frac{\sqrt{(\xi + \xi_1)^2 + (\eta + \eta_1)^2}}{2} \right) \right\} \exp \left(\frac{\eta - \eta_1}{2} \right)$$

$$\tag{1.22}$$

is Green's function for the boundary value problem for the same equation with the boundary conditions $w(\xi, 0) = 0$, $w(0, \eta) = 0$, while the function

$$w(\xi, \eta) = \int_0^\infty \int_0^\infty G(\xi, \eta, \xi_1, \eta_1) F(\xi_1, \eta_1) d\xi_1 d\eta$$

$$- \int_0^\infty \frac{\partial G}{\partial \xi_1}(\xi, \eta, 0, \eta_1) \psi(\eta_1) d\eta_1 \tag{1.23}$$

satisfies the equation

$$L_1 w = F(\xi, \eta) \qquad \text{for } 0 < \xi < \infty, \ 0 < \eta < \infty, \tag{1.24}$$

and the boundary conditions $w(0, \eta) = \psi(\eta)$, $w(\xi, 0) = 0$.

We will assume that a number of conditions are satisfied with respect to the functions $F(\xi, \eta)$ and $\psi(\eta)$, for which formula (1.23) makes sense, and which are satisfied for the problems (1.19)–(1.21). Introduce the notation $\omega = \{\xi, \eta : 0 \le \xi < \infty, \ 0 \le \eta < \infty\}$, $\omega' = \omega \backslash \{0, 0\}$ and assume that $F(\xi, \eta) \in C(\omega) \cap C^\infty(\omega')$, $\psi(\eta) \in C^\infty[0, \infty)$, $\psi(0) = 0$, and that the functions $F(\xi, \eta)$, $\psi(\eta)$ are of slow growth, i.e., that they and their derivatives do not grow at infinity faster than some power of $\xi^2 + \eta^2$. Under these assumptions, formula (1.23) easily implies that $w(\xi, \eta) \in C^1(w) \cap C^\infty(\omega')$ and is a function of slow growth.

THEOREM 1.1. *There exist functions of slow growth* $w_{2k}(\xi, \eta) \in C^1(\omega) \cap C^\infty(\omega')$ *satisfying equations and boundary conditions* (1.19)–(1.21). *The absolute values of the functions* $w_{2k}(\xi, \eta)$ *and of all their derivatives do not exceed* $M \exp(-\gamma\xi)$ *for* $\xi \geq A\eta + 1$ *for any* $A > 0$. *Here* $\gamma > 0$ *depends on* A, *while the constant* M *depends, in addition, on* k *and the order of the derivative.*

The proof follows immediately from formula (1.23) and the above remarks on this formula. The statement is valid for the function $w_2(\xi, \eta)$. Next, one checks by induction that the right-hand sides of equations (1.20) are functions of slow growth which belong to the class $C(\omega) \cap C^\infty(\omega')$, and decay exponentially for $\xi \geq A\eta + 1$, $\xi \to \infty$. Formula (1.23) now implies that the solutions $w_{2k}(\xi, \eta)$ have the same properties, and that, moreover, $w_{2k}(\xi, \eta) \in C^1(\omega)$. ∎

4.   Now, in accordance with the general scheme, one has to investigate the asymptotic behavior of the solutions $w_{2k}(\xi, \eta)$ at infinity, and then, starting from the conditions of matching the series (1.17) to the series (1.10), construct the series V.

THEOREM 1.2. *The functions* $w_{2k}(\xi, \eta)$, *i.e., the solutions of the problems* (1.19)–(1.21), *can, for* $\xi < \eta$, *be expanded in an asymptotic series*

$$w_{2k}(\xi, \eta) = \eta^k \sum_{j=0}^\infty \eta^{-j/2} \Phi_{k,j}(\theta), \qquad \eta \to \infty, \qquad (1.25)$$

where $\theta = 2^{-1}\xi\eta^{-1/2}$, *and the functions* $\Phi_{k,j}(\theta) \in C^\infty[0, \infty)$ *decay exponentially at infinity. The series* (1.25) *can be differentiated term-by-term repeatedly.*

It would seem that the way to establish relations (1.25) is straightforward: starting from the explicit formula (1.23) for the functions $w_{2k}(\xi, \eta)$, one has to obtain the asymptotic representation of these functions as $\eta \to \infty$. Unfortunately, this idea fails, so that for this purpose representation (1.23) proves useless. At the same time, it can be used in the case where the right-hand side $F(\xi, \eta)$ of (1.24) and the boundary function $\psi(\eta)$ decay rapidly enough as $\eta \to \infty$. Under these assumptions, one can obtain the asymptotic representation of the form (1.25). Therefore, the problem of investigating the asymptotics of the coefficients $w_{2k}(\xi, \eta)$ naturally falls into two steps. The first is, using the explicit formula (1.23), to obtain the asymptotic representation of the solution $w(\xi, \eta)$ in the case where the initial data and the right-hand side of the problem (1.24) decays rapidly as $\eta \to \infty$. The second step consists in reducing the boundary value problems (1.19)–(1.21) for the coefficients $w_{2k}(\xi, \eta)$ to the problems (1.24) with rapidly decaying data, i.e., constructing some asymptotic series the partial sums of which satisfy equations and conditions (1.19)–(1.21) to a sufficiently high degree of exactness as $\eta \to \infty$. This is a general approach to the investigation of the asymptotics of the coefficients of inner expansions in many problems. The first step is usually relatively simple, provided one has an integral representation for the solution of the boundary value problem, while the second requires, in each particular case, a certain degree of ingenuity, and is often, including the situation considered in this section, the main one.

We now proceed with the realization of the plan beginning with the construction of f.a.s. for the problems (1.19)–(1.21). Actually, we face the problem of constructing the common part of the asymptotics of the functions $w_{2k}(\xi, \eta)$ as $\eta \to \infty$, and the functions $v_i(\zeta, y)$ as $\zeta \to 0$, $y \to 0$. Indeed, according to the matching conditions, the equalities

$$A_{m,\zeta,y}(A_{n,\xi,\eta}W) = A_{n,\xi\eta}(A_{m,\zeta,y}V), \qquad (1.26)$$

where W and V are the series (1.17) and (1.10), must be satisfied. The left-hand side of these equalities contains segments of the asymptotic series for $w_{2k}(\xi, \eta)$ at infinity, and the right-hand side segments of the asymptotic series for $v_i(\zeta, y)$ at the origin. For large values of η, the principal part of the operator $L_1 = \Delta_{\xi,\eta} - \partial/\partial\eta$ is $\partial^2/\partial\xi^2 - \partial/\partial\eta$. The same operator (up to the scalar multiple ε^2) is the principal part of the operator $L = \partial^2/\partial\zeta^2 - a_0(y)\partial/\partial y$ in a neighborhood of the origin. Accordingly, in the next subsection we formulate some auxiliary statements and prove a number of lemmas concerning singular solutions of the heat equation.

5. Introduce the notation $L_0 = \partial^2/\partial\xi^2 - \partial/\partial\eta$. For any integer n define the function $W_n(\xi, \eta)$ as follows. If $n \geq 0$, then $W_n(\xi, \eta)$ is the solution of the boundary value problem

$$L_0 W_n = 0 \quad \text{in } \omega', \qquad W_n(0, \eta) = \eta^{n/2}, \qquad (1.27)$$
$$W_n(\xi, 0) = 0, \qquad (1.28)$$

bounded for $n = 0$, continuous in ω for $n > 0$, and slowly growing at infinity.

For $n = -(2k+1)$, $k \geq 0$, let $W_n(\xi, \eta) = \partial^{k+1}W_1(\xi, \eta)/\partial\eta^{k+1}$, and for $n = -2k$, $k > 0$, let $W_n(\xi, \eta) = \partial^k W_0(\xi, \eta)/\partial\eta^k$ so that

$$W_{-(2k+1)}(0, \eta) = \alpha_k \eta^{-(2k+1)/2} \quad \text{for } k \geq 0, \eta > 0, \alpha_k \neq 0,$$
$$W_{-2k}(0, \eta) = 0 \quad \text{for } k > 0, \eta > 0.$$

Evidently, the functions $W_n(\xi, \eta)$ satisfy equation (1.27) and condition (1.28) for all n.

It can be checked without difficulty that for any n equality $W_n(\xi, \eta) = \eta^{n/2}U_n(\theta)$ holds, where $\theta = 2^{-1}\xi\eta^{-1/2}$, and the function $U_n(\theta) \in C^\infty[0, \infty)$, and decays exponentially as $\theta \to \infty$. For $n \geq 0$ this fact is an easy consequence of the explicit formula

$$W_n(\xi, \eta) = \frac{1}{2\sqrt{\pi}} \int_0^\eta \frac{\xi}{(\eta - \tau)^{3/2}} \left[\exp\left(-\frac{\xi^2}{4(\eta - \tau)} \right) \right] \tau^{n/2} \, d\tau$$

$$= \eta^{n/2} \frac{2}{\sqrt{\pi}} \int_\theta^\infty e^{-\sigma^2} \left(1 - \frac{\theta^2}{\sigma^2} \right)^{n/2} d\sigma,$$

while for negative n it follows from the definition of $W_n(\xi, \eta)$, and the obvious relation

$$\frac{\partial}{\partial \eta} (\eta^{n/2} \psi(2^{-1}\xi\eta^{-1/2})) = \eta^{n/2-1} \left[\frac{n}{2} \psi(\theta) - \frac{\theta}{2} \psi'(\theta) \right].$$

(Here and in the rest of this section we denote $2^{-1}\xi\eta^{-1/2}$ by θ.) If one considers this relation together with the equality $(\partial^2/\partial\xi^2)(\eta^{n/2}\psi(\theta)) = \eta^{n/2-1}\psi''(\theta)/4$, it is easy to see that the equation $L_0 W_n = 0$ for the functions $W_n(\xi, \eta)$ is equivalent to the equation

$$l_n U \equiv U'' + 2\theta U' - 2nU = 0 \tag{1.29}$$

for the functions $U_n(\theta)$. (In particular, for $n = -k < 0$, the function $U_n(\theta)$ is proportional to the function $H_{k-1}(\theta)\exp(-\theta^2)$, where $H_{k-1}(\theta)$ is the Chebyshev-Hermite polynomial (see [118, p. 704]).) We also note that $U_n(\theta)$ is the unique (up to a scalar multiple) solution of equation (1.29) which exponentially decays at infinity, because the second solution which is linearly independent with respect to $U_n(\theta)$ behaves at infinity as θ^n (see [28, §6]).

The functions $W_n(\xi, \eta)$ just constructed are, as will be shown below, the terms of the asymptotic series at infinity for the solution of the homogeneous equation $L_1 w = 0$. However, since the system (1.20) includes nonhomogeneous equations, one has to examine the behavior at infinity of the solution of the equation $L_1 w = f$, where the right-hand side $f(\xi, \eta)$ is composed of the solutions of the preceding equations, $W_n(\xi, \eta)$ in particular. For this purpose it is convenient to introduce the following auxiliary classes of functions.

The class $\mathfrak{A}$ of functions $u(\xi, \eta)$ is made up in the following way:

(1) it contains the functions $W_n(\xi, \eta)$ for all integer $n \geq 0$;
(2) together with each function $u \in \mathfrak{A}$ it also contains the functions ξu and ηu;
(3) together with each function $u \in \mathfrak{A}$ it contains a function of slow growth $v(\xi, \eta)$ which is a solution of the equation $L_0 v = u$ continuous in ω, and satisfying the conditions

$$v(\xi, 0) = 0 \quad \text{for } \xi \geq 0, \qquad v(0, \eta) = 0 \quad \text{for } \eta \geq 0. \tag{1.30}$$

Functions that can be obtained by these operations form the class $\mathfrak{A}$. It is easy to see that the functions of this class are bounded at the origin.

The class $\mathfrak{B}$ is defined as the set of the linear combinations of the functions of the form $(\partial^j u/\partial \eta^j)(\xi, \eta)$, where $j \geq 0$, and $u(\xi, \eta) \in \mathfrak{A}$. The functions of this class evidently vanish for $\eta = 0$, $\xi > 0$ together with all their derivatives, but may have strong singularities at the origin. The class $\mathfrak{B}$ contains the functions $W_n(\xi, \eta)$ for all integer n. It is invariant with respect to the multiplication by ξ and η; i.e., together with each function $v(\xi, \eta)$ it also contains the functions $\xi v(\xi, \eta)$ and $\eta v(\xi, \eta)$. This follows from the evident relations

$$\xi \frac{\partial^j u}{\partial \eta^j} = \frac{\partial^j}{\partial \eta^j}(\xi u), \qquad \eta \frac{\partial^j u}{\partial \eta^j} = \frac{\partial^j}{\partial \eta^j}(\eta u) - j \frac{\partial^{j-1} u}{\partial \eta^{j-1}}.$$

One can see without difficulty that any function of the class $\mathfrak{B}$ is a sum of functions of the form $\eta^{r/2}\Phi(\theta)$, where r is an integer, and $\Phi(\theta)$ decays exponentially at infinity. A term of this form will be said to be of order $r/2$ with respect to η. Denote the subset of all functions in $\mathfrak{B}$ having a fixed order $m/2$ (where m is an integer) by $\mathfrak{B}_m$.

LEMMA 1.1. *For any integer r, and any function $v(\xi, \eta) \in \mathfrak{B}_r$, there exists a function $u(\xi, \eta) \in \mathfrak{B}_{r+2}$ satisfying the equation $L_0 u = v$ in the domain ω', and conditions (1.30) on its boundary.*

The proof follows easily from the definition of the classes $\mathfrak{A}$, $\mathfrak{B}$, $\mathfrak{B}_r$. Indeed, by the hypothesis of the lemma, $v(\xi, \eta) \in \mathfrak{B}_r \subset \mathfrak{B}$, and, consequently, $v(\xi, \eta) = (\partial^j w/\partial \eta^j)(\xi, \eta)$, where $w \in \mathfrak{A}$, $j \geq 0$. The property of the class $\mathfrak{A}$ implies that there exists a function $\tilde{w}(\xi, \eta) \in \mathfrak{A}$ satisfying conditions (1.30) such that $L_0 \tilde{w} = w(\xi, \eta)$. The function $u = (\partial^j \tilde{w}/\partial \eta^j)(\xi, \eta)$ belonging, by construction, to the class $\mathfrak{B}$, satisfies the equation $L_0 u = v$ and conditions (1.30). It is not difficult to check that this function is of order $1 + r/2$ with respect to η. . ∎

LEMMA 1.2 (On the existence of an f.a.s.). *Let $\psi(\eta)$ be a polynomial of degree n, and suppose that the function $F(\xi, \eta)$ can be expanded, as $\xi^2 + \eta^2 \to \infty$, in an asymptotic series of the form*

$$F(\xi, \eta) = \sum_{j=-2n+2}^{\infty} f_j(\xi, \eta), \tag{1.31}$$

where $f_j(\xi, \eta) \in \mathfrak{B}_{-j}$. Then, as $\xi^2 + \eta^2 \to \infty$, there exists an f.a.s. of the problem

$$L_1 v \equiv \left(\frac{\partial^2}{\partial \xi^2} + \frac{\partial^2}{\partial \eta^2} - \frac{\partial}{\partial \eta} \right) v = F(\xi, \eta), \qquad \xi > 0, \ \eta > 0, \tag{1.32}$$

$$v(0, \eta) = \psi(\eta), \quad \eta > 0, \qquad v(\xi, 0) = 0, \quad \xi > 0,$$

of the form

$$v(\xi, \eta) = \sum_{j=-2n}^{\infty} v_j(\xi, \eta), \qquad (1.33)$$

where $v_j(\xi, \eta) \in \mathfrak{B}_{-j}$.

PROOF. If $v(\xi, \eta) \in \mathfrak{B}_{-j}$ then $L_0 v \in \mathfrak{B}_{-j-2}$, and $\partial^2 v/\partial \eta^2 \in \mathfrak{B}_{-j-4}$. Therefore, an application of the operator L_1 to a term of the series (1.33) yields the relation $L_1 v_j = w_{j+2} + u_{j+4}$, where $w_{j+2} = L_0 v_j \in \mathfrak{B}_{-j-2}$, $u_{j+4} = \partial^2 v_j/\partial \eta^2 \in \mathfrak{B}_{-j-4}$. Inserting the supposed f.a.s. (1.33) into equation (1.32), one obtains the formal equality

$$\sum_{j=-2n}^{\infty} w_{j+2} + \sum_{j=-2n}^{\infty} u_{j+4} = \sum_{j=-2n+2} f_j.$$

Equating the terms of the same order results in the equalities

$$w_{-2n+2} = f_{-2n+2}, \qquad w_{-2n+3} = f_{-2n+3},$$
$$w_{-2n+4} + u_{-2n+4} = f_{-2n+4}, \dots,$$
$$w_{-2n+k} + u_{-2n+k} = f_{-2n+k}, \dots,$$

or, equivalently,

$$L_0 v_{-2n} = f_{-2n+2}, \qquad L_0 v_{-2n+1} = f_{-2n+3},$$
$$L_0 v_{-2n+k} + \frac{\partial^2}{\partial \eta^2} v_{-2n+k-2} = f_{-2n+k-2}, \qquad k > 1.$$

Now consecutively apply Lemma 1.1 to these equations with respect to $v_{-2n+k}(\xi, \eta)$, adding, for even $2n - k \geq 0$, the functions $\beta_k W_{2n-k}(\xi, \eta)$ in order to satisfy condition (1.32) for $\xi = 0$. The series (1.33) thus constructed is, owing to the properties of the functions of the class $\mathfrak{B}$, the desired f.a.s. ∎

Now there can be no doubt in the validity of Theorem 1.2. The terms of the series (1.25) are evidently the functions of the class $\mathfrak{B}$ constructed above, and the proof reduces to a consecutive application of Lemma 1.2 to each of the equations (1.19), (1.20). However, a rigorous justification requires an additional effort for the examination of the boundary value problem (1.24) with $\psi \equiv 0$ and the function $F(\xi, \eta)$ rapidly decaying at infinity.

LEMMA 1.3. *Let* $v(\xi, \eta) \in \mathfrak{B}_r$, *where* r *is an integer, and suppose that* $u(\xi, \eta)$ *is of the form* $\eta^{r/2+1} \Phi(\theta)$, *and satisfies the equation* $L_0 u = v$ *in* ω', *and that* $\Phi(\theta) \in C^\infty[0, \infty)$ *and decays exponentially as* $\theta \to \infty$. *Then* $u(\xi, \eta) \in \mathfrak{B}_{r+2}$.

PROOF. Construct, according to Lemma 1.1, the function $\tilde{u}(\xi, \eta) \in \mathfrak{B}_{r+2}$ solving the equation $L_0 \tilde{u} = v$ and satisfying condition (1.30). The function $u_1(\xi, \eta) = u(\xi, \eta) - \tilde{u}(\xi, \eta)$ satisfies the equation $L_0 u_1 = 0$ in the domain ω', and equals $\eta^{r/2+1} U(\theta)$. The function $U(\theta)$ satisfies equation (1.29) for $n = r + 2$ and decays exponentially at infinity. Therefore, $\eta^{(r+2)/2} U(\theta) = \beta W_{r+2}(\xi, \eta)$. ∎

Denote by D^k any of the differentiation operators with respect to ξ and η of order k.

LEMMA 1.4. *Suppose that the functions $\psi(\eta) \in C^\infty[0, \infty)$ and $F(\xi, \eta) \in C^\infty(\omega)$ satisfy the inequalities*

$$|D^k\psi(\eta)| \leq M_k\eta^{-N} \quad \text{for } \eta \geq 0,$$
$$|D^kF(\xi, \eta)| \leq M_k\eta^{-N} \quad \text{in } \omega, \tag{1.34}$$

$$|D^kF(\xi, \eta)| \leq M_k\exp(-\gamma\xi) \quad \text{for } \eta \leq A\xi, \tag{1.35}$$

where $k \geq 0$ is any number, N is a sufficiently large fixed number, $A > 0$ is arbitrary, and the constants γ and M_k depend on A. Then the function (1.23) solving the boundary value problem (1.24) satisfies the inequalities

$$|D^kw(\xi, \eta)| \leq \overline{M}_k\exp(-\gamma_1\xi) \quad \text{for } \eta \leq A_1\xi, \ \xi^2 + \eta^2 > 1, \ \gamma_1 > 0, \tag{1.36}$$

where $k \geq 0$ is any integer, and the relation between A_1, γ_1, and $\overline{M}_k$ is the same as between A, γ, and M_k. The function (1.23) admits the following asymptotic representation

$$D^kw(\xi, \eta) = D^k\left\{\sum_{j=0}^{N_1}\eta^{-1/2-j/2}\Phi_j(\theta)\right\} + O\left(\eta^{-N_1}\right). \tag{1.37}$$

Here $k \geq 0$ is any integer, N_1 is a sufficiently large number depending on N such that $N_1 \to \infty$ as $N \to \infty$, $\theta = 2^{-1}\xi\eta^{-1/2}$, and the functions $\Phi_j(\theta) \in C^\infty[0, \infty)$ decay exponentially at infinity.

PROOF. The function $w(\xi, \eta)$ defined by equality (1.23) is the sum of the two integrals, in the first of which $G(\xi, \eta, \xi_1, \eta_1)$ is the sum of the four functions (1.22), while in the second $(\partial G/\partial\xi_1)(\xi, \eta, 0, \eta_1)$ is the sum of the two functions. Thus, it is natural to represent the integral (1.23) as the sum of six integrals, and to prove relations (1.36), (1.37) for each of them. All these integrals are of the same type, so we consider only one of them:

$$\tilde{w}(\xi, \eta) = \int_0^\infty\int_0^\infty K_0(\sqrt{(\xi - \xi_1)^2 + (\eta - \eta_1)^2}/2)e^{\frac{\eta-\eta_1}{2}}F(\xi_1, \eta_1)\,d\xi_1\,d\eta_1.$$

The examination of the other integrals is conducted along the same lines, or is even simpler. Since $F(\xi_1, \eta_1)$ decays rapidly at infinity, the function $\tilde{w}(\xi, \eta)$ behaves at infinity, roughly speaking, as the kernel $K_0(\sqrt{(\xi - \xi_1)^2 + (\eta - \eta_1)^2}/2)e^{(\eta-\eta_1)/2}$ for fixed ξ_1, η_1. Therefore, one has first to examine the asymptotics of this kernel. It is known (see [64, p.157]) that as $r \to \infty$ the asymptotics

$$K_0(r/2) = r^{-1/2}e^{-r/2}\sum_{j=0}^\infty c_jr^{-j} \tag{1.39}$$

is valid. Denoting $r = \sqrt{t^2 + \tau^2}$, consider the function of interest $K_0(r/2)e^{\tau/2}$. For any $A > 0$, for $t > A\tau$, $r > 1$ it follows from (1.39) that

$$|K_0(r/2)e^{\tau/2}| \leq M\exp\frac{\tau - r}{2} = M\exp\left(-\frac{t^2}{2(r + \tau)}\right)$$

$$\leq M\exp\left(-\frac{t^2}{2(t/A + t\sqrt{1 + A^2/A})}\right).$$

Hence

$$|K_0(r/2)e^{\tau/2}| < Me^{-\gamma_2 t} \tag{1.40}$$

with similar estimates holding for the derivatives of the left-hand side of this inequality.

For a fixed t and $\tau \to \infty$, the kernel $K_0(r/2)e^{\tau/2}$ slowly tends to zero, and one has to find its asymptotic expansion. The main part is to rewrite the exponent in the asymptotic expansion of the functions $K_0(r/2)e^{\tau/2}$ taking into account (1.39):

$$-\frac{r}{2} + \frac{\tau}{2} = -\frac{t^2}{2(r+\tau)} = -\frac{t^2}{4\tau} + \frac{t^4}{4\tau(\tau+r)^2}.$$

Thus

$$K_0(r/2)e^{\tau/2} = e^{-\frac{t^2}{4\tau}} \left[\sum_{j=0}^{\infty} c_j r^{-\frac{1}{2}-j} \right] \exp \frac{t^4}{4\tau(r+\tau)^2}, \qquad r \to \infty. \tag{1.41}$$

This expansion and the estimate (1.40) now imply (1.36) and the asymptotic expansion (1.37) for the function $\tilde{w}(\xi, \eta)$.

In order to derive the estimate (1.36) it is sufficient to divide the domain of integration in the integral (1.38) into two parts: the domain where $\xi_1^2 + \eta_1^2 \leq 1$, and the rest of the quadrant ω, which, in its turn, is subdivided into the set where $\eta_1 \leq 2A_1\xi$, and the set where $\eta_1 > 2A_1\xi$. Taking into account the estimates (1.35), (1.40), and the fact that the kernel $K_0(r/2)\exp((\eta - \eta_1)/2)$ is a locally integrable function, one obtains the estimate (1.36) for $k = 0$. To obtain the first derivatives of $\tilde{w}(\xi, \eta)$ it is sufficient to differentiate the integrand in (1.38). Then, integrating by parts, one can move the derivatives from the kernel to the function $F(\xi_1, \eta_1)$, and then differentiate the integrand once. Proceeding in the same way, one can obtain the expressions for $D^k\tilde{w}(\xi, \eta)$ in the form of integrals of the same type as (1.38), a similar integrals along the half-axes $\xi_1 = 0$, $\eta_1 \geq 0$ and $\eta_1 = 0$, $\xi_1 \geq 0$. The estimates (1.36) are thus established for any k.

In order to derive the asymptotic expansion (1.37) it is sufficient, owing to the estimates (1.36), to consider only the case $\xi = o(\eta)$. Here the domain of integration in (1.38) must also be subdivided into several parts. The integrals over the domain $\eta_1 \geq \eta^\beta$, where $0 < \beta < 1/2$, and over the domain $\eta_1 \leq \eta^\beta$, $\xi_1 \geq \eta^\beta$ are small by virtue of the properties (1.34), (1.35) of the function $F(\xi, \eta)$. In the remaining integral one has to use the asymptotic expansion (1.41) where $t = \xi - \xi_1$, $\tau = \eta - \eta_1$ taking into account that $0 \leq \xi_1 \leq \eta^\beta$, $0 \leq \eta_1 \leq \eta^\beta$. Passing from the variables ξ, η to the variables $\theta = 2^{-1}\xi\eta^{-1/2}$, η we obtain the equalities

$$r = \sqrt{(\xi - \xi_1)^2 + (\tau - \tau_1)^2}$$
$$= \eta(1 - \eta_1\eta^{-1})\{1 + (1 - \eta_1\eta^{-1})^{-2}(2\theta\eta^{-1/2} - \xi_1\eta^{-1})^2\}^{1/2},$$

$$-\frac{t^2}{4\tau} = -\frac{(\xi - \xi_1)^2}{4(\eta - \eta_1)} = -\theta^2 + (1 - \eta_1\eta^{-1})^{-1}\{-\theta^2\eta_1\eta^{-1} + \theta\xi_1\eta^{-1/2} - \xi_1^2\eta^{-1}\},$$

$$\frac{t^4}{4\tau(r+\tau)^2} = \frac{(\xi - \xi_1)^4}{4(\eta - \eta_1)(r + \eta - \eta_1)^2}$$

$$= \frac{(2\theta - \xi_1\eta^{-1/2})^4}{4\eta(1 - \eta_1\eta^{-1})^3} \left\{ 1 + \frac{r}{\eta}\left(1 - \frac{\eta_1}{\eta}\right)^{-1} \right\}^{-2}.$$

Thus, each term of the series (1.41) is equal to

$$\eta^{-j/2}e^{-\theta^2}\varphi_j\left(\frac{\theta^2}{\sqrt{\eta}}, \frac{\eta_1}{\sqrt{\eta}}, \frac{\xi_1\theta}{\sqrt{\eta}}, \frac{\xi_1}{\sqrt{\eta}}, \frac{1}{\sqrt{\eta}}\right),$$

where the functions φ_j are smooth at the origin. Expanding these functions in Taylor series with remainders, and integrating expression (1.38) over the rest of the domain $\{\xi_1, \eta_1 : 0 \leq \xi_1 \leq \eta^\beta, 0 \leq \eta_1 \leq \eta^\beta\}$, one arrives at the asymptotic expansion (1.37) for $k = 0$. For the remaining values of k the asymptotics (1.37) is obtained by differentiating the integral (1.38) in the above manner, and applying the foregoing procedure to the resulting integrals. ∎

The proof of Theorem 1.2 follows easily from Lemmas 1.2–1.4. Construct, according to Lemma 1.2, the asymptotic series of the form (1.33) for the function $w_2(\xi, \eta)$, the solution of the problem (1.19), (1.21). The terms of this series have strong singularities at the origin. Consider, therefore, the function $w_{2,N}(\xi, \eta) = \chi(\xi, \eta)\tilde{w}_{2,N}(\xi, \eta)$, where $\tilde{w}_{2,N}(\xi, \eta)$ is the partial sum of this series, and $\chi(\xi, \eta) \in C^\infty(\omega)$ is a cut-off function vanishing in a neighborhood of the origin and equal to 1 outside some compact set. By construction, $L_1 w_{2,N} = F_{2,N}(\xi, \eta)$, where the function $F_{2,N}(\xi, \eta)$ satisfies the conditions of Lemma 1.4, and the function $w_{2,N}(0, \eta) - w_2(0, \eta)$ is compactly supported. Construct the function $w_{2,N}^*(\xi, \eta)$ solving the problem (1.24) with $F(\xi, \eta) = F_{2,N}(\xi, \eta)$, $\varphi(\eta) = w_{2,N}(0, \eta) - w_2(0, \eta)$, and apply to it Lemma 1.4. One obtains

$$w_{2,N}^*(\xi, \eta) = \sum_{j=0}^{N_1} \eta^{-1/2-j/2}\Phi_j(\theta) + O\left(\eta^{-N_1}\right).$$

The difference $w_{2,N}(\xi, \eta) - w_{2,N}^*(\xi, \eta)$ satisfies the same equation and the same boundary conditions as $w_2(\xi, \eta)$. Therefore, $w_2(\xi, \eta) = w_{2,N}(\xi, \eta) - w_{2,N}^*(\xi, \eta)$. Since N_1 is arbitrarily large, the statement of Theorem 1.2 is proved for the function $w_2(\xi, \eta)$: it can be expanded in the series (1.25), or, equivalently, in the series (1.33):

$$w_2(\xi, \eta) = \sum_{j=-2}^{\infty} v_{2,j}(\xi, \eta), \qquad \eta \to \infty,$$

where $v_{2,j}(\xi, \eta) = \eta^{-j/2}\Phi_{2,j}(\theta)$. Owing to the equation $L_1 w_2 = 0$, the functions $v_{2,j}(\xi, \eta)$ satisfy the recurrence system of equations $L_0 v_{2,-2} = 0$, $L_0 v_{2,-1} = 0$, $L_0 v_{2,j} = -\partial^2 v_{2,j-2}/\partial\eta^2$, $j > -1$. According to Lemma 1.3, one has $v_{2,j} \in \mathcal{B}_{-j}$ for all j.

The proof now proceeds by induction. If equalities (1.25) are valid for all w_{2i} for $i < n$, then the right-hand side $F_n(\xi, \eta)$ in the equation (1.20) can also be expanded in the series (1.31). Now, according to Lemma 1.2, one constructs an f.a.s. of the problem (1.20), (1.21), and the function $w_{n,N}(\xi, \eta)$, which is the product of its partial sum and the truncating function $\chi(\xi, \eta)$. The next step is the construction of the function $w_{n,N}^*$ solving the problem (1.24) with $F(\xi, \eta) = Lw_{n,N} - F_{n,N}(\xi, \eta)$, $\psi(\eta) = w_{n,N}(0, \eta) - w_{2n}(0, \eta)$. Then Lemma 1.4 is applied to the function $w_{n,N}^*$, and Lemma 1.3 to the function $w_{2n}(\xi, \eta) = w_{n,N}(\xi, \eta) - w_{n,N}^*(\xi, \eta)$. Relation (1.25) is hereby proved for $w_{2n}(\xi, \eta)$. ∎

Before we begin with the construction of the series V, let us verify that the resulting series W (see (1.17)) is an f.a.s. of the problem posed at the beginning of subsection 3 of this section. Namely, one has to verify that $A_{2N,\xi,\eta}W$

approximately satisfies (1.18), the condition $A_{2N,\xi,\eta}W|_{\xi=0}+A_{2N,x,y}U|_{x=0}=0$, and vanishes for $\eta=0$. The last equality is satisfied exactly by virtue of the first condition in (1.21). The second equality in (1.21) gives rise to the relations

$$A_{2N,\xi,\eta}W|_{\xi=0}+A_{2N,x,y}U|_{x=0}=\sum_{k=0}^{N}\varepsilon^{2k}w_{2k}(0,\eta)+\sum_{k=0}^{N}\varepsilon^{2k}u_{2k}(0,\varepsilon^2\eta)$$

$$=\sum_{k=0}^{N}\varepsilon^{2k}\left[u_{2k}(0,\varepsilon^2\eta)-\sum_{j=0}^{N-k}\frac{\varepsilon^{2j}\eta^j}{j!}\frac{\partial^j u_{2k}(0,0)}{\partial y^j}\right]$$

$$=O\left(\varepsilon^{2N+2}(\eta+1)^{N+1}\right).$$

(1.42)

Equations (1.19), (1.20) imply that

$$\mathscr{L}_\varepsilon(A_{2N,\xi,\eta}W)=\varepsilon^{-2}\left(\Delta_{\xi,\eta}(A_{2N,\xi,\eta}W)-a(\varepsilon^2\xi,\varepsilon^2\eta)\frac{\partial}{\partial\eta}A_{2N,\xi,\eta}W\right)$$

$$=\varepsilon^{-2}\sum_{k=1}^{N}\left[-a(\varepsilon^2\xi,\varepsilon^2\eta)+\sum_{j=0}^{N-k}P_j(\varepsilon^2\xi,\varepsilon^2\eta)\right]\varepsilon^{2k}\frac{\partial w_{2k}}{\partial\eta}.$$

(1.43)

We will use this equality once again below, but meanwhile it easily implies that W is an f.a.s. of the homogeneous equation for the values of η that are not very large. Indeed, the asymptotic expansion (1.25) implies, in particular, that $|\partial w_{2k}/\partial\eta|\le M(1+\eta)^{k-1}\exp(-\delta\xi/\sqrt{\eta})$. Therefore,

$$|\mathscr{L}_\varepsilon A_{2N,\xi,\eta}W|$$

$$\le\varepsilon^{-2}M\sum_{k=0}^{N}\varepsilon^{2(N-k)+2}(1+\xi^{N-k+1}+\eta^{N-k+1})\varepsilon^{2k}(1+\eta)^{k-1}\exp(-\delta\theta)$$

$$\le M\varepsilon^{2N}(1+\eta^N)\exp(-\delta\theta).$$

(1.44)

6. Now the asymptotics (1.25) of the functions $w_{2k}(\xi,\eta)$, and the matching condition (1.26) make it possible to finally determine the functions $v_k(\zeta,y)$, i.e., the solutions of the system (1.11)–(1.14). For convenience, let us again use a matching table, noting that under the change of variables $y=\varepsilon^2\eta$, $\zeta=\varepsilon\xi$, the variable θ preserves its value: $\theta=\xi/2\sqrt{\eta}=\zeta/2\sqrt{y}$.

The matching table for this problem (see Table 8, next page) is especially simple because of the absence of the logarithmic terms. There is apparently no need to explain its structure; it is built in complete analogy with the preceding tables. We only note that so far the lower halves of the squares of each horizontal row do not contain the asymptotic series for the functions

TABLE 8

W / V	$w_0(\xi, \eta)$	$\varepsilon^2 w_2(\xi, \eta)$	$\varepsilon^4 w_4(\xi, \eta)$	...
$v_0(y, \zeta)$	$\Phi_{0,0}(\theta)$	$\varepsilon^2 \eta \Phi_{2,0}(\theta)$	$\varepsilon^4 \eta^2 \Phi_{4,0}(\theta)$	...
(V_0)	$\Phi_{0,0}(\theta)$	$y\Phi_{2,0}(\theta)$	$y^2\Phi_{4,0}(\theta)$	
$\varepsilon v_1(y, \zeta)$	$\eta^{-1/2}\Phi_{0,1}(\theta)$	$\varepsilon^2 \eta^{1/2}\Phi_{2,1}(\theta)$	$\varepsilon^4 \eta^{3/2}\Phi_{4,1}(\theta)$	...
(εV_1)	$\varepsilon y^{-1/2}\Phi_{0,1}(\theta)$	$\varepsilon y^{1/2}\Phi_{2,1}(\theta)$	$\varepsilon y^{3/2}\Phi_{4,1}(\theta)$	
$\varepsilon^2 v_2(y, \zeta)$	$\eta^{-1}\Phi_{0,2}(\theta)$	$\varepsilon^2\Phi_{2,2}(\theta)$	$\varepsilon^4 \eta\Phi_{4,2}(\theta)$	...
$(\varepsilon^2 V_2)$	$\varepsilon^2 y^{-1}\Phi_{0,2}(\theta)$	$\varepsilon^2\Phi_{2,2}(\theta)$	$\varepsilon^2 y\Phi_{4,2}(\theta)$	
...	...	...	...	...

$\varepsilon^i v_i(\zeta, y)$, but the formal series $\varepsilon^i V_i(\zeta, y)$, where

$$V_i(\zeta, y) = y^{-i/2} \sum_{j=0}^{\infty} y^j \Phi_{j,i}(\theta), \qquad y \to 0. \tag{1.45}$$

THEOREM 1.3. *There exist functions $v_i(\zeta, y)$ providing the solution of the system* (1.11)–(1.14), *and having the asymptotic expansion* (1.45). *Therefore, for the series* (1.10) *and* (1.17) *thus constructed, the matching condition* (1.26) *is satisfied.*

PROOF. First, as in the preceding examples, one has to check that the series (1.45) are f.a.s. of the system (1.11)–(1.14). This can be done with the help of equality (1.43). Denote the right-hand side of this relation by $F_N(\xi, \eta, \varepsilon)$, and apply the operator $A_{N,y,\zeta}$ to the equality

$$\varepsilon^{-2}\left[\Delta_{\xi, \eta}(A_{2N, \xi, \eta}W) - a(\varepsilon^2\xi, \varepsilon^2\eta)\frac{\partial}{\partial \eta}A_{2N, \xi, \eta}W\right] = F_N(\xi, \eta, \varepsilon).$$

On the left-hand side of this equality we have to use the asymptotic expansions (1.25), and pass to the variables y, ζ. As a result, one obtains the equality

$$\varepsilon^2\Delta_{x,y}\left(\sum_{j=0}^{N-2}B_{N-j/2}V_j(\zeta, y)\varepsilon^j\right) - \sum_{j=0}^{N}\varepsilon^j\sum_{s=0}^{j}\zeta^{j-s}a_{j-s}(y)\frac{\partial}{\partial y}B_{N-s/2}V_s(\zeta, y)$$

$$= A_{N,y,\zeta}F_N(\xi, \eta, \varepsilon). \tag{1.46}$$

In order to compute $A_{N,y,\zeta}F_N$, one has to replace the functions $w_{2k}(\xi, \eta)$ with their asymptotic series (1.25) so that

$$
F_N(\xi, \eta, \varepsilon)
$$

$$
= \varepsilon^{-2} \sum_{k=1}^{N} \left[-a(\varepsilon\zeta, y) + \sum_{j=0}^{N-k} P_j(\varepsilon\zeta, y) \right] \frac{\partial}{\partial \eta} \left[\varepsilon^{2k} \eta^k \sum_{j=0}^{\infty} \eta^{-j/2} \Phi_{k,j}(\theta) \right]
$$

$$
= -\varepsilon^{-2} \sum_{k=1}^{N} \left\{ \sum_{j=0}^{N-k} (\varepsilon\zeta)^j \left[a_j(y) - \sum_{s=0}^{N-k-j} \frac{y^s}{s!} \frac{d^s a_j(0)}{dy^s} \right] + \sum_{j=N-k+1}^{\infty} (\varepsilon\zeta)^j a_j(y) \right\}
$$

$$
\times \varepsilon^2 \frac{\partial}{\partial y} \left[y^k \sum_{j=0}^{\infty} \varepsilon^j y^{-j/2} \Phi_{k,j}(\theta) \right].
$$

Hence $A_{N,\zeta,y}F_N(\xi, \eta, \varepsilon) = \sum_{p=0}^{N} \varepsilon^p \varphi_{p,N}(\zeta, y)$, where

$$
\varphi_{p,N}(\zeta, y)
$$

$$
= -\sum_{l=0}^{p} \sum_{k=1}^{N-l} \zeta^l \left(a_l(y) - \sum_{s=0}^{N-k-l} \frac{y^s}{s!} \frac{d^s a_l(0)}{dy^s} \right) \frac{\partial}{\partial y} \left(y^{k-(p-l)/2} \Phi_{k,p-l}\left(\frac{\zeta}{2\sqrt{y}} \right) \right)
$$

$$
- \sum_{l=0}^{p} \sum_{k=N-l+1}^{N} \zeta^l a_l(y) \frac{\partial}{\partial y} \left(y^{k-(p-l)/2} \Phi_{k,p-l}\left(\frac{\zeta}{2\sqrt{y}} \right) \right).
$$

The explicit form of $\varphi_{p,N}(\zeta, y)$ and the properties of the functions $\Phi_{k,j}(\theta)$ imply that

$$
\varphi_{p,N}(\zeta, y) = O\left(y^{N-p/2} \exp\left(-\gamma \frac{\zeta}{\sqrt{y}} \right) \right), \qquad \gamma > 0.
$$

Thus, it follows from (1.46) that the partial sums $B_{N-j/2}V_j(\zeta, y)$ satisfy the system of differential equations which differs from the system (1.11), (1.12) just in their right-hand parts rapidly decaying as $y \to 0$:

$$
L(B_N V_0) \equiv \frac{\partial^2}{\partial \zeta^2}(B_N V_0) - a_0(y) \frac{\partial}{\partial y}(B_N V_0) = \varphi_{0,N}(\zeta, y),
$$

$$
L(B_{N-1/2} V_1) = \zeta a_1(y) \frac{\partial}{\partial y}(B_N V_0) + \varphi_{1,N}(\zeta, y), \qquad \text{etc.}
$$

Furthermore, by construction, the functions $B_{N-j/2}V_j$ satisfy condition (1.13), while on the other part of the boundary one has $B_{N-j/2}V_j(0, y) = v_j(0, y) + O(y^{N-j/2})$. In other words, the series $V_j(\zeta, y)$ provide the f.a.s. of the system (1.11)–(1.14).

The construction of the solutions $v_i(\zeta, y)$ can now easily be carried out by applying the explicit formulas (1.15) and (1.16) to the residuals. Indeed, define, for example, the function $v_0(\zeta, y)$ as the sum of $B_N V_0$ and the solution $\tilde{v}_N(\zeta, y)$ of the problem $L\tilde{v}_N = -\varphi_{0,N}(\zeta, y)$, $\tilde{v}_N(\zeta, 0) = 0$, $\tilde{v}_N(0, y) = v_0(0, y) - B_N V_0(0, y)$. The function $\tilde{v}_N(\zeta, y)$ is constructed according to formulas (1.15), (1.16). It is clear that $B_N V_0(\zeta, y) + \tilde{v}_N(\zeta, y)$ satisfies equation (1.11) and conditions (1.13), (1.14). It is also evident that this sum does not depend on N, and is the desired function $v_0(\zeta, y)$.

The conclusion of the theorem is now obtained by induction by consecutively applying the same procedure to equations (1.12). ∎

THEOREM 1.4. *The composite asymptotic expansion*

$$T_N(x, y, \varepsilon) = A_{2N,\zeta,y}V + A_{2N,\xi,\eta}W - A_{2N,\zeta,y}(A_{2N,\xi,\eta}W), \qquad (1.47)$$

where V and W are the series (1.10) *and* (1.17), *approximately satisfies the homogeneous equation $\mathscr{L}_\varepsilon T_N = 0$ in the square Ω, and the boundary conditions $T_N(x, 0, \varepsilon) = 0$, $T_N(0, y, \varepsilon) = -U|_{x=0}$, where U is the series* (1.3). *To be more precise, there exist constants $\gamma > 0$, $\delta > 0$ such that*

$$|\mathscr{L}_\varepsilon T_N| < M\varepsilon^{\gamma N}\exp(-\delta\theta) \quad \text{for } (x, y) \in \Omega, \qquad (1.48)$$

$$T_N(x, 0, \varepsilon) = 0, \qquad (1.49)$$

$$|T_N + A_{2N,x,y}U|_{x=0} < M\varepsilon^{\gamma N} \quad \text{for } 0 \le y \le 1. \qquad (1.50)$$

PROOF. Condition (1.49) is satisfied according to the construction procedure for the functions $v_k(\zeta, y)$ and $w_{2k}(\xi, \eta)$ (see (1.13) and (1.21)). By virtue of the matching conditions for the series (1.10) and (1.17), if $y \le \varepsilon$ then

$$A_{2N,\zeta,y}V - A_{2N,\zeta,y}(A_{2N,\zeta,\eta}W) = O\left(\left(y^{N+1} + \varepsilon^{2N}y\right)e^{-\delta\theta}\right)$$
$$= O\left(\varepsilon^{N+1}e^{-\delta\theta}\right), \qquad (1.51)$$

and if $y \ge \varepsilon$ then

$$A_{2N,\xi,\eta}W - A_{2N,\zeta,y}(A_{2N,\xi,\eta}W) = O\left(\varepsilon^{2N+1}y^{-N-1/2}e^{-\delta\theta}\right)$$
$$= O\left(\varepsilon^{N+1/2}e^{-\delta\theta}\right). \qquad (1.52)$$

Similar estimates hold for the derivatives of these differences. Relations (1.51) and (1.42) imply inequality (1.50) for $y \le \varepsilon$, while condition (1.14) and the estimate (1.52) imply inequality (1.50) for $y \ge \varepsilon$.

Inequality (1.48) is proved in exactly the same way. For $y \le \varepsilon$, the system (1.19), (1.20) and the asymptotic expansion (1.25) imply that the function $A_{2N,\xi,\eta}W$ approximately satisfies the equation, while the remaining terms in the sum T_N approximately cancel each other. For $y \ge \varepsilon$, the function $A_{2N,\zeta,y}V$ approximately satisfies the equation (by virtue of the system (1.11)–(1.14), and the asymptotic expansion (1.45)), while the difference $A_{2N,\xi,\eta}W - A_{2N,\zeta,y}(A_{2N,\xi,\eta}W)$ is small. ∎

The asymptotics of the solution $u(x, y, \varepsilon)$ of the problem (1.1), (1.2) is almost constructed. We now construct the series $V^*(\varepsilon^{-1}(1 - x), y)$ near the boundary $x = 1$ in complete analogy to the series $V(\varepsilon^{-1}x, y)$, while near the vertex $(1, 0)$ the series $W^*(\varepsilon^{-2}(1 - x), \varepsilon^{-2}y)$ is constructed in complete analogy with the series $W(\varepsilon^{-2}x, \varepsilon^{-2}y)$ (see Figure 16 which shows the domains where the corresponding series provide correct asymptotics for $u(x, y, \varepsilon)$). Since the coefficients of all these series decay exponentially as $\zeta \to \infty$, $\xi \to \infty$ and, respectively, as $\zeta^* = \varepsilon^{-1}(1 - x) \to \infty$, $\xi^* = \varepsilon^{-2}(1 - x) \to \infty$, the series constructed near the boundary $x = 0$ bear no influence on the boundary condition for $x = 1$, and vice versa.

The residual in the boundary condition for $y = 1$ arising from the asymptotic expansion $U(x, y, \varepsilon)$ is already eliminated by the series S (see (1.5)). It only remains to eliminate the residuals newly arisen in the vicinity of the two vertices of the square: $(0, 1)$ and $(1, 1)$. It is sufficient to consider just the first of them.

The new residuals in the boundary conditions near the point $(0, 1)$ give rise to the following series: the series (1.5) for $x = 0$, and the series (1.10) for $y = 1$. The residual in the boundary condition at $y = 1$ equal to $-\sum_{k=0}^{\infty} \varepsilon^k v_k(1, \zeta)$ (see (1.10)) is easily eliminated, as is that due to the series (1.3), with the help of the series

$$\widetilde{S} = \sum_{k=0}^{\infty} \varepsilon^k \tilde{s}_k(\zeta, \tau), \tag{1.53}$$

where $\zeta = \varepsilon^{-1}x$, $\tau = \varepsilon^{-2}(1 - y)$.

For $\tilde{s}_k(\zeta, \tau)$ one obtains a system of ordinary differential equations which is very similar to the system (1.6) for $s_{2k}(\zeta, \tau)$. The difference is that the coefficient $a(x, y)$ has to be expanded in a series not only in y, but also in x. Let $a(\varepsilon\zeta, 1 - \varepsilon^2\tau) = \sum_{j=0}^{\infty} \varepsilon^j H_j(\zeta, \tau)$, where $H_j(\zeta, \tau)$ are polynomials of degree j, $H_0(\zeta, \tau) = a(0, 1) = \varkappa^2 > 0$. Then the equation for $\tilde{s}_k(\zeta, \tau)$ is of form

$$\frac{\partial^2 \tilde{s}_0}{\partial \tau^2} + \varkappa^2 \frac{\partial \tilde{s}_0}{\partial \tau} = 0, \qquad \frac{\partial^2 \tilde{s}_1}{\partial \tau^2} + \varkappa^2 \frac{\partial \tilde{s}_1}{\partial \tau} = -H_1(\zeta, \tau)\frac{\partial \tilde{s}_0}{\partial \tau},$$

$$\frac{\partial^2 \tilde{s}_k}{\partial \tau^2} + \varkappa^2 \frac{\partial \tilde{s}_k}{\partial \tau} = -\sum_{j=1}^{k} H_j(\zeta, \tau)\frac{\partial \tilde{s}_{k-j}}{\partial \tau} - \frac{\partial^2 \tilde{s}_{k-2}}{\partial \zeta^2}, \qquad k \geq 2, \tag{1.54}$$

and the boundary conditions are $\tilde{s}_k(\zeta, 0) = -v_k(1, \zeta)$, $\tilde{s}_k(\zeta, \tau) \to 0$ as $\tau \to \infty$. The functions $\tilde{s}_k(\zeta, \tau)$ solving this problem evidently satisfy the estimate $|\tilde{s}_k(\zeta, \tau)| < M \exp(-\delta(\zeta + \tau))$, because the functions $v_k(1, \zeta)$ satisfy a similar property.

It now remains to construct the asymptotic expansion Z which formally satisfies the homogeneous equation $\mathscr{L}_\varepsilon Z = 0$ and the boundary conditions $Z|_{x=0} = -\sum_{k=0}^{\infty} \varepsilon^{2k} s_{2k}(0, \tau) - \sum_{k=0}^{\infty} \varepsilon^k \tilde{s}_k(0, \tau)$, $Z|_{y=0} = 0$. It is exactly the same problem as that considered in Chapter I, §2 (Example 4) with the corner boundary layer arising near the vertex $(0, 1)$ (see Figure 16). The choice of the inner variables is motivated by the same argument as in similar problems above so that $\xi = \varepsilon^{-2}x = \varepsilon^{-1}\zeta$ and $\tau = \varepsilon^{-2}(1 - y)$:

$$Z = \sum_{k=0}^{\infty} \varepsilon^k z_k(\xi, \tau). \tag{1.55}$$

The equations and boundary conditions for $z_k(\xi, \tau)$ are obtained in the usual way :

$$L_2 z_0 \equiv \Delta_{\xi,\tau} z_0 + \varkappa \frac{\partial z_0}{\partial \tau} = 0,$$

$$L_2 z_k = -\sum_{j=1}^{k} Q_j(\xi,\tau) \frac{\partial z_{k-j}}{\partial \tau}, \qquad k \geq 1, \ 0 < \sigma < \infty, \ 0 < \tau < \infty, \quad (1.56)$$

$$z_{2k}(0,\tau) = -s_{2k}(0,\tau) - \tilde{s}_{2k}(0,\tau),$$

$$z_{2k+1}(0,\tau) = -\tilde{s}_{2k+1}(0,\tau).$$

Here $Q_j(\xi,\tau)$ are the polynomials of degree j obtained by expanding the function $a(x,y)$ in a Taylor series

$$a(\varepsilon^2 \xi, 1 - \varepsilon^2 \tau) = \sum_{j=0}^{\infty} \varepsilon^j Q_j(\xi,\tau),$$

$$\varkappa = Q_0(\xi,\tau) = a(0,1) > 0.$$

The problems (1.56) are very similar to the problems (1.19)–(1.21) considered above. The only difference is the sign of the coefficient at the derivative $\partial/\partial\tau$. Thus, Green's function $E(\xi,\tau,\xi_1,\tau_1)$ for the operator L_2 in the quadrant $\{\xi,\tau : 0 < \xi < \infty, 0 < \tau < \infty\}$ is of the form similar to (1.22):

$$E(\xi,\tau,\xi_1,\tau_1)$$

$$= (2\pi)^{-1} \left\{ -K_0 \left(\frac{\varkappa\sqrt{(\xi-\xi_1)^2 + (\tau-\tau_1)^2}}{2} \right) \right.$$

$$+ K_0 \left(\frac{\varkappa\sqrt{(\xi-\xi_1)^2 + (\tau+\tau_1)^2}}{2} \right)$$

$$+ K_0 \left(\frac{\varkappa\sqrt{(\xi+\xi_1)^2 + (\tau-\tau_1)^2}}{2} \right)$$

$$\left. - K_0 \left(\frac{\varkappa\sqrt{(\xi+\xi_1)^2 + (\tau+\tau_1)^2}}{2} \right) \right\} \exp\left(\varkappa \frac{\tau_1 - \tau}{2} \right).$$

The solution of the problems $L_2 z = F(\xi,\tau)$ with the boundary conditions $z(\xi,0) = 0$, $z(0,\tau) = \psi(\tau)$ is of the form

$$z(\xi,\tau) = \int_0^\infty \int_0^\infty E(\xi,\tau,\xi_1,\tau_1) F(\xi_1,\tau_1)\, d\xi_1\, d\tau_1$$

$$- \int_0^\infty \frac{\partial E}{\partial \xi_1}(\xi,\tau,0,\tau_1)\psi(\tau_1)\, d\tau_1. \quad (1.57)$$

The functions F and ψ satisfy the same smoothness conditions as in problem (1.24). But now the function $\psi(\tau)$ decays exponentially at infinity, and formula (1.57) easily implies that the solution $z(\xi,\tau)$ also has the same property provided $F(\xi, \tau) \equiv 0$. And if $|F(\xi, \tau)| < M \exp(-\delta(\xi + \tau))$, then the same inequality is satisfied by the function z and its derivative $\partial z/\partial \tau$. These properties are checked without difficulty using the explicit form (1.57), so we skip the details. A consecutive application of these properties to equation (1.56) yields the solutions $z_k(\xi, \tau)$ which decay exponentially as $\xi + \tau \to \infty$.

The asymptotics of the solution of the problem (1.1), (1.2) is now completely constructed. Near the vertex $(1, 1)$ one constructs the series $\widetilde{S}^*$ and Z^*, completely analogous to the series $\widetilde{S}$ and Z. We now provide the scheme describing the order in which the individual series realizing the asymptotic representation of the solution $u(x, y, \varepsilon)$ in different parts of the square Ω are constructed. The corresponding independent variables together with the numbers of the formulas, where the respective series are introduced and the boundary value problems for their coefficients stated, are given in the brackets:

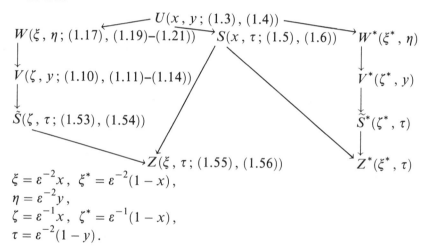

$$\xi = \varepsilon^{-2}x, \ \xi^* = \varepsilon^{-2}(1 - x),$$
$$\eta = \varepsilon^{-2}y,$$
$$\zeta = \varepsilon^{-1}x, \ \zeta^* = \varepsilon^{-1}(1 - x),$$
$$\tau = \varepsilon^{-2}(1 - y).$$

Figure 16 shows the domains where the individual asymptotic expansions are valid. As in the problems considered above, the boundaries of separate regions are shown more or less arbitrarily.

THEOREM 1.5. *Suppose that* $u(x, y, \varepsilon)$ *is a solution of the problem* (1.1), (1.2), *and let* $Y_N(x, y, \varepsilon)$ *be the following expression:*

$$Y_N(x, y, \varepsilon) = A_{2N, x, y}U + T_N + T_N^* + A_{2N, x, \tau}S$$
$$+ A_{2N, \zeta, \tau}\widetilde{S} + A_{2N, \zeta^*, \tau}\widetilde{S}^* + A_{2N, \xi, \tau}Z + A_{2N, \xi^*, \tau}Z^*,$$

where T_N *is defined by formula* (1.47), *and* T_N^* *is a similar composite asymptotic expansion obtained from the series* V^* *and* W^*. *Then there exists a constant* $\gamma > 0$ *such that the estimate*

$$|Y_N(x, y, \varepsilon) - u(x, y, \varepsilon)| < M_N \varepsilon^{\gamma N}$$

holds for any positive integer N.

The proof follows directly from Theorem 1.4, estimates (1.8), (1.9), and the similar estimates for $\tilde{s}$ and z on applying the maximum principle to equation (1.1). ∎

THEOREM 1.6. *Let* β *be a number such that* $0 < \beta < 2$. *Then the sum* $U+V+S+\tilde{S}+Z$ *is a uniform asymptotic expansion of the solution* $u(x, y, \varepsilon)$ *in the domain*

$$D_{1,\beta} = \{x, y: 0 \le x \le 1/2, \ 0 \le y \le 1\} \backslash \{x, y: 0 \le x \le \varepsilon^{\beta}, \ 0 \le y \le \varepsilon^{\beta}\},$$

and the series $U+W$ *is a uniform asymptotic expansion of* $u(x, y, \varepsilon)$ *in the domain*

$$D_{2,\beta} = \{x, y: 0 \le x \le \varepsilon^{\beta}, \ 0 \le y \le \varepsilon^{\beta}\}.$$

Similar uniform asymptotic expansions hold for $x \ge 1/2$.

PROOF. Consider the expression Y_N defined in Theorem 1.5. For $x \le 1/2$ the terms marked by an asterisk are exponentially small. It remains to prove, therefore, that $T_N(x, y, \varepsilon)$ is close to $A_{2N,\zeta,y}V$ in the domain $D_{1,\beta}$, and to $A_{2N,\xi,\eta}W$ in the domain $D_{2,\beta}$. This follows immediately from the form of T_N (see (1.47)), and inequalities (1.51), (1.52).

Note also that, since the series S, $\tilde{S}$, and Z are exponentially small far from the top side of the square, the series $U+V$ provides a uniform asymptotic expansion of the solution $u(x, y, \varepsilon)$ in that part of $D_{1,\beta}$ where $y \le 1 - \delta \, (\delta > 0)$. The precise estimate of the difference between $u(x, y, \varepsilon)$ and partial sums of the series can be easily obtained from the estimates of the functions $v_i(\zeta, y)$ and $w_{2k}(\xi, \eta)$ taking into account that N in Theorem 1.2 is an arbitrary number. ∎

§2. Asymptotics of the solution in a domain with nonsmooth boundary

In this section we continue the analysis of the problem (0.3), (0.4) in the case where the boundary contains "corner" points. In a neighborhood of the characteristics of the limit equation passing through these points (see the characteristics α_3 and α_4 in Figure 15), the asymptotics of the solution $u(x, y, \varepsilon)$ has a rather complex structure, and can be investigated by the method of matched asymptotic expansions. A similar investigation can be carried out almost without any modification, but with certain simplifications, in neighborhoods of singular characteristics passing through those points of the boundary, where the boundary function or some of its derivatives are discontinuous. We consider only the most complex of the above cases shown in Figure 17. The other typical cases of positioning a singular characteristic shown in Figure 18 are studied in a similar, but somewhat simpler way. This can be easily seen from the analysis of the following constructions.

Thus, consider the model equation

$$\mathscr{L}_{\varepsilon} u \equiv \varepsilon^2 \left(\frac{\partial^2 u}{\partial x^2} + \frac{\partial^2 u}{\partial y^2} \right) - a(x, y) \frac{\partial u}{\partial y} = f(x, y) \quad \text{for } (x, y) \in \Omega \quad (2.1)$$

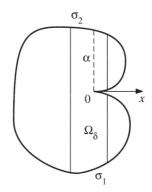

FIGURE 17

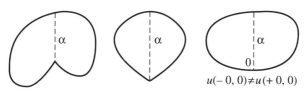

FIGURE 18

and the boundary condition

$$u(x, y, \varepsilon) = 0 \quad \text{for } (x, y) \in \partial\Omega, \tag{2.2}$$

where $a, f \in C^\infty(\overline{\Omega})$, $a(x, y) > 0$. As before, $u(x, y, \varepsilon)$ denotes a function satisfying relations (2.1) and (2.2), and continuous in $\overline{\Omega}$. (The requirement of infinite differentiability in $\overline{\Omega}$ is inappropriate here because, in general, it is not satisfied at the corner point O, but in a neighborhood of all smooth sections of the boundary $\partial\Omega$ the solution of the problem (2.1), (2.2) belongs to C^∞.) The domain Ω is shown in Figure 17. More precisely, the domain is described as follows. The point O is the only point at which the boundary $\partial\Omega$ is not smooth. In a small neighborhood of the point O, the complement of the domain Ω is the interior on an angle, i.e., the set of points where $x \geq 0$, $-Ex \leq y \leq 0$, $E > 0$. For $|x| \leq \delta$, where $\delta > 0$ is a fixed number, the boundary $\partial\Omega$ includes the sides of the angle, and two more smooth sections: $\sigma_1 = \{x, y: y = \varphi_1(x), |x| \leq \delta\}$ and $\sigma_2 = \{x, y: y = \varphi_2(x), |x| \leq \delta\}$, $\varphi_1(x) < -Ex \leq 0 < \varphi_2(x)$, $\varphi_1, \varphi_2 \in C^\infty[-\delta, \delta]$.

Thus, the subsequent investigation will be carried out just in the domain Ω_δ bounded by the straight lines $x = \pm\delta$, rays $\{y = 0, x > 0\}$, $\{y = -Ex, x > 0\}$, and curves σ_1, σ_2 (see Figure 17). In this region an f.a.s. of the equation (2.1) with the boundary condition (2.2) will be constructed. Therefore, the values of the differential operator $\mathscr{L}_\varepsilon$ applied to the difference between a partial sum of the f.a.s. and the solution $u(x, y, \varepsilon)$ will be small everywhere in Ω_δ, while the difference itself is small everywhere on the boundary $\partial\Omega_\delta$ with the exception of the lines $x = \pm\delta$. If this difference were small everywhere on $\partial\Omega_\delta$, then the justification of the f.a.s. constructed above would follow from the maximum principle. It turns out that the requirement that the residual must be small on the lines $x = \pm\delta$ is not necessary. For small ε a more powerful maximum principle is valid. It is formulated as follows. If v is sufficiently smooth in Ω_δ and continuous in $\overline{\Omega}_\delta$, $\mathscr{L}_\varepsilon v$ is small in Ω_δ, the values of v are bounded on the lines $x = \pm\delta$ and small on the rest of the boundary $\partial\Omega_\delta$, then v is small in the domain Ω_{δ_1}, where $\delta_1 < \delta$. The precise statement together with the proof of the corresponding lemma will be given at the end of the chapter (Lemma 5.1). This lemma is of an auxiliary character concerning just the justification of the asymptotics, and makes it possible to localize the procedure of constructing an f.a.s. in the vicinity of singular characteristics. It relates the construction of an f.a.s. with the justification of the asymptotics not only in this section, but also in the two succeeding ones. Therefore, our attention in these sections is limited just to the construction of an f.a.s.

Denote by $p_j(\xi, \eta)$, $q_j(\xi)$ polynomials of degree not greater than j, and by $P_j(x, y)$, $Q_j(x, y)$ homogeneous polynomials of degree j.

1. The outer expansion of the solution $u(x, y, \varepsilon)$ is constructed, as usual, in the form

$$U = \sum_{k=0}^\infty \varepsilon^{2k} u_{2k}(x, y). \tag{2.3}$$

The equations for $u_{2k}(x, y)$ are also obtained in the usual way which we have already used many times:

$$a(x, y)\frac{\partial u_0}{\partial y} = -f(x, y), \qquad a(x, y)\frac{\partial u_k}{\partial y} = \Delta u_{2k-2}, \quad k \geq 1. \tag{2.4}$$

We have also repeatedly discussed that the correct definition of boundary data for the functions $u_k(x, y)$ is equivalent to the fact that conditions (2.2) are satisfied at the lower end of each segment $x = \text{const}$. Thus,

$$u_{2k}(x, \varphi_1(x)) = 0 \quad \text{for } |x| \leq \delta,$$
$$u_{2k}(x, 0) = 0 \quad \text{for } x \geq 0. \tag{2.5}$$

The problems (2.4), (2.5) can be solved explicitly by integrating over y either from the point $y = \varphi_1(x)$, or from the point $y = 0$. Thus, all the functions $u_{2k}(x, y)$ are defined everywhere in Ω_δ. They are evidently infinitely differentiable everywhere in Ω_δ with the exception of the segment $\alpha = \{x, y: x = 0, 0 \leq y \leq \varphi_2(0)\}$. On the segment α all the functions $u_{2k}(x, y)$ have, in general, a discontinuity of the first kind, because to the left of the segment the corresponding function is obtained by integrating equations (2.4) from the point $y = \varphi_1(x)$, and to the right of it from the point $x = 0$ (see Figure 17). However, both to the left and to the right of α, each of the functions $u_{2k}(x, y)$, after being extended by continuity to α, becomes infinitely differentiable in the corresponding closed domain containing α.

2. The outer expansion does not satisfy condition (2.2) on the boundary $y = -Ex$, but this residual can be eliminated with the help of the exponential boundary layer considered in Chapter I. The independent variables in this layer are x and $\tau = \varepsilon^{-2}(-y - Ex)$. The f.a.s. of the boundary value problem (2.1), (2.2) in the domain $y \le -Ex$, $x > 0$ is the sum of the series (2.3) and the series

$$S = \sum_{k=0}^{\infty} \varepsilon^{2k} s_{2k}(x, \tau). \tag{2.6}$$

On inserting the series (2.6) into the homogeneous equation (2.1), and the sum $U + S$ into the boundary condition (2.2) (for $y = -Ex$), one obtains the recurrence system of boundary value problems

$$(1 + E^2)\frac{\partial^2 s_{2k}}{\partial \tau^2} + a(x, -Ex)\frac{\partial s_{2k}}{\partial \tau} = g_k(x, \tau),$$

where

$$g_k(x, \tau) = -2E\frac{\partial^2 s_{2(k-1)}}{\partial x \partial \tau} - \frac{\partial^2 s_{2(k-2)}}{\partial x^2} - \sum_{j=1}^{k} \frac{(-1)^j}{j!}\frac{\partial^j a}{\partial y^j}(x, -Ex)\frac{\partial s_{2(k-1)}}{\partial \tau},$$

$$s_{2k}(x, 0) = -u_{2k}(x, -Ex).$$

One takes for $s_{2k}(x, \tau)$ the solutions of this system which are infinitely differentiable for $\tau \ge 0$, and decay exponentially as $\tau \to \infty$. Evidently, $|s_{2k}(x, \tau)| \le M_k e^{-\gamma \tau}$, where $0 < \gamma < (1 + E^2)^{-1}\min_{\overline{\Omega}} a(x, y)$ with the same estimates holding for the derivatives of the functions $s_{2k}(x, \tau)$.

3. We now proceed with the construction of the asymptotic expansion in a neighborhood of the singular characteristic $\alpha = \{x, y: x = 0, 0 \le y \le \varphi_2(0)\}$. As in §1, subsection 2, introduce new independent variables $\zeta = \varepsilon^{-1}x$, y to equate the orders of the two principal terms in the left-hand side of (2.1), i.e., $\varepsilon^2 \partial^2/\partial x^2$ and $\partial/\partial y$. The inner expansion is of the form

$$V = \sum_{i=0}^{\infty} \varepsilon^i v_i(\zeta, y). \tag{2.7}$$

The system of recurrence equations for the coefficients $v_i(\zeta, y)$ is similar to system (1.12). The difference is that, while in §1 the asymptotics of the solution $u(x, y, \varepsilon)$ in a neighborhood of the characteristic $x = 0$ is given by the sum of the series U and V, now, in view of the discontinuity of the coefficients $u_k(x, y)$, it is more convenient to look for the asymptotics of the solution in a neighborhood of the characteristic α in the form of the series V. The equation for $v_k(\zeta, y)$ includes, therefore, the terms of the Taylor expansion of the right-hand side $f(x, y)$. Thus, the equations are of the form

$$Lv_0 = f(0, y),$$

$$Lv_i = -\frac{\partial^2 v_{i-2}}{\partial y^2} + \sum_{j=1}^{i} a_j(y)\zeta^j \frac{\partial u_{i-j}}{\partial y} + \frac{\zeta^i}{i!}\frac{\partial^i f}{\partial x^i}(0, y), \qquad i \geq 1, \qquad (2.8)$$

$$v_{-1}(\zeta, y) \equiv 0,$$

where $a(x, y) = \sum_{j=0}^{\infty} a_j(y)x^j$, $x \to 0$, $L = \partial^2/\partial\zeta^2 - a_0(y)\partial/\partial y$.

We now have to solve equation (2.8) not only for $\zeta \geq 0$, as was the case in §1, but for all ζ, i.e., in the entire strip $\{\zeta, y: -\infty < \zeta < \infty, 0 \leq y \leq \varphi_2(0)\}$. For $\zeta > 0$ the initial conditions for $v_k(\zeta, y)$ are evidently induced by condition (2.2):

$$v_k(\zeta, 0) = 0 \quad \text{for } \zeta > 0, \qquad (2.9)$$

while for $\zeta < 0$ the initial conditions are obtained by matching the series V to the series U. Replacing x in the series (2.3) with $\varepsilon\zeta$, and expanding each of the functions $u_{2k}(\varepsilon\zeta, 0)$ in a Taylor series, one obtains the formal series $\sum_{i=0}^{\infty} \varepsilon^i q_i(\zeta)$, where $q_i(\zeta) = \sum_{j+2k=i} \zeta^i(1/j!)(\partial^j u_{2k}(0, 0)/\partial x^j)$. It is now clear that the polynomials $q_i(\zeta)$ have to be taken for the initial conditions for the functions $v_i(\zeta, y)$:

$$v_i(\zeta, 0) = q_i(\zeta) \quad \text{for } \zeta < 0. \qquad (2.10)$$

One can easily see that the functions $v_i(\zeta, 0)$ are, in general, discontinuous at the origin because the initial data for $\zeta < 0$ are defined (through the coefficients of the outer expansion $u_{2k}(x, y)$) by the boundary conditions of the original problem on σ_1, while for $\zeta > 0$ they are defined by the boundary conditions on the x-axis. Therefore, the functions $v_i(\zeta, y)$ have singularities at the origin, the order of which grows with i. Naturally, one cannot use these functions for approximating the solution $u(x, y, \varepsilon)$ everywhere in the domain Ω_δ.

Furthermore, as in §1, neither equations (2.8) and initial conditions (2.9), (2.10), nor a given order of the growth of the function $v_i(\zeta, y)$ at the origin define these functions uniquely. The situation repeats that of the preceding section where it was discussed in subsection 2. As in §1, in order to define the functions $v_i(\zeta, y)$ uniquely, one more inner expansion is introduced.

In a neighborhood of the origin, the inner expansion of the solution $u(x, y, \varepsilon)$ is of the form

$$W = \sum_{k=0}^{\infty} \varepsilon^{2k} w_{2k}(\xi, \eta), \qquad (2.11)$$

where $\xi = \varepsilon^{-2}x$, $\eta = \varepsilon^{-2}y$. Assume, for simplicity, that $a(0, 0) = 1$. The equations for the coefficients $w_{2k}(\xi, \eta)$ are almost the same as in §1 (see (1.19), (1.20)) with the only difference being that, due to nonhomogeneity,

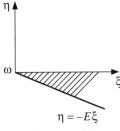

$$\eta = -E\xi$$

FIGURE 19

additional terms appear in their right-hand sides (as in equations (2.8)):

$$L_1 w_0 = 0, \tag{2.12}$$

$$L_1 w_{2k} = \sum_{j=1}^{k} P_j(\xi, \eta) \frac{\partial w_{2k-2j}}{\partial \eta} + Q_{k-1}(\xi, \eta), \qquad k > 0. \tag{2.13}$$

Here $L_1 = \Delta_{\xi, \eta} - \partial/\partial \eta$, $a(x, y) = \sum_{j=0}^{\infty} P_j(x, y)$, $x \to 0$, $y \to 0$, $f(x, y) = \sum_{i=0}^{\infty} Q_i(x, y)$, $x \to 0$, $y \to 0$. However, the domain of definition for the functions $w_{2k}(\xi, \eta)$ and the boundary conditions for them are now different. By ω we mean the entire (ξ, η)-plane with the exception of those points for which $-E\xi < \eta < 0$ (see Figure 19), while $\omega' = \omega \backslash (0, 0)$.

Thus, solutions of equations (2.12) and (2.13) are to be constructed in ω, while the boundary conditions for them follow from (2.2):

$$\begin{aligned} w_{2k}(\xi, 0) = 0 \quad \text{for } \xi \geq 0, \\ w_{2k}(\xi, -E\xi) = 0 \quad \text{for } \xi \geq 0. \end{aligned} \tag{2.14}$$

Since the right-hand sides of equations (2.13) exhibit polynomial growth, it is natural to look for the solutions $w_{2k}(\xi, \eta)$ in the class of functions growing not faster than some power of $|\xi| + |\eta|$. Solutions of the problems (2.12)–(2.14) are not defined in this class of functions uniquely. To make them unique one has to match the series (2.3) to (2.11).

Passing in the series U from the variables x, y to the variables ξ, η, and expanding each function $u_{2k}(\varepsilon^2 \xi, \varepsilon^2 \eta)$ in a Taylor series, one obtains the formal series $\sum_{i=0}^{\infty} \varepsilon^{2i} p_i(\xi, \eta)$, where

$$p_i(\xi, \eta) = \sum_{j+l+k=i} \xi^j \eta^l \frac{1}{j! l!} \frac{\partial^{j+l} u_{2k}}{\partial x^j \partial y^l}(0, 0).$$

It is natural to consider such solutions of the problems (2.12)–(2.14) which for $\xi^2 + \eta^2 \to \infty$, $\eta \leq 0$ are close to the polynomials $p_k(\xi, \eta)$. It turns out that one can construct solutions which are exponentially close to such polynomials not only as $\eta \to -\infty$, but also as $\xi \to -\infty$.

4. We now proceed with the construction of the solutions $w_{2k}(\xi, \eta)$ which will be constructed in the form

$$w_{2k}(\xi, \eta) = p_k(\xi, \eta) + z_{2k}(\xi, \eta). \tag{2.15}$$

Note that, by construction, the polynomials $p_k(\xi, \eta)$ satisfy the same system of recurrence relations (2.12), (2.13) as the desired functions w_{2k}. As at the end of §1, subsection 5, this is easily proved in view of the fact that the formal series U is an f.a.s. of equation (2.1). One obtains, therefore, the following homogeneous system of equations for the functions $z_{2k}(\xi, \eta)$:

$$L_1 z_0 = 0, \qquad (\xi, \eta) \in \omega',$$

$$L_1 z_{2k} = \sum_{j=1}^{k} P_j(\xi, \eta) \frac{\partial z_{2k-2j}}{\partial \eta}, \qquad k \geq 1, \ (\xi, \eta) \in \omega'. \tag{2.16}$$

The boundary conditions for the functions $z_{2k}(\xi, \eta)$ follow from (2.14), (2.15):

$$z_{2k}(\xi, 0) = -p_k(\xi, 0) \quad \text{for } k \geq 0,$$
$$z_{2k}(\xi, -E\xi) = -p_k(\xi, -E\xi) \quad \text{for } \xi \geq 0. \tag{2.17}$$

The solutions of the problems (2.16), (2.17) will actually be constructed in the class of functions decaying exponentially for $\eta \to -\infty$, $\xi \geq 0$, as well as for $|\xi| + |\eta| \to \infty$, $\xi \leq 0$, $\eta \leq 0$ and growing not faster than some power of $|\xi| + |\eta|$ for $|\xi| + |\eta| \to \infty$, $\eta > 0$. However, it is more convenient to consider a wider class of functions which is naturally associated with the operator L_1. This class is defined as follows. Let D be an angle in the (ξ, η)-plane, $D = \{\rho, \varphi: \rho \geq 0, \varphi_0 \leq \varphi \leq \varphi_1\}$, ρ, φ are polar coordinates, $D' = D \backslash \{0, 0\}$. A function $u(\xi, \eta)$ will be said to belong to the class $\mathfrak{U}$ in the domain D ($u(\xi, \eta) \in \mathfrak{U}(D)$) if it is infinitely differentiable in D', bounded on any bounded subdomain of D, and satisfies the condition

$$\frac{u(\xi, \eta)}{I_0(\sqrt{\xi^2 + \eta^2}/2) e^{\eta/2}} \xrightarrow[|\xi|+|\eta| \to \infty]{} 0, \qquad (\xi, \eta) \in D \tag{2.18}$$

($I_0(\rho)$ is the Bessel function of pure imaginary argument). For the domain D one can take the angle ω defined above, although in what follows some other values of φ_0 and φ_1 will also be used.

THEOREM 2.1. *There exist unique solutions $z_{2k}(\xi, \eta)$ of the problems* (2.16), (2.17) *such that $z_{2k}(\xi, \eta) \in \mathfrak{U}(\omega)$, and for some positive α_k, β_k*

$$|z_{2k}(\xi, \eta)| < M_k \exp(\alpha_k \xi + \beta_k \eta) \quad \text{for } (\xi, \eta) \in \omega, \tag{2.19}$$

$$\left| \frac{\partial^l z_{2k}(\xi, \eta)}{\partial \xi^{l_1} \partial \eta^{l_2}} \right| < M_{k,l} \exp(\alpha_k \xi + \beta_k \eta) \quad \text{for } (\xi, \eta) \in \omega, \tag{2.20}$$

$$|\xi| + |\eta| \geq 1.$$

The proof of the theorem will be preceded by a few simple lemmas.

LEMMA 2.1. *If $u(\xi, \eta) \in \mathfrak{U}(D)$, $L_1 u = 0$ for $(\xi, \eta) \in D'$, $u|_{\varphi=\varphi_0} = u|_{\varphi=\varphi_1} = 0$, then $u(\xi, \eta) \equiv 0$ in D.*

PROOF. It is not difficult to check that

$$L_1\left(I_0\left(\frac{\sqrt{\xi^2 + \eta^2}}{2}\right) \exp \frac{\eta}{2}\right) = 0.$$

Then the statement of the lemma follows easily from the maximum principle, and relation (2.18). ∎

LEMMA 2.2. *There exist constants* $\beta > 0$ *and* $\mu > 0$ *such that* $\Phi(\xi, \eta) = \exp[(E\beta + \mu)\xi + \beta\eta] \in \mathfrak{U}(\omega)$, $L_1\Phi < 0$ *for* $(\xi, \eta) \in \omega'$.

PROOF. With the use of the asymptotic representation of the function $I_0(\rho/2)e^{\eta/2}$ as $\rho \to \infty$ (see, for example, [64, §5.7]), one can easily conclude that condition (2.18) is satisfied for the function $\Phi(\xi, \eta)$ if $(E\beta + \mu)\xi + \beta\eta < \sigma(\rho/2 + \eta/2)$, where $\sigma < 1$. In its turn, the condition $L_1\Phi < 0$ requires that $(E\beta + \mu)^2 + \beta^2 - \beta < 0$. For the first of these inequalities to be satisfied, it is sufficient that the relation $(E\beta + \mu)^2 + (\sigma/2 - \beta)^2 < \sigma^2/4$ holds. Choose β so small that this relation and the second inequality above are satisfied for $\mu = 0$ and $\sigma = 1$. Then one can choose μ so small, and σ close to 1 such that these inequalities remain valid. ∎

LEMMA 2.3. *Let* $q_1(\xi)$, $q_2(\xi)$ *be polynomials, and* $F(\xi, \eta) \in C(\omega) \cap C^\infty(\omega')$. *Suppose that the inequality* $|F(\xi, \eta)| \leq M\Phi(\xi, \eta)$, *where* $\Phi(\xi, \eta)$ *is the function defined in the preceding lemma, holds everywhere in* ω. *Then there exists a unique solution of the equation* $L_1u = F(\xi, \eta)$ *belonging to the class* $\mathfrak{U}$ *in* ω, *and satisfying the conditions*

$$u(\xi, 0) = q_1(\xi), \quad u(\xi, -E\xi) = q_2(\xi) \quad for \ \xi > 0. \tag{2.21}$$

Furthermore, the relation

$$|u(\xi, \eta)| \leq M_l\Phi(\xi, \eta) \tag{2.22}$$

holds everywhere in ω, *and for* $(\xi, \eta) \in \omega$ *and* $\xi^2 + \eta^2 \geq 1$ *one has*

$$\left|\frac{\partial^l u(\xi, \eta)}{\partial \xi^{l_1} \partial \eta^{l_2}}\right| \leq M_l\Phi(\xi, \eta). \tag{2.23}$$

PROOF. Fix a positive number R, and consider the domain $\omega_R = \{\xi, \eta: (\xi, \eta) \in \omega, 0 < \rho < R\}$. Denote by $u_R(\xi, \eta)$ the solution of the equation $L_1u_R = F(\xi, \eta)$ in the domain ω_R which is continuous everywhere in $\overline{\omega}_R$, with the exception of the corner points, satisfies conditions (2.21) on the straight-line sections of the boundary of this domain, and vanishes on the rest of the boundary. Such a solution exists, is infinitely differentiable in the domain ω_R and bounded. Using the maximum principle, one can obtain without difficulty that the inequality $|u_R(\xi, \eta)| \leq M_1\Phi(\xi, \eta)$ holds everywhere in ω_R.

The fact that the functions $u_R(\xi, \eta)$ are uniformly bounded implies that they are compact together with all derivatives on any compact belonging to ω_R (see [61, Chapter 3]). Therefore, one can choose a subsequence of functions $u_{R_i}(\xi, \eta)$ converging in this sense to a function $u(\xi, \eta)$. This limit function is evidently infinitely differentiable in the domain ω', satisfies the equation $L_1u = F$ in this domain, condition (2.21) on its boundary, and relation (2.22) everywhere in ω'. Then the a priori estimates for the derivatives easily imply that it also satisfies relations (2.23). Evidently, this limit function also satisfies relation (2.18), i.e., $u(\xi, \eta) \in \mathfrak{U}(\omega)$.

This proves the existence of the solution $u(\xi, \eta)$ satisfying the requirements of this lemma. The uniqueness follows from Lemma 2.1. ∎

We now proceed with the proof of Theorem 2.1. It is easy to see that it reduces to the consecutive application of Lemma 2.3 to the equations of system (2.16). Consider the first of these equations: $L_1z_0 = 0$. According to Lemma 2.3, there exists a function $z_0(\xi, \eta)$ satisfying this equation and conditions (2.17) such that $z_0(\xi, \eta) \in \mathfrak{U}(\omega)$. The estimates (2.19), (2.20) for this function follow easily from the estimates (2.22), (2.23) of Lemma 2.3.

We now begin with the construction of the function $z_2(\xi, \eta)$. Consider the right-hand side of equation (2.16): $F(\xi, \eta) = P_1(\xi, \eta)(\partial z_0/\partial \eta)$. One can show without difficulty that this function satisfies the conditions of Lemma 2.3. Indeed, the estimate (2.23) for the derivative $\partial z_0/\partial \eta$ implies that an estimate similar to (2.22) holds for the function $F(\xi, \eta)$ outside any fixed neighborhood of the origin. At the origin, the derivatives of the function $z_0(\xi, \eta)$, in general, have singularities. However, using the asymptotic representation for $\partial z_0/\partial \eta$ as $|\xi| + |\eta| \to 0$ (see, for example, [59]) together with the fact that the polynomial $P_1(\xi, \eta)$ is homogeneous, it is not difficult to obtain that $|F(\xi, \eta)| = |P_1(\xi, \eta)(\partial z_0/\partial \eta)| \le M$ as $|\xi| + |\eta| \to 0$ as well.

Therefore, according to Lemmas 2.1 and 2.3 there is a unique solution of the equation $L_1 z_2 = P_1(\xi, \eta)(\partial z_0/\partial \eta)$ satisfying conditions (2.17) such that $z_2(\xi, \eta) \in \mathfrak{U}(\omega)$. The function $z_2(\xi, \eta)$ is constructed, and, evidently, satisfies estimates (2.22), (2.23).

The construction of the functions $z_{2k}(\xi, \eta)$ for $k \ge 2$ is carried out in the same way. ∎

Let us now check that the inner expansion W is matched to the expansion $U + S$. Consider the series (2.6) for $x \to +0$, i.e., expand each of the functions $s_{2k}(x, \tau)$ in a Taylor series in x. One obtains the formal series

$$\sum_{k=0}^{\infty} \varepsilon^{2k} s_{2k}(x, \tau) = \sum_{k=0}^{\infty} \varepsilon^{2k} \sum_{i=0}^{\infty} b_{i,k}(\tau) x^i, \tag{2.24}$$

where the properties of the functions $s_{2k}(x, \tau)$ imply that the relations

$$|b_{i,k}(\tau)| \le M \exp(-\alpha_{i,k}\tau), \qquad \alpha_{i,k} > 0, \tag{2.25}$$

hold for $b_{i,k}(\tau)$ and that similar relations hold for the derivatives of $b_{i,k}(\tau)$.

Taking into account that $\tau = \varepsilon^{-2}(-y - Ex) = -\eta - E\xi$, and passing in the formal series (2.24) to the variables ξ, η one can rewrite it in the form

$$\sum_{k=0}^{\infty} \varepsilon^{2k} h_{2k}(\xi, \eta), \tag{2.26}$$

where $h_{2k}(\xi, \eta) = \sum_{i=0}^{k} b_{k-i,i}(-\eta - E\xi)\xi^{k-i}$.

By construction (in view of the fact that the formal series S is an f.a.s. of the homogeneous equation (2.1)), the coefficients $h_{2k}(\xi, \eta)$ of the series (2.26) satisfy the same homogeneous sequence of equations (2.16) as the functions $z_{2k}(\xi, \eta)$. For the same reasons (namely, by virtue of the fact that $S|_{\tau=0} = -U|_{y=-Ex}$), the functions $h_{2k}(\xi, \tau)$ satisfy conditions (2.17) on the boundary $\eta = -E\xi$:

$$h_{2k}(\xi, -E\xi) = -p_k(\xi, -E\xi). \tag{2.27}$$

THEOREM 2.2. *The functions* $z_{2k}(\xi, \eta)$ *admit the following representation in the domain* $\{\xi, \eta\{: \xi \ge 1, \eta \le -E\xi\}$:

$$z_{2k}(\xi, \eta) = h_{2k}(\xi, \eta) + \sigma_{2k}(\xi, \eta), \tag{2.28}$$

where the functions $h_{2k}(\xi, \eta)$ *are defined above, and the functions* $\sigma_{2k}(\xi, \eta)$
are infinitely differentiable in this domain and satisfy the estimate

$$\left| \frac{\partial^l \sigma_{2k}(\xi, \eta)}{\partial \xi^{l_1} \partial \eta^{l_2}} \right| \le M_l \exp(-\alpha_k \xi + \beta_k \eta), \qquad \alpha_k > 0, \, \beta_k > 0. \tag{2.29}$$

PROOF. Represent the function $z_{2k}(\xi, \eta)$ in the domain $D_0 = \{\xi, \eta : \xi \ge 0, \, \eta \le -E\xi\}$ in the form (2.28). As mentioned above, the functions $h_{2k}(\xi, \eta)$ satisfy the same system of equations in the domain D_0 as the functions $z_{2k}(\xi, \eta)$ and also satisfy (2.17). In addition, the estimate (2.25) implies that the functions $h_{2k}(\xi, \eta)$ satisfy the condition (2.18) everywhere in D_0.

Therefore, the functions $\sigma_{2k}(\xi, \eta)$ are, in their turn, the unique solutions of the problems

$$L_1 \sigma_{2k} = \sum_{j=1}^{k} P_j(\xi, \eta) \frac{\partial \sigma_{2k-2j}}{\partial \eta} \quad \text{for } \xi > 0, \, \eta < -E\xi, \tag{2.30}$$

$$\sigma_{2k}(\xi, -E\xi) = 0, \quad \sigma_{2k}(0, \eta) = z_{2k}(0, \eta - h_{2k}(0, \eta)) \quad \text{for } \eta < 0,$$

which belong to the class $\mathfrak{u}$ in this domain. One can check without difficulty that

$$|\sigma_{2k}(0, \eta)| \le M_k \exp(\beta_k \eta) \quad \text{for } \eta < 0, \text{ where } 0 < \beta < 1/2.$$

Consider the problem (2.30) for $k = 0$. Proceeding along the same lines as in the proof of Lemma 2.3, but taking $M \exp(-\alpha \xi + \gamma \eta)$ where $0 < \gamma < \beta_0$, $\alpha > 0$, $\alpha^2 < \gamma - \gamma^2$ for the barrier function, construct the function $\tilde{\sigma}_0(\xi, \eta)$ which is a solution of this problem such that $|\tilde{\sigma}_0(\xi, \eta)| \le M \exp(-\alpha \xi + \gamma \eta)$ for $(\xi, \eta) \in D_0$, and $|\partial^l \tilde{\sigma}_0 / \partial \xi^{l_1} \partial \eta^{l_2}| \le M_l \exp(-\alpha \xi + \gamma \eta)$ for $(\xi, \eta) \in D_0$, $\xi^2 + \eta^2 \ge 1$. Evidently, $\tilde{\sigma}_0(\xi, \eta) \in \mathfrak{u}(D_0)$. Owing to the uniqueness (see Lemma 2.1), $\sigma_0(\xi, \eta) \equiv \tilde{\sigma}_0(\xi, \eta)$ everywhere in D_0. Thus, the statement of the theorem is proved for the function $z_0(\xi, \eta)$. For the remaining functions $z_{2k}(\xi, \eta)$ ($k \ge 1$) the proof is carried out in the same way. ∎

5. Recall that we have not yet constructed the series V in a neighborhood of the singular characteristic α. The general approach requires the examination of the coefficients of the constructed inner expansion W as $\eta \to \infty$, leading, with the use of the matching condition, to the final determination of the function $v_k(\zeta, y)$.

According to formula (2.15), one has to establish the asymptotics of the function $z_{2k}(\xi, \eta)$ as $\eta \to \infty$. Consider the values of $z_{2k}(\xi, \eta)$ for $\eta = 0$. The estimate (2.19) implies that $z_{2k}(\xi, 0)$ decay exponentially as $\xi \to -\infty$. It follows from (2.17) that $z_{2k}(\xi, 0) + p_k(\xi, 0) = 0$ for $\xi > 0$. Such initial data and equations for $z_{2k}(\xi, \eta)$ give rise (as will be shown below) to the following behavior of these functions in a neighborhood of the ξ-axis: the functions $z_{2k}(\xi, \eta)$ are exponentially small for $\xi + C\eta < 0$, $\xi \to -\infty$, $\eta > 0$ $\forall C > 0$, while for $\xi - C\eta > 0$, $\xi \to \infty$ $\forall C > 0$ the sum $z_{2k}(\xi, \eta) + p_k(\xi, \eta)$ is exponentially small.

It will be shown below that the functions $w_{2k}(\xi, \eta)$ and $z_{2k}(\xi, \eta)$ have, as $\eta \to \infty$, asymptotic expansions of the same form as the functions $w_{2k}(\xi, \eta)$

in §1 (see (1.25)):

$$z_{2k}(\xi, \eta) = \eta^k \sum_{j=0}^{\infty} \eta^{-j/2} \Phi_{k,j}(\theta), \qquad \eta \to \infty, \tag{2.31}$$

where $\theta = 2^{-1}\xi\eta^{-1/2}$. According to the remarks on the behavior of the functions $z_{2k}(\xi, \eta)$ for $\xi \to \pm\infty$, it is natural to expect that the functions $\Phi_{k,j}(\theta)$ decay exponentially as $\theta \to -\infty$, while as $\theta \to \infty$ the series (2.31) must be asymptotically equal to $-p_k(\xi, \eta)$. This polynomial can be represented in the following form:

$$-p_k(\xi, \eta) = \eta^k \sum_{j=0}^{2k} \eta^{-j/2} q_{k,j}(\theta),$$

where $q_{k,j}$ are polynomials. Therefore the following are natural conditions on $\Phi_{k,j}(\theta)$:

$$\Phi_{k,j}(\theta) = O\left(\exp(\nu_{k,j}\theta)\right), \qquad \theta \to -\infty,$$

$$\Phi_{k,j}(\theta) = q_{k,j}(\theta) + O\left(\exp(-\nu_{k,j}\theta)\right), \qquad \theta \to \infty, \tag{2.32}$$

$$\nu_{k,j} > 0, \qquad q_{k,j}(\theta) \equiv 0 \quad \text{for } j > 2k.$$

THEOREM 2.3. *The functions* $z_{2k}(\xi, \eta)$ *solving the problems* (2.16), (2.17) *can be expanded, as* $\eta \to \infty$, *in the asymptotic series* (2.31) *uniformly with respect to* ξ, *where* $\theta = 2^{-1}\xi\eta^{-1/2}$, *and the functions* $\Phi_{k,j}(\theta) \in C^{\infty}(\mathbf{R}^1)$ *satisfy conditions* (2.32). *Moreover, for the remainders of the series* (2.31) *the following sharpened estimates hold*:

$$|z_{2k}(\xi, \eta) - B_N z_{2k}| \leq \begin{cases} M\eta^{-N} & \text{for } |\xi| < \eta, \\ M\exp(-\alpha_k\xi) & \text{for } |\xi| \geq \eta \end{cases} \tag{2.33}$$

for $\alpha_k > 0$ *and* N *sufficiently large. Similar estimates are valid for the derivatives of* $z_{2k}(\xi, \eta)$.

The situation arising in the investigation of the asymptotics of the solutions $z_{2k}(\xi, \eta)$ basically repeats that discussed in §1 (subsections 4,5, Theorem 1.2, Lemmas 1.1–1.4). Accordingly, the proof of Theorem 2.3 will be made as brief as possible. We will direct our attention just to the differences between the analogous statements in this and in the preceding section.

The functions $z_{2k}(\xi, \eta)$ in the upper half-plane $\eta \geq 0$ are the solutions of the problem

$$L_1 z_{2k} = \sum_{j=1}^{k} P_j(\xi, \eta) \frac{\partial z_{2k-2j}}{\partial \eta}, \tag{2.34}$$

$$\begin{aligned} z_{2k}(\xi, 0) &= -p_k(\xi, 0) \quad \text{for } \xi > 0, \\ z_{2k}(\xi, 0) &= \psi_k(\xi) \quad \quad \text{for } \xi < 0 \end{aligned} \tag{2.35}$$

and belong in this half-plane to the class $\mathfrak{u}$. The estimates (2.19) imply that the boundary function $\psi_k(\xi)$ satisfies the estimate

$$|\psi_k(\xi)| \leq M_k \exp(\alpha_k\xi), \qquad \alpha_k > 0,$$

with similar estimates holding for the derivatives of this function. One can write out an integral representation for the function $z_{2k}(\xi, \eta)$ similar to (1.23) as follows:

$$z_{2k}(\xi, \eta) = \int_0^\infty \int_0^\infty G(\xi, \eta; \xi_1, \eta_1) F_k(\xi_1, \eta_1) \, d\xi_1 \, d\eta_1$$

$$- \int_{-\infty}^\infty \frac{\partial G}{\partial \eta_1}(\xi, \eta; \xi_1, 0) z_{2k}(\xi_1, 0) \, d\xi_1, \tag{2.36}$$

where $F_k(\xi, \eta)$ is the right-hand side of equality (2.34), and

$$G(\xi, \eta; \xi_1, \eta_1) = (2\pi)^{-1} \left\{ -K_0 \left(\frac{\sqrt{(\xi - \xi_1)^2 + (\eta - \eta_1)^2}}{2} \right) \right.$$

$$\left. + K_0 \left(\frac{\sqrt{(\xi - \xi_1)^2 + (\eta + \eta_1)^2}}{2} \right) \right\} \exp \frac{\eta - \eta_1}{2}.$$

As in §1, the investigation of the asymptotics of the functions $z_{2k}(\xi, \eta)$ falls into two steps. On the first step an f.a.s. of the problem (2.34), (2.35) is constructed for $\eta \to \infty$, and on the second step, using formula (2.36), one obtains the asymptotic expansion for the difference between $z_{2k}(\xi, \eta)$ and the partial sum of the constructed f.a.s.

The definition of the classes of functions in which formal asymptotics are constructed has to be modified. The principal role is still played by the operator $L_0 = \partial^2/\partial\xi^2 - \partial/\partial\eta$, but for the domain of definition of the functions one should consider the half-plane $\eta > 0$ instead of the quadrant $\xi > 0, \eta > 0$.

Thus, for integer $n \geq 0$, the function $W_n(\xi, \eta)$ is the solution of the boundary value problem

$$L_0 W_n = 0 \quad \text{for } \eta > 0,$$

$$W_n(\xi, 0) = 0 \quad \text{for } \xi < 0,$$

$$W_n(\xi, 0) = \xi^n \quad \text{for } \xi > 0,$$

bounded for $n = 0$, continuous for $\eta \geq 0$ if $n > 0$, and slowly growing at infinity. One can easily write out the explicit formula for $W_n(\xi, \eta)$:

$$W_n(\xi, \eta) = \int_0^\infty G_0(\xi - \xi_1, \eta) \xi_1^n \, d\xi_1,$$

where $G_0(\xi, \eta) = (1/2\sqrt{\pi\eta}) \exp(-\xi^2/4\eta)$ is the fundamental solution of the heat equation.

For each $n = -k$, where $k > 0$, we set $W_n(\xi, \eta) = \partial^k W_0/\partial\xi^k$ so that $(\partial^l W_n/\partial\xi^{l_1}\partial\eta^{l_2})(\xi, \eta) = \alpha_{n,l_1,l_2} W_{n-l-2l_2}(\xi, \eta)$ for any integer n. In particular, $W_{-1}(\xi, \eta) = G_0(\xi, \eta)$. Evidently, $L_0 W_n = 0$ for $\eta > 0$ and all integers n, while $W_{-k}(\xi, 0) = 0$ for $k > 0$, $\xi \neq 0$.

For any integer n, one has $W_n(\xi, \eta) = \eta^{n/2} U_n(\theta)$, where $\theta = 2^{-1}\xi\eta^{-1/2}$, and the function $U_n(\theta)$ is a nontrivial solution of the equation (1.29):

$$l_n U \equiv U'' + 2\theta U' - 2nU = 0.$$

For all integer n the functions $U_n(\theta)$ decay exponentially as $\theta \to -\infty$. But for $\theta \to \infty$ the functions $U_n(\theta)$ behave differently: for $n < 0$ they still decay exponentially, while for $n \geq 0$ they are exponentially close to a polynomial.

All these statements about $W_n(\xi, \eta)$ (with the exception of the last one) follow from the explicit form of $W_n(\xi, \eta)$ for $n \geq 0$, and are verified as in §1. The last statement is implied by the following lemma.

LEMMA 2.4. *For* $n \geq 0$ *the functions* $W_n(\xi, \eta)$ *have the following asymptotic representation: for all* $0 \leq \eta \leq \xi$

$$W_n(\xi, \eta) = X_n(\xi, \eta) + O\left(\exp(-\alpha\xi)\right), \qquad \alpha > 0, \tag{2.37}$$

where $X_n(\xi, \eta) = \sum_{i,l} \alpha_{i,l} \xi^l \eta^l$, $i + 2l = n$.

The proof of the lemma reduces to the construction of a polynomial $X_n(\xi, \eta)$ that would satisfy the equation $L_0 X_n = 0$ and the condition $X_n(\xi, 0) = \xi^n$. Such a polynomial is easily constructed explicitly. The estimate (2.37) follows from the maximum principle applied to the difference $W_n - X_n$ with the help of the barrier function $M \exp(-\alpha\xi + \beta\eta)$. ∎

The functional classes corresponding to the classes $\mathfrak{A}$ and $\mathfrak{B}$ of §1 now have a simpler form. The class $\mathfrak{A}$ is not introduced at all, while the class $\mathfrak{B}$ is the set of linear combinations of the functions $\xi^k \eta^l W_n(\xi, \eta)$, where n is any integer, and k and l are any nonnegative integers. The definition of the order of the functions $u(\xi, \eta) \in \mathfrak{B}_m$ remains the same—a function is of order $m/2$ if $u(\xi, \eta) = \eta^{m/2} \Phi(\theta)$. Lemma 2.4 now implies the following properties of the function $\Phi(\theta)$ in this representation:

$$\begin{aligned}
\Phi(\theta) &= O\left(\exp(\nu\theta)\right), & \theta &\to -\infty, \\
\Phi(\theta) &= q(\theta) + O\left(\exp(-\nu\theta)\right), & \theta &\to \infty,
\end{aligned} \tag{2.38}$$

where $\nu > 0$, and $q(\theta)$ is a polynomial of degree not greater than m.

In order to construct an f.a.s. of the problems (2.34), (2.35) one should replace Lemma 1.1 by the following lemma.

LEMMA 2.5. *For any integer* r, *and any function* $v(\xi, \eta) \in \mathfrak{B}_r$ *there exists a function* $u(\xi, \eta) \in \mathfrak{B}_{r+2}$ *satisfying the equation* $L_0 u = v$ *for* $\eta > 0$ *and the condition* $u(\xi, 0) = 0$ *for* $\xi \neq 0$.

PROOF. It is sufficient to prove the statement of the lemma for the function $v(\xi, \eta) = \xi^k \eta^l W_n(\xi, \eta)$, where k, l, and n are integers such that $k \geq 0$, $l \geq 0$, $k+2l+n = r$. One can easily verify that

$$\begin{aligned}
L_0(\xi^k &\eta^{l+1} W_n(\xi, \eta)) \\
&= k(k-1)\xi^{k-2}\eta^{l+1} W_n(\xi, \eta) + 2k\xi^{k-1}\eta^{l+1}\frac{\partial W_n}{\partial \xi}(\xi, \eta) \\
&\quad + \xi^k\eta^{l+1} L_0 W_n(\xi, \eta) - (l+1)\xi^k\eta^l W_n(\xi, \eta) \\
&= k(k-1)\xi^{k-2}\eta^{l+1} W_n(\xi, \eta) + 2\alpha k\xi^{k-1}\eta^{l+1} W_{n-1}(\xi, \eta) \\
&\quad - (l+1)\xi^k\eta^l W_n(\xi, \eta).
\end{aligned}$$

Hence, choosing $u_0(\xi, \eta) = -(l+1)^{-1}\xi^k \eta^{l+1} W_n(\xi, \eta) = -(l+1)^{-1}\eta v$, one has

$$v - L_0 u_0 = \eta^{l+1}[a_{1,1}\xi^{k-1}W_{n-1}(\xi, \eta) + a_{2,1}\xi^{k-2}W_n(\xi, \eta)] \equiv v_1(\xi, \eta).$$

On the next step let $u_1 = -(l+2)^{-1}\eta v_1$. Then

$$v - L_0(u_0 + u_1) = \eta^{l+2}[a_{1,2}\xi^{k-2}W_{n-2}(\xi, \eta) + a_{2,2}\xi^{k-3}W_{n-1}(\xi, \eta)$$
$$+ a_{3,2}\xi^{k-4}W_n(\xi, \eta)] \equiv v_2(\xi, \eta).$$

Now let $u_2 = -(l+3)^{-1}\eta v_2$, etc. Since k is a nonnegative integer, the process terminates at the $(k+1)$th step yielding $L_0(u_0 + u_1 + \cdots + u_k) = v$. One can easily see that $u_i(\xi, \eta) \in \mathfrak{B}_{r+2}$ for all i.

The construction of the f.a.s. of the problem $L_0 v = F$, $v(\xi, 0) = 0$, with $\xi < 0$, $v(\xi, 0) = \psi(\xi)$ for $\xi > 0$, where $\psi(\xi)$ is a polynomial, and the function $F(\xi, \eta)$ is the same as in formula (1.31), almost repeats that of Lemma 1.2.

The modifications in Lemmas 1.3 and 1.4 for this problem are absolutely inessential. The function $\Phi(\theta)$ in Lemma 1.3 is now defined not only for $[0, \infty)$, but everywhere on $\mathbf{R}^1$, and decays exponentially as $|\theta| \to \infty$. In Lemma 1.4 one now has to examine formula (2.36) instead of formula (1.23).

The proof of Theorem 2.3 is only slightly different from that of Theorem 1.2. The difference is that the functions in Theorem 1.2 equal polynomials on the positive half of the η-axis, while in Theorem 2.3 they equal polynomials on the positive half of the ξ-axis. Therefore, the functions $z_{2k}(\xi, \eta)$, in general, do not decay exponentially as $\xi \to \infty$, but are asymptotically equal to polynomials. The estimates (2.32) and (2.33) follow from relations (2.38) for functions of the class $\mathfrak{B}_{m/2}$. ∎

The asymptotic expansions of the functions $w_{2k}(\xi, \eta) = p_k(\xi, \eta) + z_{2k}(\xi, \eta)$ (see formula (2.15)) are of the same form as those of $z_{2k}(\xi, \eta)$. Starting with these asymptotic expansions, as in §1 (see Table 8), one constructs the formal series $V_i(\zeta, y)$ (formula (1.45)).

THEOREM 2.4. *There exist functions* $v_i(\zeta, y)$ *solving the system* (2.8)–(2.10) *and admitting asymptotic expansions* (1.45). *Therefore, for the series* (2.7) *and* (2.11) *thus constructed the matching condition* (1.26) *holds.*

The proof repeats that of Theorem 1.3.

The composite asymptotic expansion in the problem considered here is virtually the same as in the preceding section (see formula (1.47) and Theorem 1.5) but slightly more cumbersome. Besides the boundary layer series (2.6), one has to introduce a similar boundary layer series near the boundary σ_2 (see Figure 17). These series almost coincide with the series S and $\widetilde{S}$ of

§1, and are of the form:

$$S^* = \sum_{k=0}^{\infty} \varepsilon^{2k} s_{2k}^*(x, \tau^*), \quad \text{for } \tau^* = \varepsilon^{-2}[\varphi_2(x) - y],$$

$$s_{2k}^*(x, 0) = -u_{2k}(x, \varphi_2(x)),$$

$$\widetilde{S} = \sum_{k=0}^{\infty} \varepsilon^2 \tilde{s}_k(\zeta, \tilde{\tau}), \quad \text{where } \zeta = \varepsilon^{-1} x, \ \tilde{\tau} = \varepsilon^{-2}[\varphi_2(0) - y].$$

The boundary values for $\tilde{s}_k(\zeta, 0)$ are obtained from the condition that the sum of the series $\widetilde{S}$ and V vanishes on the boundary σ_2. The equations for the functions $s_{2k}^*(x, \tau^*)$ and $\tilde{s}_k(\zeta, \tilde{\tau})$ are obtained in the usual way. These functions decay exponentially as τ^* and $\tilde{\tau}$ tend to infinity.

The coefficients of all the series just constructed with the exception of the functions $s_{2k}(x, \tau)$ are defined everywhere in Ω_δ. The functions $s_{2k}(x, -\varepsilon^{-2}(y + Ex))$ are defined only for $x > 0$, $y < -Ex$. Extend them by zero to the rest of Ω_δ, i.e., to the points where $x < 0$, $\varphi_1(x) \leq y \leq \varphi_2(x)$ and $x \geq 0$, $0 \leq y \leq \varphi_2(x)$.

The next difference from §1 is that the functions $w_{2k}(\xi, \eta) = p_k(\xi, \eta) + z_{2k}(\xi, \eta)$ have different asymptotic representations in different parts of the domain ω at infinity. In the neighborhood of the boundary $\eta + E\xi = 0$ the functions $z_{2k}(\xi, \eta)$ are, by virtue of (2.28), (2.29), close to $h_{2k}(\xi, \eta)$; for $0 \leq \eta < \xi$ they are close to $-p_k(\xi, \eta)$; and in the domain $D = \{\xi, \eta : \eta + c_1\xi < -E\xi, \ \eta \in \mathbf{R}^1\}$, where c_1 is any positive constant, the functions $z_{2k}(\xi, \eta)$ are exponentially small. As $\eta \to \infty$ the functions $w_{2k}(\xi, \eta)$ can be expanded in the asymptotic series (2.31) (these series are written out for $z_{2k}(\xi, \eta)$, but for $w_{2k}(\xi, \eta)$ they have exactly the same form but with other functions $\Phi_{k, j}(\theta)$).

Thus, in contrast to the problems of Chapter III, and from §1 of Chapter IV, the functions of the inner expansion have different asymptotic representations at infinity. Introduce the corresponding notation

$$[w_{2k}(\xi, \eta)]_1 = p_k(\xi, \eta), \qquad [w_{2k}(\xi, \eta)]_2 = h_{2k}(\xi, \eta)$$

$$[w_{2k}(\xi, \eta)]_3 = \eta^k \sum_{j=0}^{\infty} \eta^{-j/2} \Phi_{k, j}(\theta).$$

Similar notation is used for linear combinations of $w_{2k}(\xi, \eta)$, and, in particular, for $A_{N, \xi, \eta} W$. It follows from the matching conditions proved above that for any positive integers N_1 and N_2, one has

$$A_{N_2, \xi, \eta} A_{N_1, x, y} U = A_{N_1, x, y} [A_{N_2, \xi, \eta} W]_1,$$

$$A_{N_2, \xi, \eta} A_{N_1, x, \tau} S = A_{N_1, x, \tau} [A_{N_2, \xi, \eta} W]_2,$$

$$A_{N_2, \xi, \eta} A_{N_1, \zeta, y} V = A_{N_1, \zeta, y} [A_{N_2, \xi, \eta} W]_3.$$

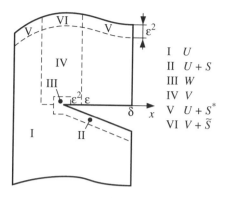

FIGURE 20

Now define the composite asymptotic expansion

$$Y_N(x, y, \varepsilon) = A_{2N,x,y}U + A_{2N,x,\tau}S + A_{2N,x,\tau^*}S^* + A_{2N,\zeta,y}\widetilde{S}$$
$$+ A_{2N,\xi,\eta}W + A_{2N,\zeta,y}V - A_{2N,\xi,\eta}A_{2N,x,y}U$$
$$- A_{2N,x,y}A_{2N,\xi,y}V - A_{2N,\xi,\eta}A_{2N,x,\tau}S - A_{2N,\xi,\eta}A_{2N,\zeta,y}V.$$

THEOREM 2.5. *There exists a constant* $\gamma > 0$ *such that for any positive integer* N *the following estimates are valid in the domain* Ω_δ:

$$|\mathscr{L}_\varepsilon Y_N(x, y, \varepsilon) - f(x, y)| < M\varepsilon^{\gamma N},$$
$$|Y_N(x, y, \varepsilon)| < M\varepsilon^{\gamma N} \quad \text{for } (x, y) \in \partial D \cap \overline{D}_\delta,$$
$$|Y_N(x, y, \varepsilon) < M \quad \text{for } |x| = \delta.$$

THEOREM 2.6. *In the domain* Ω_δ *the following estimate is valid:*

$$|Y_N(x, y, \varepsilon) - u(x, y, \varepsilon)| < M\varepsilon^{\gamma N},$$

where $u(x, y, \varepsilon)$ *is the solution of the problem* (2.1), (2.2), *and* Y_N *and* γ *are the same as in the preceding theorem.*

THEOREM 2.7. *Let* β *be a number such that* $0 < \beta < 1$. *Then the series* W *is a uniform asymptotic expansion of the solution* $u(x, y, \varepsilon)$ *in the domain* $\Omega_\delta \cap \{x, y: |x| < \varepsilon^\beta, |y| < \varepsilon^{2\beta}\}$, *the sum* $U + S^* + \widetilde{S}$ *is such an expansion in the domain* $\{x, y: |x| < \varepsilon^\beta, \varepsilon^{2\beta} < y \le \varphi_2(x)\}$, *and the sum* $U + S + S^*$ *in the rest of* Ω_δ.

The proof of Theorem 2.5 almost repeats that of Theorem 1.4. The conclusion of Theorem 2.6 follows from Theorem 2.5 and Lemma 5.1, while the proof of Theorem 2.7 virtually coincides with that of Theorem 1.6. ■

Figure 20 shows, marked by Roman numerals, the subdomains of the domain Ω_δ, in which the individual asymptotic expansions hold (the corresponding series are written out to the right of the figure). As before, the boundaries of the domains are shown more or less arbitrarily.

§3. The case of a singular characteristic tangent
to the boundary of the domain from the outside

Here we consider the asymptotics of the solution (0.3), (0.4) in a neighborhood of the point c_2 (see Figure 15). At this point the characteristic of the limit equation is tangent to the boundary on the outside, i.e., in a small neighborhood of the point c_2 the tangent characteristic has no other points in common with the closed domain $\overline{\Omega}$ but the point c_2. The case of an inner tangent characteristic to the boundary will be considered in the next section. Both sections consider just the case of general position which is equivalent to tangency of the characteristic to the boundary being of the first order.

Suppose that the tangency point is at the origin O, and the equation has a special form

$$\mathscr{L}_\varepsilon \equiv \varepsilon^3 \left(\frac{\partial^2 u}{\partial x^2} + \frac{\partial^2 u}{\partial y^2} \right) - \frac{\partial u}{\partial y} = f(x, y), \qquad (3.1)$$

with the boundary condition

$$u(x, y, \varepsilon) = 0 \quad \text{for } (x, y) \in \partial\Omega, \qquad (3.2)$$

and the boundary $\partial\Omega$ in a neighborhood of O is of the form $\{x, y: x = y^2\}$ (for convenience, the small parameter before the Laplace operator is denoted by ε^3). Such simplifications are not crucial. If the singular characteristic is an outer tangent of the first order, then all constructions and proofs for the second order elliptic equation (0.3) differ from the following constructions only in minor details.

Following the remark made at the beginning of §2, we limit our attention just to the construction of an f.a.s. in the domain $\Omega_\delta = \{x, y: y^2 < x < \delta\}$, where $\delta > 0$ is a fixed number (see Figure 21). After an f.a.s. of the problem (3.1), (3.2) is constructed, it follows from Lemma 5.1 that this f.a.s. is an asymptotic expansion of the solution $u(x, y, \varepsilon)$ in a domain Ω_{δ_1} for $\delta_1 < \delta$.

1. We seek the outer expansion in the usual form

$$U = \sum_{k=0}^{\infty} \varepsilon^{3k} u_{3k}(x, y). \qquad (3.3)$$

The equations and boundary conditions for $u_{3k}(x, y)$ are the following:

$$\frac{\partial u_0}{\partial y} = -f(x, y), \quad \frac{\partial u_{3k}}{\partial y} = \Delta_{x, y} u_{3k-3}, \quad k \geq 1, \qquad (3.4)$$

$$u(x, -\sqrt{x}) = 0, \qquad 0 < x \leq \delta.$$

The origin of relations (3.4) has been repeatedly discussed in similar situations in Chapter I, Chapter II, and the preceding sections of Chapter IV, and there is no need to go into details.

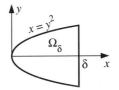

FIGURE 21

The solutions of the problems (3.4) are expressed in quadratures:

$$u_0(x, y) = -\int_{-\sqrt{x}}^{y} f(x, y_1)\, dy_1,$$

$$u_{3k}(x, y) = \int_{-\sqrt{x}}^{y} \Delta_{x, y_1} u_{3k-3}(x, y_1)\, dy_1. \tag{3.5}$$

These functions are defined and infinitely differentiable in the domain $\overline{\Omega}_\delta \backslash 0$.

The residual in the boundary condition on the top part of the boundary of the domain $y = \sqrt{x}$ is eliminated with the help of the boundary layer series

$$S = \sum_{k=0}^{\infty} \varepsilon^{3k} s_{3k}(x, \tau), \tag{3.6}$$

where $\tau = \varepsilon^{-3}(\sqrt{x} - y)$. Inserting the series (3.6) into the homogeneous equation (3.1), we obtain the recurrence system of equations

$$L_1 s_0 \equiv (1 + (4x)^{-1})\frac{\partial^2 s_0}{\partial \tau^2} + \frac{\partial s_0}{\partial \tau} = 0,$$

$$L_1 s_3 = -\frac{1}{\sqrt{x}}\frac{\partial^2 s_0}{\partial x \partial \tau} + \frac{1}{4x^{3/2}}\frac{\partial s_0}{\partial \tau}, \tag{3.7}$$

$$L_1 s_{3k} = -\frac{1}{\sqrt{x}}\frac{\partial^2 s_{3k-3}}{\partial x \partial \tau} + \frac{1}{4x^{3/2}}\frac{\partial s_{3k-3}}{\partial \tau} - \frac{\partial^2 s_{3k-6}}{\partial x^2}, \qquad k \geq 2.$$

The boundary conditions for the functions $s_{2k}(x, \tau)$ are

$$s_{2k}(x, 0) = -u_{3k}(x, \sqrt{x}),$$

$$s_{3k}(x, \tau) \to 0 \quad \text{for } \tau \to \infty. \tag{3.8}$$

The solutions of the problems (3.7), (3.8) are also easily written out explicitly: they are polynomials in τ of degree $2k$ multiplied by $\exp(-(1 + (4x)^{-1})\tau)$. For $x > 0$ the functions $s_{3k}(x, \tau)$, as well as $u_{3k}(x, y)$, smoothly depend on x, but both sequences have growing singularities as $x \to 0$. Thus, the problem (3.1), (3.2) is a bisingular one.

Before we begin with the construction of the inner expansion, it is useful to examine the structure of singularities of the already obtained functions $u_{3k}(x, y)$ and $s_{3k}(x, \tau)$ as $x \to 0$.

LEMMA 3.1. *The functions* $u_{3k}(x, y)$ *defined by relations* (3.5) *can, as* $x \to 0$, *be expanded in the asymptotic series*

$$u_{3k}(x, y) = x^{(1-3k)/2} \sum_{j=0}^{\infty} x^{j/2} \varphi_{j,k}(yx^{-1/2}), \qquad (3.9)$$

where $\varphi_{j,k}(\theta) \in C^{\infty}[-1, 1]$, *and the series can be repeatedly differentiated term-by-term.*

PROOF. Expanding the function $f(x, y_1)$ in the integrand of (3.5) in a Taylor series in y_1, and integrating, one obtains the expressions

$$\psi(x)[y^k - (-x^{1/2})^k] = x^{k/2}\psi(x)[(y/x^{1/2})^k - (-1)^k], \qquad k \geq 1,$$

for each term of the series. On expanding the functions $\psi(x)$ in a Taylor series in x, and estimating the remainder one obtains the asymptotics (3.9) for $k = 0$. Since the derivatives of $u_0(x, y)$ are either integrals of the same type, or simply smooth functions, they also admit similar asymptotic expansions. Therefore, the series for $u_0(x, y)$ can be differentiated term-by-term any number of times. In particular, $\Delta_{x,y}u_0(x, y)$ can be expanded in an asymptotic series of the form (3.9) with the coefficient $x^{-3/2}$ appearing before the sum. After the same procedure as was applied to $f(x, y)$, one obtains the asymptotic expansion (3.9) for $u_3(x, y)$. Relation (3.9) is now easily verified for all $k \geq 0$ by induction. ∎

LEMMA 3.2. *The functions* $s_{3k}(x, \tau)$ *solving the problems* (3.7), (3.8) *admit the asymptotic expansions*

$$s_{3k}(x, \tau) = \left[\exp\left(-\frac{4x}{1+4x}\tau\right)\right] x^{(1-3k)/2} \sum_{l=0}^{2k}(x\tau)^l \sum_{j=0}^{\infty} c_{k,l,j} x^{j/2}, \qquad x \to 0.$$

$$(3.10)$$

These series can be infinitely differentiated term-by-term.

The proof follows from the analysis of the explicit solutions of the system (3.7), (3.8). Indeed, $s_0(x, \tau) = -u_0(x, \sqrt{x})\exp(-4x\tau/(1+4x))$. By the preceding lemma, equality (3.10) is therefore satisfied for $k = 0$. Assuming, by induction, that (3.10) is satisfied for $k \leq n$, write out the right-hand side of equation (3.7) for $k = n + 1$:

$$\frac{\partial s_{3n}}{\partial \tau} = \left[\exp\left(-\frac{4x}{1+4x}\tau\right)\right] x^{3/2-3n/2} \sum_{l=0}^{2n}(x\tau)^l \sigma_{n,l,1}(x),$$

$$\frac{\partial^2 s_{3n}}{\partial x \partial \tau} = \left[\exp\left(-\frac{4x}{1+4x}\tau\right)\right] x^{1/2-3n/2} \sum_{l=0}^{2n+1}(x\tau)^l \sigma_{n,l,2}(x),$$

$$\frac{\partial^2 s_{3n-3}}{\partial x^2} = \left[\exp\left(-\frac{4x}{1+4x}\tau\right)\right] x^{-3n/2} \sum_{l=0}^{2n}(x\tau)^l \sigma_{n,l,3}(x).$$

Here $\sigma(x)$ with various subscripts stands for the asymptotic series of the form $\sum_{j=0}^{\infty} c_j x^{j/2}$, $x \to 0$. In what follows the subscripts in $\sigma(x)$ will be omitted. Thus

$$L_1 s_{3(n+1)} = \left[\exp\left(-\frac{4x}{1+4x}\tau \right) \right] x^{-3n/2} \sum_{l=0}^{2n+1} (x\tau)^l \sigma(x). \tag{3.11}$$

Taking into account that, by (3.8), (3.9), $s_{3n+3}(x, 0) = -u_{3n+3}(x, \sqrt{x}) = x^{(-2-3n)/2}\sigma(x)$, the explicit form of the solution of equation (3.11) implies that $s_{3n+3}(x, \tau)$ is of the form (3.10). ∎

2. The asymptotics of the coefficients of the series (3.3), (3.6) is hereby established. It suggests the following form for the inner variables in a neighborhood of the point O. The principal terms in the coefficients of the series (3.3), (3.6) for $x \to 0$ are of the form $\varepsilon^{3k} x^{(1-3k)/2}$. It is, therefore, natural to take $x = \varepsilon^2\xi$. Similarly, in order to make the principal term of the exponent $4x\tau/(1+4x) \approx 4x(\sqrt{x} - y)/\varepsilon^3$ independent of ε, it is reasonable to set $y = \varepsilon\eta$.

The choice of the new variables can be approached from another standpoint. Let

$$x = \varepsilon^\alpha\xi, \quad y = \varepsilon^\beta\eta, \quad \text{where } \alpha > 0, \ \beta > 0. \tag{3.12}$$

As mentioned above, after such a change of variables, equation (3.1) must include at least two highest order terms. The terms of equation (3.1) contain the following exponents of ε: $3 - 2\alpha$, $3 - 2\beta$, and $-\beta$. Therefore, there are only three possibilities: (1) $\alpha = \beta$, $\beta \geq 3$; (2) $\beta = 2\alpha - 3$, $0 < \beta \leq 3$; (3) $\beta = 3$, $0 < \alpha < 3$. It is also necessary to take into account the fact that the change of variables (3.12) either preserves the parabola $x = y^2$ (for $\alpha = 2\beta$), or transforms it into a curve close to the straight line $\xi = 0$ (for $\alpha < 2\beta$), or the cut $\eta = 0$, $\xi \geq 0$ (for $\alpha > 2\beta$). Clearly, only the first two cases are acceptable. The corresponding straight-line segments are shown in Figure 22 (next page). All points (α, β) lying on these segments satisfy the above requirements. However, as mentioned above, practical considerations show that the correct change of variables is obtained if one takes the extreme values of α and β. In this case there are two such values: $\alpha = 2$, $\beta = 1$, and $\alpha = \beta = 3$. For the first of them the highest order terms in equation (3.1) are $\varepsilon^3 \partial^2/\partial y$ and $\partial/\partial y$; for the second all the three terms in the left-hand side of the equation are of the same order. It is the first change of variables that we make:

$$x = \varepsilon^2\xi, \qquad y = \varepsilon\eta. \tag{3.13}$$

It is unclear why does the second possibility remain unused, but it turns out that the coefficients of the inner expansion

$$W = \sum_{i=1}^{\infty} \varepsilon^i w_i(\xi, \eta), \tag{3.14}$$

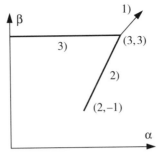

FIGURE 22

i.e. the functions $w_i(\xi, \eta)$ are smooth at the point $(0, 0)$. Thus, if one introduces another set of inner variables $\xi_1 = \varepsilon^{-3}x = \varepsilon^{-1}\xi$, $\eta_1 = \varepsilon^{-3}y = \varepsilon^{-2}\eta$, then the new "inner expansion" is simply a series obtained from W by expanding the functions $w_i(\varepsilon\xi_1, \varepsilon^2\eta_1)$ in Taylor series. There is clearly no need in such additional "inner expansion."

Thus, consider the inner expansion (3.14), where the variables ξ and η are defined by formulas (3.13). The functions $w_i(\xi, \eta)$ will be sought inside the parabola $D = \{\xi, \eta : \xi > \eta^2, \eta \in \mathbf{R}^1\}$. (The series (3.14) starts with the term $i = 1$ owing to condition (3.2). If the boundary condition were inhomogeneous, the series would begin with the term $i = 0$.) Equations for $w_i(\xi, \eta)$, and the boundary conditions are obtained in the usual way:

$$L_0 w_1 \equiv \frac{\partial^2 w_1}{\partial \xi^2} - \frac{\partial w_1}{\partial \eta} = p_0(\xi, \eta), \qquad L_0 w_2 = p_1(\xi, \eta),$$

$$L_0 w_i = -\frac{\partial^2 w_{i-2}}{\partial \eta^2} + p_{i-1}(\xi, \eta), \qquad i \geq 3,$$

(3.15)

$$w_i(\eta^2, \eta) = 0 \quad \text{for } \eta \in \mathbf{R}^1, \tag{3.16}$$

where the polynomials $p_i(\xi, \eta)$ are coefficients of the Taylor expansion of the function $f(x, y)$

$$f(\varepsilon^2 \xi, \varepsilon\eta) = \sum_{i=0}^{\infty} \varepsilon^i p_i(\xi, \eta).$$

The functions $w_i(\xi, \eta)$ are evidently unbounded at infinity, and it is natural to seek them in the class of functions growing not faster that some power of ξ. The solution of the problems (3.15), (3.16) can be investigated independently of the results obtained in subsection 1, but it is more convenient to use the matching condition and construct the functions $w_i(\xi, \eta)$ starting from the already constructed series U and S.

Make the change of variables (3.13) in each of the series (3.3), (3.6). Then

$$yx^{-1/2} = \eta\xi^{-1/2}, \quad x\tau = \xi(\sqrt{\xi} - \eta), \text{ and}$$

$$\exp\left(-\frac{4x\tau}{1 + 4x}\right) = \exp(-4x\tau)\exp\left(4x\frac{4x\tau}{1 + 4x}\right)$$

$$= \exp(-4\xi(\sqrt{\xi} - \eta))\sum_{l=0}^{\infty}\varepsilon^{2l}[\xi^2(\sqrt{\xi} - \eta)]^l\sum_{j=0}^{\infty}a_{j,l}\varepsilon^{2j}\xi^j.$$

Replace the functions $u_{3k}(x, y)$ and $s_{3k}(x, \tau)$ with their asymptotic expansions. After these formal transformations the sum $U + S$ takes the form $\sum_{i=1}^{\infty}\varepsilon^i H_i$, where

$$H_i = \xi^{i/2}\sum_{j=0}^{\infty}\xi^{-3j/2}\left\{\varphi_{i-1,j}(\eta\xi^{-1/2})\right.$$

$$\left. + \exp(-4\xi(\sqrt{\xi} - \eta))\sum_{l=0}^{2j+i}b_{i,j,l}[\xi(\sqrt{\xi} - \eta)]^l\right\}. \quad (3.17)$$

Since $U + S$ is an f.a.s. of the problem (3.1), (3.2) for $x > 0$, one can easily show, as in subsection 5 of §1, that the series H_i are f.a.s. of the problems (3.15), (3.16) as $\xi \to \infty$.

THEOREM 3.1. *There exist functions $w_i(\xi, \eta)$ in $C^{\infty}(\overline{D})$ that are solutions of the problems* (3.15), (3.16) *and have the asymptotic expansions H_i as $\xi \to \infty$.*

The proof of this theorem is achieved in the same standard way used above in the proof of Theorems 1.1 and 2.1 of Chapter II, Theorem 1.2 and Lemma 4.1 of Chapter III, Theorem 1.2 of Chapter IV. The outline of the proof is as follows. If the series H is an f.a.s. of the equation $Lh = g$, denote by h_N the partial sum $B_N H$ smoothed in any reasonable way. Then h_N satisfies the equation $Lh_N = g_N$, where the difference $g - g_N$ is asymptotically small. Then one has to construct the solution h_N^* of the equation $Lh_N^* = g - g_N$, and to show that it is also asymptotically small. The sum $h_N + h_N^*$ is the desired solution, and it remains to prove that this sum does not depend on N. In any particular realization of this scheme, one has to make sure that the solutions of the auxiliary problems satisfy (either exactly or approximately) the corresponding boundary conditions.

Going back to problems (3.15), (3.16) we prove two auxiliary lemmas.

LEMMA 3.3. *Suppose that the function $\psi(\xi, \eta) \in C^{\infty}(D)$ satisfies the estimates*

$$\left|\frac{\partial^{i+j}\psi}{\partial\xi^i\partial\eta^j}(\xi, \eta)\right| \leq M_{i,j}\xi^{-N+n_i+l_j} \quad \forall i, j \geq 0 \quad (3.18)$$

for some sufficiently large N, where n_i and l_i are numbers independent of N (not necessarily positive). Then there exist a function $h(\xi, \eta) \in C^{\infty}(\overline{D})$ vanishing on ∂D (i.e., for $\xi = \eta^2$), and satisfying (in D) the equation $L_0 h \equiv \partial^2 h/\partial\xi^2 - \partial h/\partial\eta = \psi(\xi, \eta)$, and the estimates

$$\left|\partial^{i+j}h\partial\xi^i\partial\eta^j(\xi, \eta)\right| \leq \overline{M}_{i,j}\xi^{-N+\overline{n}_i+\overline{l}_j} \quad \forall i, j \geq 0, \quad (3.19)$$

where $\overline{n}_i$ and $\overline{l}_j$ are also some numbers independent of N.

PROOF. Let $\chi(z) \in C^\infty(\mathbf{R}^1)$, $\chi(x) \equiv 1$ for $z \le 0$, $\chi(z) \equiv 0$ for $z \ge 1$. Denote $\psi_m(\xi, \eta) = \psi(\xi, \eta)\chi(\xi - m)$. Construct the function $h_m(\xi, \eta)$ solving the problem

$$L_0 h_m = \psi_m(\xi, \eta) \quad \text{for } \eta^2 \le \xi \le m + 3, \ -\sqrt{m + 2} \le \eta \le \sqrt{m + 2},$$

$$h_m(\xi, -\sqrt{m + 2}) = 0, \qquad h_m(\eta^2, \eta) = 0, \qquad h_m(m + 3, \eta) = 0.$$

Such a solution evidently exists, is infinitely differentiable, and vanishes identically for $-\sqrt{m + 2} \le \eta \le -\sqrt{m + 1}$ (see [30, Chapter 4]). The central point of the proof is the derivation of an estimate for the function $h_m(\xi, \eta)$ that is uniform with respect to $h_m(\xi, \eta)$. For that purpose we consider the auxiliary function $\Phi(\xi, \eta) = (2\sqrt{\xi + d} - \eta)^{-2N+1}$, where the constant $d > 0$ will be chosen later. Since $\xi \ge \eta^2$, one has

$$\Phi(\xi, \eta) > 0, \tag{3.20}$$

in D, and

$$L_0 \Phi = (2N - 1)(2\sqrt{\xi + d} - \eta)^{-2N-1}\left(\frac{2N}{\xi + d} + \frac{2\sqrt{\xi + d} - \eta}{2(\xi + d)^{3/2}} - (2\sqrt{\xi + d} - \eta)\right)$$

$$< -(2N - 1)(2\sqrt{\xi + d} - \eta)^{-2N}\left(1 - \frac{4N + 1}{2d^{3/2}}\right).$$

Choose d such that $L_0\Phi < -(2\sqrt{\xi + d} - \eta)^{-2N}$. Hence $L_0\Phi < -\mu\xi^{-N}$, $\mu > 0$. This inequality, inequality (3.20), the estimate (3.18) for $i = j = 0$, and the maximum principle imply the estimate

$$|h_m(\xi, \eta)| \le \mu_1(\xi + d)^{-N+1}, \tag{3.21}$$

where the constants μ_1 and $d > 0$ do not depend on m. Thus, the functions $h_m(\xi, \eta)$ are uniformly bounded.

It follows from the a priori estimates for solutions of a parabolic equation (see [30, Chapter 4]) that the functions $h_m(\xi, \eta)$ are compact in C^2 on any compact set in $\overline{D}$. Passing to the limit as $m \to \infty$, one obtains the desired solution for which (by virtue of (3.21)) the estimate (3.19) holds for $i = j = 0$. The other estimates in (3.19) follow from the a priori estimates of the derivatives for solutions of a parabolic equation. However, one has to bear in mind that in estimating the derivatives near the boundary ∂D, a section of the boundary has to be straightened causing the appearance of variable coefficients in the equation (in general, growing). Since in this case the boundary is a parabola, the coefficients (and their derivatives) grow at most polynomially. Therefore, the estimates of the derivatives may contain some powers of ξ, which is reflected in (3.19). ∎

REMARK. A more detailed analysis can provide more precise form of the exponents $\overline{n}_i$ and $\overline{l}_j$, but this is absolutely unnecessary for the construction of the functions $w_i(\xi, \eta)$. Moreover, for $w_i(\xi, \eta)$ themselves, the precise (as to the order) estimates of their derivatives will follow (after Theorem 3.1 is proved) from the asymptotic expansions of $w_i(\xi, \eta)$.

LEMMA 3.4. *Suppose that the function $w(\xi, \eta) \in C^\infty(\overline{D})$ vanishes on ∂D, and satisfies (in $\overline{D}$) the equation $L_0 w = 0$, and the estimate $|w(\xi, \eta)| \le M\xi^{-N}$, where N is a positive number. Then $w(\xi, \eta) = 0$.*

PROOF. For any $\nu > 0$ the functions $\nu \pm w(\xi, \eta)$ are positive on the boundary of the domain $\{\xi, \eta: \eta^2 < \xi < \lambda, |\eta| < \lambda^{1/2}\}$ for λ sufficiently large. The maximum principle implies that $|w(\xi, \eta)| \le \nu$. ∎

The proof of Theorem 3.1 follows the above scheme. First, we prove the existence of the solution $w_1(\xi, \eta)$. Denote by $w_{1,N}(\xi, \eta)$ the product $\chi(2 - \xi)B_N H_1$ where χ is the truncating function defined in the proof of Lemma 3.3, H_1 is the series (3.17), and $B_N H_1$ its partial sum. Let $L_0 w_{1,N} = g_N(\xi, \eta)$. Since H_1 is an f.a.s. of equation (3.15), then $p_0(\xi, \eta) - g_N(\xi, \eta) = O\left(\xi^{-N_1}\right)$, where $N_1 \to \infty$ as $N \to \infty$. According to Lemma 3.4, one can construct a function $w_{1,N}^*(\xi, \eta) = 0$ solving the problem $L_0 w_{1,N}^* = p_0(\xi, \eta) - g_N(\xi, \eta)$, $w_{1,N}^*(\xi, \eta) = 0$ on ∂D. By Lemma 3.4, the function $w_{1,N}(\xi, \eta) + w_{1,N}^*(\xi, \eta)$ does not depend on N and is the desired function $w_1(\xi, \eta)$.

The existence of the function $w_2(\xi, \eta)$ is proved in the same way. The proof of the existence of the other functions $w_i(\xi, \eta)$ solving the problems (3.15), (3.16) is achieved by induction similarly to the proof of the existence of $w_1(\xi, \eta)$. ∎

3. Thus, the series (3.14) which is matched to the sum of the series (3.3) and (3.6) is constructed. One can see from the construction of the functions $w_i(\xi, \eta)$ that the matching condition here is of a somewhat different form as compared to that of §1. For $\xi \to \infty$ each of the functions $w_i(\xi, \eta)$ can be represented as the sum of two series corresponding to the two sums in formula (3.17) so that the first series is matched to the series U, and the second to S. Denote these series by $[w_i(\xi, \eta)]_1$ and $[w_i(\xi, \eta)]_2$. We will use the same notation for a partial sum of the series W, i.e., a linear combination of these functions: $[A_{N,\xi,\eta}W]_1$ and $[A_{N,\xi,\eta}W]_2$. Thus, the matching condition is of the form

$$A_{N_1,x,y}[A_{N_2,\xi,\eta}W]_1 + A_{N_1,x,\tau}[A_{n_2,\xi,\eta}W]_2$$
$$= A_{N_2,\xi,\eta}A_{N_1,x,y}U + A_{N_2,\xi,\eta}A_{N_1,x,\tau}S \qquad \forall N_1, N_2.$$

As before, introduce the following notation for the composite asymptotic expansion:

$$T_N(x, y, \varepsilon) = A_{N,x,y}U + A_{N,x,\tau}S + A_{N,\xi,\eta}W \\ - A_{N,\xi,\eta}A_{N,x,y}U - A_{N,\xi,\eta}A_{N,x,\tau}S. \tag{3.22}$$

THEOREM 3.2. *The function $T_N(x, y, \varepsilon)$ satisfies the following estimates in the domain Ω_δ:*

$$|\mathscr{L}_\varepsilon T_N(x, y, \varepsilon)| < M\varepsilon^{N_1}, \quad |T_N(\delta, y, \varepsilon)| < M, \quad T_N(x, y, \varepsilon) = 0$$

on $\partial\Omega$, where $N_1 \to \infty$ as $N \to \infty$, and the constant M depends only on N and δ.

The proof virtually repeats that of Theorem 1.4.

THEOREM 3.3. *The series (3.14) is a uniform asymptotic expansion of the solution of the problem (3.1), (3.4) in the domain $\{x, y : y^2 \le x \le \varepsilon^\gamma\}$, while the sum of the series (3.3) and (3.6) is a uniform asymptotic expansion of the solution of the same problem in the domain $\Omega_\delta \cap \{x, y : x \ge \varepsilon^\gamma\}$, where γ is any number such that $0 < \gamma < 2$.*

PROOF. It follows from Theorem 3.2 and Lemma 5.1 that

$$|T_N(x, y, \varepsilon) - u(x, y, \varepsilon)| < M\varepsilon^{N_2},$$

where $T_N(x, y, \varepsilon)$ is defined by formula (3.22), and $N_2 \to \infty$ as $N \to \infty$. Now the assertion of the theorem is proved along the same lines as that of Theorem 1.6. ■

§4. The case of a characteristic tangent
to the boundary of the domain from the inside

In this section we consider the asymptotics of the solution of the problem (0.3), (0.4) in a neighborhood of the singular characteristic α_2 (see Figure 15). As in the preceding section, we consider a model equation and boundary, corresponding to the case of general position.

Thus, let the equation be of the form

$$\mathscr{L}_\varepsilon u = \varepsilon^3 \left(\frac{\partial^2 u}{\partial x^2} + \frac{\partial^2 u}{\partial y^2} \right) - a(x, y) \frac{\partial u}{\partial y} = f(x, y), \qquad a(x, y) > 0, \quad (4.1)$$

where $a, f \in C^\infty(\overline{D})$, $a(0, 0) = 1$, the boundary in a neighborhood of the point O is of the form $\{x, y : x = y^2\}$ and the boundary condition is

$$u(x, y, \varepsilon) = 0 \quad \text{for } (x, y) \in \partial\Omega. \quad (4.2)$$

But now the domain Ω lies to the left of the parabola $\{x, y : x = y^2\}$ so that the y-axis lies inside Ω in some neighborhood of the origin. The f.a.s. will again be constructed in the domain $\Omega_\delta = \Omega \cap \{x, y : |x| \le \delta\}$. We assume the domain Ω_δ to be bounded by two smooth curves: $\sigma_1 = \{x, y : y = \varphi_1(x)\}$ and $\sigma_2 = \{x, y : y = \varphi_2(x)\}$, the straight line $|x| = \delta$, and the parabola mentioned above (see Figure 23).

The position of the boundary of the domain and the singular characteristic $\alpha = \{x, y : x = 0, 0 \le y \le \varphi_2(0)\}$ is topologically equivalent to the situation considered in §2 (see Figure 17). On the other hand, the local behavior of the solution in a neighborhood of the point O has much in common with that of the solution studied in §3—the same inner variables, the same structure of singularities of the exponential boundary layer functions, etc.

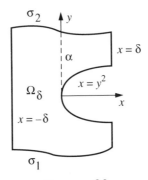

FIGURE 23

The investigation of the problem (4.1), (4.2) is, therefore, a combination of similar investigations of the two preceding sections. Since many aspects of the analysis of the problem (4.1), (4.2) have been considered above, the exposition in this section will be made more concise.

1. The outer expansion of the solution is of the same form as in §3 (formula (3.3)), but here, as in §2, the coefficients of the outer expansion are discontinuous on the singular characteristic (see Figure 23). It is, therefore, convenient to represent the outer expansion as the sum of the two series, the first of which has smooth coefficients everywhere in $\overline{\Omega}_\delta$, while the second differs from zero only to the right of the singular characteristic α. We can assume, without any loss of generality, that the function $f(x, y)$ is smoothly extended to the entire plane. Define the coefficients $u_{3k}(x, y)$ everywhere in the strip $|x| \leq \delta$ as the solutions of the problems

$$a(x, y)\frac{\partial u_0}{\partial y} = -f(x, y),$$

$$\frac{\partial u_{3k}}{\partial y} = \frac{\Delta u_{3k-3}}{a(x, y)}, \quad k \geq 1, \quad u_{3k}(x, \varphi_1(x)) = 0.$$

Evidently, $u_{3k}(x, y) \in C^\infty(\overline{\Omega}_\delta)$.

The outer expansion will be looked for in the form $U + Z$, where

$$U = \sum_{k=0}^\infty \varepsilon^{3k} u_{3k}(x, y), \tag{4.3}$$

$$Z = \sum_{k=0}^\infty \varepsilon^{3k} z_{3k}(x, y), \tag{4.4}$$

$u_{3k}(x, y)$ are the functions just defined, and the functions $z_{3k}(x, y)$ are different from zero only to the right of α, i.e., in the domain $\{x, y: 0 < x \leq \delta, \sqrt{x} \leq y \leq \varphi_2(x)\}$. These functions are the solutions of the problems

$$\frac{\partial z_0}{\partial y} = 0, \quad \frac{\partial z_{3k}}{\partial y} = \frac{\Delta z_{3k-3}}{a(x, y)}, \quad k \geq 1, \quad z_{3k}(x, \sqrt{x}) = -u_{3k}(x, \sqrt{x}). \tag{4.5}$$

The series $U + Z$ thus constructed is evidently an f.a.s. of equation (4.1) and the boundary condition (4.2) on σ_1 and the arc $\{x, y: 0 < x \leq \delta, y = \sqrt{x}\}$. Clearly, $z_0(x, y) = -u_0(x, \sqrt{x})$.

LEMMA 4.1. *For* $y \geq \sqrt{x}$, $k \geq 1$, $x \to 0$ *the functions* $z_{3k}(x, y)$ *defined by relations* (4.5) *can be expanded in the asymptotic series*

$$z_{3k}(x, y) = x^{1/2-2k} \sum_{j=0}^\infty a_{j, k}(y)x^{j/2}, \quad a_{j, k}(y) \in C^\infty[0, \sigma_2(0)].$$

These series can be differentiated term-by-term to any order.

The proof is easily achieved by induction as in Lemma 3.1. ∎

The residual in the boundary condition on the arc $\{x, y: 0 \leq x \leq \delta, y = -\sqrt{x}\}$ is eliminated by adding the following series of boundary layer functions

$$S = \sum_{k=0}^{\infty} \varepsilon^{3k} s_{3k}(x, \tau), \qquad (4.6)$$

where

$$\tau = \varepsilon^{-3}(-y - \sqrt{x}). \qquad (4.7)$$

The equations for $s_{3k}(x, \tau)$ almost coincide with equations (3.7), while the boundary conditions are

$$s_{3k}(x, 0) = -u_{3k}(x, -\sqrt{x}), \qquad s_{3k}(x, \tau) \to 0 \quad \text{for } \tau \to \infty.$$

LEMMA 4.2. *For the functions* $s_{3k}(x, \tau)$ *the asymptotic expansions*

$$s_{3k}(x, \tau) = \left[\exp\left(-\frac{4xa(x, -\sqrt{x})\tau}{1 + 4x}\right)\right] x^{-3k/2} \sum_{l=0}^{2k}(x, \tau)^l \sum_{j=0}^{\infty} c_{k,l,j} x^{j/2},$$

$$x \to 0.$$

are valid.

The proof almost repeats that of Lemma 3.2. The only difference is that in Lemma 3.2 the principal term of the asymptotics equals $x^{(1-3k)/2}$, while in this lemma it is of the form $x^{-3k/2}$. The reason is the difference in the boundary conditions. In §3 the boundary condition for $s_0(x, \tau)$ equals zero for $x = 0$, while in the problem considered in this section, in general, $s_0(0, 0) = -u_0(0, 0) \neq 0$. The inessential factor $a(x, -\sqrt{x})$ appears in the exponent due to the fact that the coefficient of $\partial u/\partial y$ is now variable. ∎

2. As in §3, introduce inner variables in a neighborhood of the point O

$$x = \varepsilon^2 \xi, \qquad y = \varepsilon \eta \qquad (4.8)$$

looking for the asymptotics of the solution $u(x, y, \varepsilon)$ in the form of the sum of the series U and the inner expansion

$$W = \sum_{i=0}^{\infty} \varepsilon^i w_i(\xi, \eta). \qquad (4.9)$$

Now, naturally, we will mean by D the exterior region of the parabola, i.e., the domain $\{\xi, \eta: \xi < \eta^2, \eta \in \mathbf{R}^1\}$. Otherwise the form of the equations and boundary conditions for $w_i(\xi, \eta)$ is scarcely different from those in §3. Since the series W must be an f.a.s. of the homogeneous equation, one has

$$L_0 w_0 \equiv \frac{\partial^2 w_0}{\partial \xi^2} - \frac{\partial w_0}{\partial \eta} = 0, \qquad L_0 w_1 = 0,$$

$$L_0 w_i = -\frac{\partial^2 w_{i-2}}{\partial \eta^2} + \sum_{\substack{s<i \\ 2j+l+s=i}} \frac{1}{j!l!} \frac{\partial^{j+l} a}{\partial x^j \partial y^l}(0, 0) \frac{\partial w_s}{\partial \eta} \xi^j \eta^l, \qquad i \geq 2. \qquad (4.10)$$

The boundary conditions for $w_i(\xi, \eta)$ follow from (4.2). Making the change of variables (4.8) in the series (4.3), one obtains the series $\sum_{i=0}^{\infty} \varepsilon^i q_i(\xi, \eta)$, where $q_i(\xi, \eta)$ are polynomials of degree not greater than i. Therefore,

$$w_i(\eta^2, \eta) = -q_i(\eta^2, \eta) \quad \text{for } \eta \in \mathbf{R}^1. \tag{4.11}$$

Here the functions $w_i(\xi, \eta)$ must tend to zero as $\eta \to -\infty$ and as $\xi \to -\infty$, while for $\xi \to \infty$, $\eta < 0$ in a neighborhood of the parabola ∂D the series (4.9) must be matched to the series S.

Equations (4.10) are parabolic, whereby it is sufficient to prove the existence of solutions for the problems (4.10), (4.11) for $\eta < -\eta_0$, where η_0 is some number. The solutions can then be easily extended to $\eta \geq -\eta_0$ with the help of known theorems on the existence of a solution of the boundary value problem for the heat equation.

Passing in the series S to the variables ξ, η according to formulas (4.7), (4.8) with the use of Lemma 4.2, one obtains the formal series $\sum_{i=0}^{\infty} \varepsilon^i H_i$, where

$$H_i = \xi^{i/2} \exp(4\xi(\sqrt{\xi} + \eta)) \sum_{j=0}^{\infty} \xi^{-j/2} \sum_{l=0}^{2j+i} b_{i,j,l} [\xi(\sqrt{\xi} + \eta)]^l \tag{4.12}$$

(cf. formula (3.17)). These series provide f.a.s. for the problems (4.10), (4.11) for $\eta \leq -\sqrt{\xi}$, $\xi \to \infty$, and the functions $w_i(\xi, \eta)$ must have asymptotic expansions H_i in this domain. The terms of the series H_i are exponentially small at infinity everywhere except a small neighborhood of the lower branch of the parabola, for example, for $\eta < -\sqrt{\xi} - \xi^{\beta}$, where $\beta < 1$. The shaded area in Figure 24 indicates the region where the terms of the series H_i are not exponentially small. One can assume that the region where the functions $w_i(\xi, \eta)$ must have asymptotics (4.12) is substantially wider so that the boundary of this domain has a simpler form. As in §2, take for such a boundary the straight line $\xi = -\eta$. However, in §2, the coefficients of the inner expansion are solutions of elliptic problems, while here they are solutions of nonhomogeneous heat equations whereby they are constructed somewhat more easily. Introduce the following notation for the

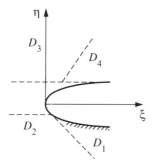

FIGURE 24

subdomains of the domain D. Let $D_1 = \{\xi, \eta: \eta \le -1, -\eta \le \xi < \eta^2\}$, $D_2 = \{\xi, \eta: \eta \le -1, \xi < -\eta\}$ (see Figure 24).

THEOREM 4.1. *There exist functions* $w_i(\xi, \eta) \in C^\infty(\overline{D})$ *solving the problems* (4.10), (4.11), *asymptotically equal to* H_i *in* D_1, *where* H_i *are the series* (4.12), *and exponentially small in* D_2. *To be more precise, for any positive integer* N

$$\left| w_i(\xi, \eta) - \xi^{i/2} \exp(4\xi(\sqrt{\xi} + \eta)) \sum_{j=0}^{2N} \xi^{-j/2} \sum_{l=0}^{2j+i} b_{i,j,l} [\xi(\sqrt{\xi} + \eta)]^l \right|$$

$$\le M\xi^{-N+i/2} \exp(\beta(\xi - \eta^2)) \text{ in } D_1, \tag{4.13}$$

$$|w_i(\xi, \eta))| < M \exp(\beta(\xi - \eta^2)) \text{ in } D_2,$$

where $\beta > 0$ *is a fixed number, and the coefficients* $b_{i,j,l}$ *are the same as in formula* (4.12). *Similar estimates hold for all derivatives of* $w_i(\xi, \eta)$.

For any $\eta_1 > 0$ *the following estimates are valid:*

$$|w_i(\xi, \eta)| \le M \exp(\beta\xi) \quad \text{for } \eta \le \eta_1, \tag{4.14}$$

where the constant M *depends just on* i *and* η_1, *with similar estimates holding for all derivatives of* $w_i(\xi, \eta)$.

The proof follows the common scheme repeatedly used above and described in subsection 2 of §3.

LEMMA 4.3. *Suppose that the function* $\psi(\xi) \in C^\infty(\overline{D})$ *satisfies the following estimates: for all* $i \ge 0$, $j \ge 0$

$$\left| \frac{\partial^{i+j} \psi(\xi\eta)}{\partial\xi^i \partial\eta^j} \right| \le \begin{cases} M_{i,j}|\eta|^{-2N+n_i+l_j} \exp(\beta(\xi - \eta^2)) & \text{in } D_1, \\ M_{i,j} \exp(\beta(\xi - \eta^2)) & \text{in } D_2. \end{cases} \tag{4.15}$$

Then there exists a function $h(\xi, \eta) \in C^\infty(\overline{D})$ *vanishing on* ∂D, *satisfying the equation* $L_0 h \equiv \frac{\partial^2 h}{\partial\xi^2} - \frac{\partial h}{\partial\eta} = \psi(\xi, \eta)$ *in* D, *and the estimates*

$$\left| \frac{\partial^{i+j} h(\xi\eta)}{\partial\xi^i \partial\eta^j} \right| \le \begin{cases} \overline{M}_{i,j}|\eta|^{-2N+\overline{n}_i+\overline{l}_j} \exp(\beta_1(\xi - \eta^2)) & \text{in } D_1, \\ \overline{M}_{i,j} \exp(\beta_1(\xi - \eta^2)) & \text{in } D_2, \end{cases} \tag{4.16}$$

for all $i \ge 0$, $j \ge 0$. *Here* $M_{i,j}$, $\overline{M}_{i,j}$, n_i, l_j, $\overline{n}_i$, $\overline{l}_j$ *are the same as in Lemma* 3.3, β *is a positive number, and* $\beta_1 > 0$ *is any number smaller than* β.

The proof is achieved along the same lines as that of Lemma 3.3. Suppose that the functions $\psi_m(\xi, \eta) \in C^\infty(\overline{D})$ are obtained by multiplying the function $\psi(\xi, \eta)$ by truncating functions so that $\psi_m(\xi, \eta) \equiv \psi(\xi, \eta)$ for $\eta > -m$, $\psi_m(\xi, \eta) \equiv 0$

for $\eta \leq -m - 1$, while $h_m(\xi, \eta)$ are the solutions of the equations $L_0 h_m = \psi_m$ in $D \cap \{\xi, \eta : \eta \geq -m - 2\}$ vanishing on ∂D and for $\eta = -m - 2$. Such solutions exist (see [30, Chapter 4]), and the main point now is to find their uniform bound.

The functions $\psi_m(\xi, \eta)$ evidently satisfy the same estimates (4.15) as the function $\psi(\xi, \eta)$ (possibly, with other constants $M_{i,j}$). In order to estimate $h_m(\xi, \eta)$ one can use the auxiliary function $\Phi(\xi, \eta) = \eta^{-2N} \exp(\gamma(\xi - \eta^2))$. Since $L_0 \Phi = \eta^{-2N}(2N\eta^{-1} + 2\gamma\eta + \gamma^2) \exp(\gamma(\xi - \eta^2)) < -\eta^{-2N} \exp(\gamma(\xi - \eta^2))$ for $\eta < -\eta_0$, where η_0 is a sufficiently large fixed number, the function $M\Phi(\xi, \eta)$ is a majorant for the functions $\pm h_m(\xi, \eta)$ for any $\gamma < \beta$, $\gamma > 0$.

Next, as in Lemma 3.3, one proves the existence of the solution $h(\xi, \eta)$ and the estimates (4.16) for $\eta < -\eta_0$. For $\eta \geq -\eta_0$ the function $h(\xi, \eta)$ is extended as the solution of the equation $L_0 h = \psi$ vanishing on ∂D with the initial function defined at $\eta = -\eta_0$. Since this function $h(\xi, -\eta_0)$ decays exponentially together with all its derivatives, the solution $h(\xi, \eta)$ has the same property for $\eta_0 \leq \eta \leq \eta_1$ for any fixed η_1. This fact is easily verified with the help of the maximum principle and the auxiliary function $\exp(\gamma\xi + \mu\eta)$, where $\mu > 0$. ∎

LEMMA 4.4. *Suppose that the function* $v(\xi, \eta) \in C^\infty(\overline{D})$ *vanishes on* ∂D, *satisfies the equation* $L_0 v = 0$ *in* D, *grows not faster than some power of* $|\xi|$ *as* $\xi \to -\infty$, *and tends to zero as* $\eta \to -\infty$ *uniformly with respect to* ξ. *Then* $v(\xi, \eta) \equiv 0$.

The proof is an immediate consequence of the maximum principle for the heat equation. ∎

The proof of Theorem 4.1 follows the general scheme described above. To prove the existence of the function $w_0(\xi, \eta)$ we first find the function $w_{0,N}(\xi, \eta)$ constructed from the asymptotic series H_0. This can be achieved by replacing the series in formula (4.12) by its partial sum multiplied by a truncating function vanishing for $\xi \leq -1$ and for $\xi \leq -\eta$. The proof is then based on Lemmas 4.3, 4.4 and repeats that of Theorem 3.1. The existence of the other $w_i(\xi, \eta)$ for $\eta \leq -1$ and the estimates (4.13) are proved along the same lines by induction. For $\eta \geq -1$ the functions $w_i(\xi, \eta)$ are extended as the solutions of the boundary value problems (4.10), (4.11). The estimates (4.14) follow from the maximum principle with the help of the auxiliary function $\exp(\gamma\xi + \mu\eta)$, and from the a priori estimates for solutions of parabolic equations. ∎

3.   The series (4.9) is thus constructed, but this is where the similarity with the problem considered in §3 ends. The subsequent investigation is close to that conducted in §2 (cf. Figure 17, and Figure 23). One has to find the asymptotics of the solutions $w_i(\xi, \eta)$ as $\eta \to \infty$, and then, using these asymptotics and the matching condition, construct the series V—the inner expansion in a neighborhood of the singular characteristic α. The analysis of the asymptotics of the solutions $w_i(\xi, \eta)$ in this section is somewhat different from that in §2: here $w_i(\xi, \eta)$ are solutions of nonhomogeneous heat equations, while in §2 $w_i(\xi, \eta)$ are solutions of elliptic equations. However, the result turns out to be very similar. This comes as no surprise because for $\eta \to \infty$ the operator $\partial^2/\partial\xi^2 + \partial^2/\partial\eta^2 - \partial/\partial\eta$ is, in a certain sense, asymptotically close to the operator $\partial^2/\partial\xi^2 - \partial/\partial\eta$. The difference in the asymptotics as $\eta \to \infty$ is due to the difference in the behavior of the functions $w_i(\xi, \eta)$

as $\xi \to \infty$ in this problem and in §2, rather than to the different form of the operators.

In this subsection we study the functions $w_i(\xi, \eta)$ solving the problems (4.10), (4.11) in the domain $D^+ = \{\xi, \eta : \eta \geq 2, -\infty < \xi < \eta^2\}$. Theorem 4.1 implies that $w_i(\xi, \eta)$ decay exponentially together with their derivatives for $\eta = 2$, $\xi \to -\infty$.

As opposed to §2, the problem considered here admits no explicit formula for the solutions $w_i(\xi, \eta)$ similar to formula (2.36), i.e., a formula expressing the solution of the nonhomogeneous heat equation in domain D^+ through the right-hand side of the equation and the values of the solution on ∂D^+. However, using the fundamental solution of the heat equation

$$G(\xi, \eta) = \frac{1}{2\sqrt{\pi\eta}} \exp\left(-\frac{\xi^2}{\eta}\right), \tag{4.17}$$

one can easily write out Green's formula in the domain D^+. Indeed, suppose that the function $w(\xi, \eta) \in C^\infty(D^+)$ satisfies the equation

$$L_0 w \equiv \frac{\partial^2 w}{\partial \xi^2} - \frac{\partial w}{\partial \eta} = -F(\xi, \eta) \tag{4.18}$$

and grows at infinity no faster that some power of $|\xi| + |\eta|$. Then, multiplying (4.18) expressed in terms of the variables ξ_1, η_1 by $G(\xi - \xi_1, \eta - \eta_1)$, and integrating over the domain $D^+ \cap \{\xi_1, \eta_1 : 2 < \eta_1 < \eta\}$, one arrives at Green's formula

$$w(\xi, \eta) = I_1(\xi, \eta) + I_2(\xi, \eta) + I_3(\xi, \eta), \tag{4.19}$$

where

$$I_1(\xi, \eta) = \int_{-\infty}^{4} G(\xi - \xi_1, \eta - 2)g(\xi_1)\, d\xi_1, \tag{4.20}$$

$$\begin{aligned} I_2(\xi, \eta) = \int_{2}^{\eta} \Big\{ & g_2(\eta_1)G(\xi - \eta_1^2, \eta - \eta_1) \\ & + g_3(\eta_1)\frac{\partial G}{\partial \xi}(\xi - \eta_1^2, \eta - \eta_1) \\ & + 2\eta_1 g_3(\eta_1)G(\xi - \eta_1^2, \eta - \eta_1) \Big\}\, d\eta_1, \end{aligned} \tag{4.21}$$

$$I_3(\xi, \eta) = \iint_{D^+} G(\xi - \xi_1, \eta - \eta_1)F(\xi_1, \eta_1)\, d\xi_1\, d\eta_1, \tag{4.22}$$

$$g_1(\xi) = w(\xi, 2), \quad g_2(\eta) = \frac{\partial w}{\partial \xi}(\eta^2, \eta), \quad g_3(\eta) = w(\eta^2, \eta). \tag{4.23}$$

This formula yields, without difficulty, the asymptotics of $w(\xi, \eta)$ as $\eta \to \infty$ in the case where the right-hand side $F(\xi, \eta)$ and the values of w and $\partial w/\partial \xi$ on ∂D^+ decay sufficiently rapidly at infinity. The study of the asymptotics of $w_i(\xi, \eta)$ as $\eta \to \infty$ will follow the same scheme as in §§1 and 2: for a fixed i one first constructs an f.a.s. of the problems (4.10), (4.11), and then applies formulas (4.19)–(4.23) to the resulting residual.

We begin with the investigation of the asymptotics of the integrals (4.20)–(4.22) with the densities in the integrands decaying rapidly at infinity. Although the asymptotics of these integrals as $\eta \to \infty$ is of the same form everywhere in D^+, it is more convenient to consider this asymptotics separately in some subdomains of D^+. Introduce the following notation:

$$D_3 = \{\xi, \eta: \xi \leq \eta, \eta \geq 2\}, \qquad D_4 = \{\xi, \eta: \eta \leq \xi \leq \eta^2, \eta \geq 2\}$$

(see Figure 24). We also note that in reality the functions $w_i(\xi, \eta)$ decay exponentially as $\xi \to -\infty$ (albeit nonuniformly with respect to η). However, the proof of this is rather troublesome, and in order to achieve our main objective, i.e., the construction and justification of the asymptotics of the solution $u(x, y, \varepsilon)$, it is sufficient to verify that $w_i(\xi, \eta)$ tend to zero as $\xi \to -\infty$ faster than any power of $|\xi|^{-1}$. Therefore, in the investigation of the integrals $I_1(\xi, \eta)$, $I_2(\xi, \eta)$, $I_3(\xi, \eta)$ our attention will be limited only to power-type estimates.

LEMMA 4.5. *Suppose that the function $g_1(\xi)$ satisfies the inequality $|g_1(\xi)| < M(1 + \xi^2)^{-2N}$ for $\xi \leq 4$, where N is a sufficiently large number. Then the integral $I_1(\xi, \eta)$ defined by formula (4.20) admits the following asymptotic representation:*

$$\frac{\partial^{i+l}}{\partial \xi^i \partial \eta^l} I_1(\xi, \eta) = \frac{\partial^{i+l}}{\partial \xi^i \partial \eta^l} \sum_{j=0}^{N_1} c_{j,1} \frac{\partial^j}{\partial \xi^j} G(\xi, \eta) + \rho_{i,l,N,1} \qquad (4.24)$$

in D^+ for $\eta \geq 3$, where $i + l \leq N_2$,

$$|\rho_{i,l,N,1}(\xi, \eta)| \leq M(\xi^2 + \eta^2)^{-N_3} \qquad \text{in } D^+ \text{ for } \eta \geq 3, \qquad (4.25)$$

and N_1, N_2, N_3 are some numbers which tend to infinity as $N \to \infty$.

PROOF. Represent the function $G(\xi - \xi_1, \eta)$ in the integrand of $I_1(\xi, \eta)$ in the following form:

$$\sum_{j=0}^{2N} \frac{(-\xi_1)^j}{j!} \frac{\partial^j}{\partial \xi^j} G(\xi, \eta) - \frac{\xi_1^{2N+1}}{(2N+1)!} \frac{\partial^{2N+1}}{\partial \xi^{2N+1}} G(\xi - \gamma \xi_1, \eta),$$

where $0 < \gamma < 1$. Inserting this expression into the integral $I_1(\xi, \eta)$, and integrating, one obtains the first terms of the series (4.24) for $i = l = 0$ plus the remainder. Thus, it only remains to estimate this remainder. The explicit form of the function G makes it clear that its absolute value does not exceed

$$M \int_{-\infty}^{4} (1 + \xi_1^2)^{-N+1} \eta^{-N-1} \exp\left(-\frac{(\xi - \gamma \xi_1)^2}{8\eta}\right) d\xi_1. \qquad (4.26)$$

For $|\xi| \leq \eta$ this integral evidently does not exceed the right-hand side of inequality (4.25). For $|\xi| \geq \eta$ we split the integral (4.26) into two integrals: one over the domain where $|\xi_1| < \frac{1}{2}|\xi|$, and the other over the domain where $|\xi_1| \geq \frac{1}{2}|\xi| \geq \frac{1}{2}|\eta|$. Estimating each of the resulting integrals, one arrives at inequality (4.25).

Clearly, the derivatives of $I_1(\xi, \eta)$ for $\eta \geq 3$ are of the form similar to that of $I_1(\xi, \eta)$, and, consequently, can be expanded into a series similar to (4.24) for $i = l = 0$. This implies representation (4.24) for the remaining values of i and l.
∎

We note that the terms of the asymptotic expansion (4.24) are of the form $\eta^\beta \Phi(\xi \eta^{-1/2})$, where the function $\Phi(\theta)$ decays exponentially at infinity. Thus, for $|\xi| < C\sqrt{\eta}$ the terms of the series have power-type asymptotics, while for $|\xi| > \eta^{1/2+\mu}$ ($\mu > 0$) the only essential term in the asymptotic representation is the remainder $\rho_{i,l,N,1}$.

LEMMA 4.6. *Suppose that the functions* $g_2(\eta)$ *and* $g_3(\eta)$ *satisfy the estimates* $|g_k(\eta)| < M\eta^{-8N}$, *where* N *is sufficiently large. Then the integral* $I_2(\xi, \eta)$ *defined by formula* (4.21) *satisfies the following relations:*

$$\frac{\partial^{i+l}}{\partial \xi^i \partial \eta^l} I_2(\xi, \eta) = \frac{\partial^{i+l}}{\partial \xi^i \partial \eta^l} \sum_{j=0}^{N_1} c_{j,2} \frac{\partial^j G}{\partial \xi^j}(\xi, \eta) + \rho_{i,l,N,2}(\xi, \eta) \qquad (4.27)$$

in the domain D_3, *where* $i + l \leq N_2$, *and*

$$|\rho_{i,l,N,2}(\xi, \eta)| \leq M(\xi^2 + \eta^2)^{-N_3}. \qquad (4.28)$$

Here N_1, N_2, N_3 *are numbers tending to infinity as* $N \to \infty$.

The proof of the asymptotics for the integral $I_2(\xi, \eta)$ reduces to the proof of the asymptotics for all three integrals constituting $I_2(\xi, \eta)$. All three are examined along the same lines. Accordingly, we consider only the integral

$$I_{2,1}(\xi, \eta) = \int_2^\eta g_2(\eta_1) G(\xi - \eta_1^2, \eta - \eta_1) \, d\eta_1.$$

Expanding the function $G(\xi - \eta_1^2, \eta - \eta_1)$ in a Taylor series in the variable η_1, and taking into account the equality $\partial G/\partial \eta = \partial^2 G/\partial \xi^2$, one obtains

$$G(\xi - \eta_1^2, \eta - \eta_1) = \sum_{j=0}^{2N} P_{2j}(\eta_1) \frac{\partial^j G}{\partial \xi^j}(\xi, \eta) + \Lambda_{N,2}(\xi, \eta, \eta_1). \qquad (4.29)$$

Here P_{2j} is a polynomial of degree not greater than $2j$. The remainder $\Lambda_{N,2}(\xi, \eta, \eta_1)$ is the sum of functions of the form $P_m(\eta_1)(\partial^n G/\partial \xi^n)(\xi - \gamma^2 \eta_1^2, \eta - \gamma \eta_1)$, where $0 < \gamma < 1$, $m \leq 4N + 2$, $2N + 1 \leq n \leq 4N + 2$. Inserting expression (4.29) for $G(\xi - \eta_1^2, \eta - \eta_1)$ into the integral $I_{2,1}(\xi, \eta)$, one obtains formula (4.27) for $i = l = 0$, where

$$\rho_{0,0,N,2}(\xi, \eta) = -\sum_{j=0}^{2N} \frac{\partial^j G}{\partial \xi^j}(\xi, \eta) \int_\eta^\infty P_{2j}(\eta_1) g_2(\eta_1) \, d\eta_1$$

$$+ \int_2^\eta g_2(\eta_1) \Lambda_{N,2}(\xi, \eta, \eta_1) \, d\eta_1.$$

The condition imposed on the function $g_2(\eta)$ and the explicit form (4.17) of the function $G(\xi, \eta)$ imply that each summand appearing under the summation sign satisfies the estimate (4.28). The absolute value of the last summand does not exceed

$$M \int_2^\eta \eta_1^{-4N+2} (\eta - \gamma\eta_1)^{-(n+1)/2} \exp\left(-\frac{(\xi - \gamma^2\eta_1^2)^2}{8(\eta - \gamma\eta_1)}\right) d\eta_1$$

Represent this integral as the sum of two integrals: one over the interval with the endpoints 2 and $1 + \eta/2$, and the other over the interval with the endpoints $1 + \eta/2$ and η. For the first of them the estimate (4.28) is evident. Since for $z \geq 0$ inequality $\exp(-z/16) < M_i(1 + z)^{-i} \ \forall i$ holds, the second integral does not exceed

$$M \int_{1+\eta/2}^\eta \eta_1^{-4N+2} \left[(\xi - \gamma^2\eta_1^2)^2 + \eta - \gamma\eta_1\right]^{-(n+1)/2} \exp\left(-\frac{(\xi - \gamma^2\eta_1^2)^2}{16(\eta - \gamma\eta_1)}\right) d\eta_1.$$

The factor $\left[(\xi - \gamma^2\eta_1^2)^2 + \eta - \gamma\eta_1\right]^{-(n+1)/2}$ is uniformly bounded in D_3. This is easily verified if one considers separately the case where $\gamma\eta_1$ is close η, and the case where $\gamma\eta_1$ is far from η. Here one has to take into account that $\gamma\eta_1 < \eta$, $\xi \leq \eta$, and $\eta \geq 2$. It is now clear that the integral from $1 + \eta/2$ to η also satisfies estimate (4.28) for $|\xi| \leq \eta$. For $\xi \leq -\eta$ one has

$$\left[\left(\xi - \gamma^2\eta_1^2\right)^2 + \eta - \gamma\eta_1\right]^{-(n+1)/2} \leq |\xi|^{-n-1}$$

and, consequently, the estimate (4.28) is also valid.

Relation (4.27) is hereby proved for $i = l = 0$. For the derivatives of the function $I_2(\xi, \eta)$ the remark made at the end of the proof of Lemma 4.5 is valid. ∎

LEMMA 4.7. *Suppose that the function* $F(\xi, \eta) \in C^\infty(D^+)$ *satisfies the estimates*

$$\left|\frac{\partial^{i+l}F}{\partial\xi^i\partial\eta^l}(\xi, \eta)\right| \leq M\left(\xi^2 + \eta^2\right)^{-4N} \quad \text{in } D^+, \tag{4.30}$$

where $i + l \leq N$, *and* N *is a sufficiently large number. Then the integral* $I_3(\xi, \eta)$ *defined by formula (4.22) satisfies the relations*

$$\frac{\partial^{i+l}I_3}{\partial\xi^i\partial\eta^l}(\xi, \eta) = \sum_{j=0}^{N_1} c_{j,3}\frac{\partial^j G}{\partial\xi^j}(\xi, \eta) + \rho_{i,l,N,3}(\xi, \eta), \tag{4.31}$$

where $i + l \leq N_2$,

$$\left|\rho_{i,l,N,3}(\xi, \eta)\right| \leq M\left(\xi^2 + \eta^2\right)^{-N_3} \quad \text{in } D_3, \tag{4.32}$$

and N_1, N_2, *and* N_3 *are numbers tending to infinity as* $N \to \infty$.

PROOF. Represent the integral $I_3(\xi, \eta)$ as the sum of the two integrals: $I_{3,1}(\xi, \eta)$ over the domain where $2 \leq \eta_1 \leq 1 + \eta/2$, and $I_{3,2}(\xi, \eta)$ over the domain where $1 + \eta/2 \leq \eta_1 \leq \eta$. Expand the function $G(\xi - \xi_1, \eta - \eta_1)$ in the first integral in a Taylor series with remainder, taking into account that $\partial G/\partial\eta = \partial^2 G/\partial\xi^2$. One has

$$G(\xi - \xi_1, \eta - \eta_1) = \sum_{j=0}^{2N} P_j(\xi_1, \eta_1)\frac{\partial^i G}{\partial\xi^j}(\xi, \eta) + \Lambda_{N,3}(\xi, \eta, \xi_1, \eta_1),$$

where P_j are polynomials of degree not greater than j. Here $\Lambda_{N,3}(\xi, \eta, \xi_1, \eta_1)$

is the sum of functions of the form $P_m(\xi_1, \eta_1)(\partial^n G/\partial \xi^n)(\xi - \gamma \xi_1, \eta - \gamma \eta_1)$, where $0 < \gamma < 1$, $m \leq 2N + 2$, $2N + 1 \leq n \leq 4N + 2$. Inserting this expression for $G(\xi - \xi_1, \eta - \eta_1)$ into the integral $I_{3,1}(\xi, \eta)$, one gets formula (4.31) for $i = l = 0$, where

$$\rho_{0,0,N,3}(\xi, \eta) = -\sum_{j=0}^{2N} \frac{\partial^j G}{\partial \xi^j}(\xi, \eta) \int_{1+\eta/2}^{\infty} \int_{-\infty}^{\eta_1^2} P_j(\xi_1, \eta_1) F(\xi_1, \eta_1) \, d\xi_1 \, d\eta_1$$

$$+ \int_{2}^{1+\eta/2} \int_{-\infty}^{\eta_1^2} F(\xi_1, \eta_1) \Lambda_{N,3}(\xi, \eta, \xi_1, \eta_1) \, d\xi_1 \, d\eta_1.$$

The conditions imposed on the function $F(\xi, \eta)$ and the explicit form of the functions $G(\xi, \eta)$ imply that the summands under the summation sign satisfy the estimates (4.32). The last summand does not exceed the quantity

$$M \int_{2}^{1+\eta/2} \int_{-\infty}^{\eta_1^2} (\xi_1^2 + \eta_1^2)^{-2N} \eta^{-N} \exp\left(-\beta \frac{(\xi - \gamma \xi_1)^2}{\eta}\right) d\xi_1 \, d\eta_1,$$

where $\beta > 0$. Considering the cases $|\xi| \leq \eta$ and $\xi \leq -\eta < 0$ separately, we conclude that this integral also satisfies estimates (4.32).

It remains to estimate the integral $I_{3,2}(\xi, \eta)$. By hypothesis, it does not exceed

$$M \int_{1+\eta/2}^{\eta} \int_{-\infty}^{\eta_1^2} \left(\xi_1^2 + \eta_1^2\right)^{-4N} G(\xi - \xi_1, \eta - \eta_1) \, d\xi_1 \, d\eta_1.$$

One can easily see that this integral also does not exceed $M(\xi^2 + \eta^2)^{-N_3}$. Relation (4.31) is hereby proved for $i = l = 0$.

The function $I_3(\xi, \eta)$ cannot, in general, be differentiated under the integral sign with respect to ξ and η a large number of times. Since $\partial I_3/\partial \eta = \partial^2 I_3/\partial \xi^2 + cF(\xi, \eta)$, it is sufficient to investigate the derivatives with respect to ξ. After a single differentiation of $I_3(\xi, \eta)$ under the integral sign, the resulting integral can be transformed via integration by parts. As a result, it turns out that $(\partial I_3/\partial \xi)(\xi, \eta)$ is a sum of integrals of type $I_3(\xi, \eta)$ and $I_2(\xi, \eta)$. Therefore, the asymptotic representation (4.31) holds for $(\partial I_3/\partial \xi)(\xi, \eta)$. The same representation is then proved for other derivatives by induction. ∎

LEMMA 4.8. *Suppose that* $w(\xi, \eta) \in C^\infty(D^+)$, $L_0 w = F(\xi, \eta)$, *the estimates* (4.30) *hold for the function* $F(\xi, \eta)$, *and the boundary function* $g_3(\eta) = w(\eta^2, \eta)$ *satisfies the estimates*

$$\left|\frac{d^i}{d\eta^i} g_3(\eta)\right| \leq M\eta^{-4N} \quad \text{for } i \leq N, \tag{4.33}$$

where N *is a sufficiently large number. Then the estimates*

$$\left|\frac{\partial^{i+l} w}{\partial \xi^i \partial \eta^l}(\xi, \eta)\right| < M\eta^{-N_3} \tag{4.34}$$

hold in the domain $D_4 = D^+ \cap \{\xi, \eta: \xi \geq \eta \geq 2\}$ *for* $i + l \leq N_2$, *where* N_2 *and* N_3 *are numbers tending to infinity as* $N \to \infty$.

PROOF. First, we prove the estimate (4.34) for $i = l = 0$. Consider the auxiliary functions $\Phi(\xi, \eta) = \overline{M} \eta^N \xi^{-2N} \pm w(\xi, \eta)$ in the domain $\{\xi, \eta: 4N\sqrt{\eta} < \xi < \eta^2, \eta > N\}$. Since the maximum principle implies that the solution $w(\xi, \eta)$ is bounded, the functions $\Phi(\xi, \eta)$ are positive on the "parabolic" boundary of this domain, i.e., for $\eta = N$, $4N^{3/2} < \xi < N^2$, for $\xi = 4N\sqrt{\eta}$, $\eta \geq N$ and for $\xi = \eta^2$, $\eta \geq N$, provided $\overline{M}$ is sufficiently large. Since

$$L_0\Phi = \overline{M}\eta^{N-1}\xi^{-2N}(-N + 2N(2N+1)\eta\xi^{-2}) \pm F(\xi, \eta)$$

$$< -\overline{M}\frac{N}{2}(\xi^2 + \eta^2)^{-N/2} + |F(\xi, \eta)|,$$

the maximum principle implies that $\Phi(\xi, \eta) > 0$ in the domain under consideration. This yields the estimate (4.34) for $i = l = 0$. The remaining estimates follow from conditions (4.33), the estimate of $w(\xi, \eta)$ just obtained, and the a priori estimates for derivatives of solutions of parabolic equations. ∎

THEOREM 4.2. *Suppose that* $w(\xi, \eta) \in C^\infty(D^+)$, $L_0 w = F(\xi, \eta)$, *and the following estimates are satisfied*:

$$\left|\frac{\partial^{i+l} F}{\partial\xi^i\partial\eta^l}(\xi, \eta)\right| \leq M(\xi^2 + \eta^2)^{-4N} \quad in \ D^+,$$

$$\left|\frac{\partial^i w}{\partial\xi^i}(\xi, 2)\right| \leq M(1 + \xi^2)^{-4N} \quad for \ \xi \leq 4, \tag{4.35}$$

$$\left|\frac{d^i g_3}{d\eta^i}(\eta)\right| \leq M\eta^{-4N} \quad for \ \eta \geq 2,$$

where $g_3(\eta) = w(\eta^2, \eta)$, $i + l \leq N$, N *is a sufficiently large number. Then the following asymptotic representations are valid for the function* $w(\xi, \eta)$:

$$\frac{\partial^{i+l} w}{\partial\xi^i\partial\eta^l}(\xi, \eta) = \frac{\partial^{i+l}}{\partial\xi^i\partial\eta^l}\sum_{j=0}^{N_1} c_j\frac{\partial^j G}{\partial\xi^j}(\xi, \eta) + \rho_{i,l,N}(\xi, \eta) \tag{4.36}$$

in D^+ *for* $i + l \leq N_2$, *where* $G(\xi, \eta)$ *is the function* (4.17),

$$|\rho_{i,l,N}(\xi, \eta)| \leq M(\xi^2 + \eta^2)^{-N_3} \quad in \ D^+, \tag{4.37}$$

and N_1, N_2, *and* N_3 *are numbers tending to infinity as* $N \to \infty$.

The proof follows from formulas (4.19)–(4.23) for the solution of the equation $L_0 w = F$ and Lemmas 4.5–4.8. Indeed, the a priori estimates for derivatives of a solution of a parabolic equation imply that $w(\xi, \eta)$ rapidly tends to zero as $\xi \to -\infty$, $2 \leq \eta \leq \eta_1$, while for $\eta \to \infty$ not only $g_3(\eta) = w(\eta^2, \eta)$ but also $g_2(\eta) = (\partial w/\partial\xi)(\eta^2, \eta)$ rapidly tends to zero. Relation (4.36) now follow from Lemmas 4.5–4.7 in the domain D_3, and from Lemma 4.8 in the domain D_4 (in this domain (4.36) reduces to the estimates (4.37)). ∎

4. The functions $w_i(\xi, \eta)$ solving the problems (4.10), (4.11) rapidly tend to zero for a fixed η and $\xi \to -\infty$, i.e., satisfy one of conditions (4.35) of Theorem 4.2. However, for $\xi = \eta^2$ they do not tend to zero, but satisfy condition (4.11). The right-hand sides of the equations for $w_i(\xi, \eta)$ vanish only for $i = 0$ and $i = 1$. For $i \geq 2$ the right-hand sides of the equations include derivatives of the preceding $w_i(\xi, \eta)$. These derivatives, as we have clarified in the preceding subsection, do not rapidly tend to zero at infinity, and can only be expanded in series of the form (4.36). Hence, in order to be able to apply Theorem 4.2 to the functions $w_i(\xi, \eta)$, one first has to construct f.a.s. of the problems (4.10), (4.11) as $\eta \to \infty$. This construction is the main subject of the present subsection.

As in §§1 and 2, the f.a.s. has to be looked for in the class of functions of the form $\eta^\beta \Phi(\theta)$, where $\theta = 2^{-1}\xi\eta^{-1/2}$. However, the set of functions here is wider than in the classes considered in §2. This can be explained as follows. The class of functions has to be chosen in such a way that conditions (4.11) are satisfied. The condition on the parabola $\xi = \eta^2$ as $\xi \to \infty$ asymptotically coincides in the first term with the condition for $\eta = 0$, $\xi > 0$. The functions should, therefore, assume the values $\eta^k = \xi^{k/2}$, as opposed to just integer powers of ξ in §2. The differentiation naturally gives rise to negative powers. In order to obtain ξ^{-1}, one has to take $\ln\xi$ for $\xi > 0$, and then to differentiate the resulting solution with respect to ξ. With this in mind, we introduce in this section another class $\mathfrak{B}$, which is wider than that of §2.

Denote by $W_{k/2}(\xi, \eta)$, where $k \geq 0$ is an integer, the solution of the boundary value problem

$$L_0 W_{k/2} = 0 \qquad \text{for } \eta > 0, \tag{4.38}$$

$$W_{k/2}(\xi, 0) = 0 \qquad \text{for } \xi < 0,$$
$$\tag{4.39}$$
$$W_{k/2}(\xi, 0) = \xi^{k/2} \quad \text{for } \xi > 0.$$

If k is an odd negative integer equal to $-(2n+1)$, we set

$$W_{k/2}(\xi, \eta) = \gamma_k \frac{\partial^n}{\partial \xi^n} W_{1/2}(\xi, \eta), \tag{4.40}$$

where the constants γ_k are chosen in such a way that equalities (4.39) are satisfied. If k is an even negative integer equal to $-2n$, we set

$$W_{k/2}(\xi, \eta) = \gamma_k \frac{\partial^n}{\partial \xi^n} W^*(\xi, \eta), \tag{4.41}$$

where $W^*(\xi, \eta)$ is the solution of the problem

$$L_0 W^* = 0 \qquad \text{for } \eta > 0,$$
$$W^*(\xi, 0) = 0 \qquad \text{for } \xi < 0,$$
$$W^*(\xi, 0) = \ln\xi \quad \text{for } \xi > 0.$$

The constants γ_k are also chosen to satisfy equalities (4.39). Under a solution of the problem for $W^*(\xi, \eta)$, and solutions of the problems (4.38), (4.39) we mean the integral

$$\int_0^\infty G(\xi - \xi_1, \eta)\varphi(\xi_1)\,d\xi_1, \tag{4.42}$$

where $G(\xi, \eta)$ is the fundamental solution (4.17), and $\varphi(\xi_1)$ the corresponding initial function. Thus, relations (4.38), (4.39) are valid for all integer k, although for $k < 1$ the solutions cannot be written in the form of the integral (4.42) understood in the usual sense (the integral diverges).

It follows from the explicit form (4.42) and formulas (4.40), (4.41) for the remaining k that

$$W_{k/2}(\xi, \eta) = \eta^{k/4}(\Phi_k(\theta) + \Psi_k(\theta) \ln \eta), \qquad (4.43)$$

where $\theta = 2^{-1}\xi\eta^{-1/2}$, and $\Psi_k(\theta)$ is not identically zero only for negative even values of k. It also follows from (4.40)–(4.42) that the functions $\Phi_k(\theta)$ and $\Psi_k(\theta)$ decay exponentially as $\theta \to -\infty$, and the function $\Psi_k(\theta)$ also decays exponentially for $\theta \to \infty$. Their asymptotics as $\theta \to \infty$ can be deduced from the explicit form of the functions $W_k(\xi, \eta)$, but it is more convenient to obtain it in another way. This will be done a bit later, in Lemma 4.9, after we introduce auxiliary classes of functions.

Denote by $\mathfrak{B}$ the linear span of the functions $P(\xi, \eta, D_\xi, D_\eta)W_{k/2}(\xi, \eta)$ for all integer k and polynomial P. The order of a function $u(\xi, \eta) \in \mathfrak{B}$ is defined as in §§1 and 2: if this function is of the form

$$u(\xi, \eta) = \eta^{k/4}(\Phi(\theta) + \Psi(\theta) \ln \eta), \qquad (4.44)$$

then its order equals $k/4$. The set of all elements in $\mathfrak{B}$ of order $k/4$ is denoted $\mathfrak{B}_{k/2}$. Clearly, if $u(\xi, \eta) \in \mathfrak{B}_{k/2}$ then $(\partial u/\partial \xi)(\xi, \eta) \in \mathfrak{B}_{k/2-1}$, $\partial u/\partial \eta \in \mathfrak{B}_{k/2-2}$, $\xi u \in \mathfrak{B}_{k/2+1}$, $\eta u \in \mathfrak{B}_{k/2+2}$. Therefore, any function in $\mathfrak{B}$ is a finite sum of functions each of which belongs to some $\mathfrak{B}_{k/2}$. We also note that since $(\partial W_0/\partial \xi)(\xi, \eta) = G(\xi, \eta)$, $(\partial^k G/\partial \xi^k)(\xi, \eta) \in \mathfrak{B}_{-1-k}$.

LEMMA 4.9. *For $\xi > \eta > 1$, $\theta \to \infty$ any function $u(\xi, \eta) \in \mathfrak{B}_{k/2}$ admits an asymptotic expansion*

$$u(\xi, \eta) = \sum_{j=0}^{\infty} a_j \xi^{k/2-2j} \eta^j. \qquad (4.45)$$

This series can be repeatedly differentiated term-by-term.

LEMMA 4.10. *For any function $z(\xi, \eta) \in \mathfrak{B}_{k/2}$ there exists a function $u(\xi, \eta) \in \mathfrak{B}_{k/2+2}$ such that*

$$L_0 u = z \quad \text{for } \eta > 0, \qquad u(\xi, 0) = 0 \quad \text{for } \xi \neq 0. \qquad (4.46)$$

LEMMA 4.11. *For any integer k, there exists an f.a.s. of the problem*

$$L_0 u = 0 \quad \text{in } D^+, \qquad (4.47)$$

$$u(\xi, 2) = 0 \quad \text{for } \xi \leq 4,$$

$$u(\xi, \sqrt{\xi}) = \xi^{k/2} \quad \text{for } \xi > 4 \qquad (4.48)$$

as $\xi^2 + \eta^2 \to \infty$, which is of the form

$$\sum_{j=0}^{\infty} v_j(\xi, \eta), \qquad v_j(\xi, \eta) \in \mathfrak{B}_{k/2-3j/2}.$$

LEMMA 4.12. *For any function* $z(\xi, \eta) \in \mathfrak{B}_{k/2}$ *there exists an f.a.s. of the problem*

$$L_0 u = z(\xi, \eta) \quad \text{for } (\xi, \eta) \in D^+,$$

$$u(\xi, 2) = 0 \cdot \text{for } \xi \leq 4, \qquad u(\xi, \sqrt{\xi}) = 0 \quad \text{for } \xi \geq 4$$

as $\xi^2 + \eta^2 \to \infty$, *which is of the form*

$$\sum_{j=0}^{\infty} v_j(\xi, \eta), \qquad v_j(\xi, \eta) \in \mathfrak{B}_{k/2+2-3j/2}. \tag{4.50}$$

PROOF OF LEMMA 4.9. The definition of the class $\mathfrak{B}$ clearly implies that it is sufficient to prove the asymptotics (4.45) for $u(\xi, \eta) = W_{k/2}$. First, construct an f.a.s. of the problem (4.38), (4.39) for $\xi \geq 1$, $\theta \to \infty$. This is easily achieved by considering the formal series $\sum_{j=0}^{\infty} v_j(\xi, \eta)$, and letting $v_0(\xi, \eta) = \xi^{k/2}$, $v_{j+1}(\xi, \eta) = \int_0^{\eta} (\partial^2 v_j / \partial \xi^2)(\xi, \eta_1) \, d\eta_1$.

As a result, one obtains the series (4.45). Now, as usual, consider the function $z_N(\xi, \eta) = \chi(\xi)[W_{k/2} - B_N W]$, where $\chi(\xi)$ is a truncating function vanishing for $\xi \leq 1/2$, and equal to 1 for $\xi \geq 1$. The construction implies that $z_N(\xi, 0) = 0$, $L_0 z_N = 0$ for $\xi \leq 1/2$, and $L_0 z_N = O(\theta^{-N_1} \xi^{k/2})$ for $\xi \geq 1/2$. The explicit representation of the solution of the Cauchy problem in the form of convolution of the fundamental solution $G(\xi, \eta)$ with the right-hand side $L_0 z_N$ easily yields $z_N(\xi, \eta) = O(\theta^{-N_2})$ for $\xi > \eta > 1$. ∎

PROOF OF LEMMA 4.10. It is evidently sufficient to prove the lemma in the case $z(\xi, \eta) = \xi^p \eta^l w(\xi, \eta)$, where p and l are nonnegative integers, and the function $w(\xi, \eta) \in \mathfrak{B}_{k/2}$ satisfies the equation $L_0 w = 0$. The proof is achieved by induction in p. If $p = 0$, one can take $u(\xi, \eta) = -(l+1)^{-1} \eta^{l+1} w(\xi, \eta)$ for the solution of equation (4.46). If $p > 0$, we look for the solution of this equation in the form $-(l+1)^{-1} \xi^p \eta^{l+1} w(\xi, \eta) + \tilde{u}(\xi, \eta)$. The resulting equation for the function $\tilde{u}(\xi, \eta)$ is

$$L_0 \tilde{u} = (l+1)^{-1} \eta^{l+1} \left(p(p-1) \xi^{p-2} w + 2p \xi^{p-1} \frac{\partial w}{\partial \xi} \right).$$

The problem for $\tilde{u}$ is hereby reduced to those already solved by the induction hypothesis. The explicit form of the solution makes it clear that $u(\xi, \eta) \in \mathfrak{B}_{k/2+2}$. ∎

PROOF OF LEMMA 4.11. Clearly, one has to set $v_0(\xi, \eta) = W_{k/2}(\xi, \eta)$. Equation (4.47) is then satisfied exactly, and condition (4.48) asymptotically. By Lemma 4.9,

$$v_0(\xi, \sqrt{\xi}) = \xi^{k/2} + \sum_{j=1}^{\infty} a_j \xi^{k/2-3j/2}, \qquad \xi \to \infty.$$

Thus, on the parabola $\xi = \eta^2$ appears a residual of smaller order for $\xi \to \infty$. Eliminating the residual $-a_1 \xi^{k/2-3/2}$, we set

$$v_1(\xi, \eta) = -a_1 W_{k/2-3/2}(\xi, \eta),$$

and the process then proceeds by induction. ∎

PROOF OF LEMMA 4.12. Take the solution of the problem (4.46) for $v_0(\xi, \eta)$ (the first term in (4.50)). This satisfies all relations in (4.49) except the last one. By Lemma 4.9, the values of the function $v_0(\xi, \eta)$ for $\eta = \sqrt{\xi}$, $\xi \to \infty$ are given by the series (4.45) with k replaced by $k + 4$ because $v_0(\xi, \eta) \in \mathfrak{B}_{k/2+2}$. Thus,

$v_0(\xi, \sqrt{\xi}) = \sum_{j=0}^{\infty} a_j \xi^{k/2+2-3j/2}$, $\xi \to \infty$. The residuals appearing in the boundary condition are eliminated according to Lemma 4.11 yielding the series (4.50). ∎

THEOREM 4.3. *For the functions* $w_i(\xi, \eta)$ *solving the problems* (4.10), (4.11) *the following asymptotic expansions are valid in the domain* D^+ *as* $\xi^2 + \eta^2 \to \infty$:

$$w_i(\xi, \eta) = \sum_{j=0}^{\infty} \psi_{j,i}(\xi, \eta), \tag{4.51}$$

where $\psi_{j,i}(\xi, \eta) \in \mathfrak{B}_{2i-j/2}$. *This series can be repeatedly differentiated term-by-term. Relation* (4.51) *has the following meaning: for any* $i \geq 0$ *for all sufficiently large* N , *the following estimate holds in* D^+ :

$$\left| w_i(\xi, \eta) - \sum_{j=0}^{N} \psi_{j,i}(\xi, \eta) \right| < M(\xi^2 + \eta^2)^{-N_1}, \tag{4.52}$$

where $N_1 \to \infty$ *as* $N \to \infty$. *Similar estimates are also valid for the derivatives of* $w_i(\xi, \eta)$.

The proof is achieved by induction using Lemmas 4.9–4.12 and Theorem 4.2. Note that the definitions of the polynomials $q_i(\xi, \eta)$ imply the equalities

$$q_i(\xi, \sqrt{\xi}) = a_i \xi^{i/2}. \tag{4.53}$$

By Lemma 4.11, construct an f.a.s. for the problem $L_0 w_0 = 0$, $w_0(\xi, \sqrt{\xi}) = -q_0$ in the form (4.51). The difference between $w_0(\xi, \eta)$ and the partial sum of the constructed series is a solution of the problem considered in Theorem 4.2. According to this theorem, the difference and its derivatives admit the asymptotic expansions (4.36) with the estimates of the remainder given by (4.52). The terms of the asymptotic expansion (4.36) belong to $\mathfrak{B}_{-k}$ for various values of k , where k are positive integers. Thus, relation (4.51) is proved for $i = 0$. Relations (4.51), (4.52) for $w_1(\xi, \eta)$ are proved in the same way. Starting with $i = 2$, the functions $w_i(\xi, \eta)$ satisfy nonhomogeneous equations.

Suppose that relations (4.51), (4.52) are proved for all $i \leq n-1$. It follows from (4.10) that the function $w_n(\xi, \eta)$ satisfies the following equation in D^+ :

$$L_0 w_n = -\frac{\partial^2 w_{n-2}}{\partial \eta^2} + \sum_{\substack{s<n \\ 2j+l+s=n}} \frac{1}{j!l!} \frac{\partial^{j+l} a}{\partial x^j \partial y^l}(0, 0) \frac{\partial w_s}{\partial \eta}(\xi, \eta) \xi^j \eta^l.$$

By the inductive hypothesis, each term in the right-hand side of this equation admits, as $\xi^2 + \eta^2 \to \infty$, the asymptotic expansion $\sum_{j=0}^{\infty} \psi_{n,j}^*(\xi, \eta)$, where $\psi_{n,j}^*(\xi, \eta) \in \mathfrak{B}_{m-j/2}$. For the first term one has $m = 2(n-2) - 4$, and for each term of the sum $m = j + 2l + 2s - 2 = 2n - 3j - 2 \leq 2n - 2$.

Thus, the right-hand side admits the asymptotic expansion $\sum_{j=0}^{\infty} \psi_{n,j}(\xi, \eta)$, where $\psi_{n,j}(\xi, \eta) \in \mathfrak{B}_{2n-2-2j}$. By Lemma 4.12, construct for each term of this series the f.a.s. of the problem $L_0 u = \psi_{n,j}$, $u(\xi, 2) = 0$, $u(\xi, \sqrt{\xi}) = 0$ in the domain D^+. Each such solution belongs to $\mathfrak{B}_{2n-j/2}$. The f.a.s. of the nonhomogeneous equation with zero boundary conditions is hereby constructed. By adding to it the f.a.s. of the homogeneous equation subject to the condition $w_n(\xi, \sqrt{\xi}) = -q_n(\xi, \sqrt{\xi})$, constructed according to Lemma 4.11, one obtains the f.a.s. of the complete problem for $w_n(\xi, \eta)$. Now an application of Theorem 4.2, as in the case of $w_0(\xi, \eta)$, yields relations (4.51), (4.52) for $w_n(\xi, \eta)$. ∎

REMARK. A careful examination of the subscripts j in formula (4.51) reveals (see (4.53) and Lemmas 4.11, 4.12) that the f.a.s. include only those terms $\psi_{i,j}(\xi, \eta)$ for which j is a multiple of 3. The addition of the series (4.36) to it gives rise to the terms $\psi_{i,j}(\xi, \eta)$ for even j. There are no other terms in the series (4.51). Moreover, for $w_0(\xi, \eta)$ the series consist only of even j. All terms of the f.a.s. with the numbers $j \equiv 0 \mod 3$ are easily computed explicitly through the coefficients of the outer expansion (4.3), or, to be more precise, through their terms at the origin. The remaining terms of the series (4.51) (appearing, according to Theorem 4.2, due to the solutions of the problems with the data rapidly decaying at infinity) are found with much more difficulty. The explicit formulas for these terms contain not only the Taylor coefficients of the functions u_{3k}, z_{3k}, but also absolute constants obtained from the solutions of some special problems. The problem of deriving explicit formulas for these constants is not considered here.

5.   The study of the asymptotics of the functions $w_i(\xi, \eta)$ at infinity is complete, making it possible to construct the asymptotic expansion of the solution $u(x, y, \varepsilon)$ in a neighborhood of the singular characteristic α in the same way as in §2. The only difference is that, due to representation (4.43), the change of variables gives rise to terms containing $\ln \varepsilon$. The exponents in the powers of ε are also altered, because the coefficient before the Laplace operator is denoted by ε^2 in §2, and by ε^3 in this section. Finally, since the series (4.3) has no singularities anywhere in Ω_δ, it is more convenient to look for the inner expansion of the solution $u(x, y, \varepsilon)$ in a neighborhood of the straight-line segment $\alpha = \{x, y: x = 0, 0 \leq y \leq \varphi_2(0)\}$ in the form of the sum of the series (4.3) and the series

$$V = \sum_{k=0}^{\infty} \varepsilon^{k/4} v_k(\zeta, y) + \ln \varepsilon \sum_{k=1}^{\infty} \varepsilon^{k/4} v_{k,1}(\zeta, y), \qquad (4.54)$$

where $\zeta = \varepsilon^{-3/2} x$. Thus, the series V has to be matched to the series (4.4) as $\zeta \to \infty$, to zero as $\zeta \to -\infty$, and, most important, to the series (4.9) as $y \to 0$. The equations and initial conditions for $v_k(\zeta, y)$ and $v_{k,1}(\zeta, y)$

are obtained in the usual way:

$$L_3 v_0 \equiv \frac{\partial^2 v_0}{\partial \zeta^2} - a(0, y)\frac{\partial v_0}{\partial y} = 0,$$

$$L_3 v_k = -\frac{\partial^2 v_{k-12}}{\partial y^2} + \sum_{j \geq 1} \frac{\zeta^j}{j!}\frac{\partial^j a}{\partial x^j}(0, y)\frac{\partial v_{k-6j}}{\partial y},$$

$$v_k(\zeta, y) \equiv 0 \quad \text{and} \quad v_{k,1}(\zeta, y) \equiv 0 \quad \text{for } k < 0,$$

$$v_{k,1}(\zeta, 0) = 0, \qquad v_k(\zeta, 0) = 0 \quad \text{for } \zeta < 0,$$

$$v_k(\zeta, 0) = - \sum_{6i+3j+12l=k} \frac{1}{i!j!}\frac{\partial^{i+j} u_{3l}}{\partial x^i \partial y^j}(0, 0)\zeta^{i+j/2} \quad \text{for } \zeta > 0.$$

(4.55)

The last relation is obtained from the formal equality

$$V = -\sum_{k=0}^{\infty} \varepsilon^{3k} u_{3k}(x, y) \quad \text{for} \quad y = \sqrt{x}, \quad x = \varepsilon^{3/2}\zeta.$$

The same equality implies the boundary condition (4.11) for $w_i(\xi, \eta)$ so that for $y = 0$, $\zeta \to 0$ the series (4.54) is automatically matched to the series (4.9) as $\xi = \eta^2 \to \infty$.

The functions $v_k(\xi, \eta)$, $v_{k,1}(\zeta, y)$ are constructed as in §§1 and 2. Rewrite the series (4.9) replacing the functions $w_i(\xi, \eta)$ with their asymptotic expansions (4.51), the functions $\psi_{i,j}(\xi, \eta)$ in this expansion with their representations (4.44), and, finally, making the change of variables by the formulas $\xi = \varepsilon^{-1/2}\zeta$, $\eta = \varepsilon^{-1}y$, $\theta = 2^{-1}\xi\eta^{-1/2} = 2^{-1}\zeta y^{-1/2}$:

$$W = \sum_{j=0}^{\infty} \varepsilon^i \sum_{j=0}^{\infty} \eta^{i-j/4}\left(\Phi_{i,j}(\theta) + \Psi_{i,j}(\theta)\ln \eta\right) = \sum_{k=0}^{\infty} \varepsilon^{k/4}(V_k + V_{k,1}\ln \varepsilon).$$

Here V_k and $V_{k,1}$ are the formal series

$$V_k = y^{-k/4}\sum_{i=0}^{\infty} y^i\left(\Phi_{i,k}(\theta) + \Psi_{i,k}(\theta)\ln y\right), \qquad (4.56)$$

$$V_{k,1} = -y^{-k/4}\sum_{i=0}^{\infty} y^i\Psi_{i,k}(\theta). \qquad (4.57)$$

THEOREM 4.4. *There exist functions* $v_k(\zeta, y)$, $v_{k,1}(\zeta, y)$ *solving the problem* (4.55) *and having the asymptotic expansions* (4.56), (4.57) *as* $y \to 0$. *The series* (4.54) *and* (4.9) *satisfy the matching condition.*

The proof of the theorem virtually coincides with that of Theorem 1.3.

TABLE 9

V ╲ W	$w_0(\xi,\eta)$	$\varepsilon w_1(\xi,\eta)$	$\varepsilon^2 w_2(\xi,\eta)$	...
$v_0(\zeta,y)$	$\Phi_{0,0}(\theta)$ ----- $\Phi_{0,0}(\theta)$	$\varepsilon\eta\Phi_{1,0}(\theta)$ ----- $y\Phi_{1,0}(\theta)$	$\varepsilon^2\eta^2\Phi_{2,0}(\theta)$ ----- $y^2\Phi_{2,0}(\theta)$	...
$\varepsilon^{1/2}v_2(\zeta,y)$	$\eta^{-1/2}\Phi_{0,2}(\theta)$ ----- $\varepsilon^{1/2}y^{-1/2}\Phi_{0,2}(\theta)$	$\varepsilon\eta^{1/2}\Phi_{1,2}(\theta)$ ----- $\varepsilon^{1/2}y^{1/2}\Phi_{1,2}(\theta)$	$\varepsilon^2\eta^{3/2}\Phi_{2,2}(\theta)$ ----- $\varepsilon^{1/2}y^{3/2}\Phi_{2,2}(\theta)$	...
$\varepsilon^{3/4}v_3(\zeta,y)$		$\varepsilon\eta^{1/4}\Phi_{1,3}(\theta)$ ----- $\varepsilon^{3/4}y^{1/4}\Phi_{1,3}(\theta)$	$\varepsilon^2\eta^{5/4}\Phi_{2,3}(\theta)$ ----- $\varepsilon^{3/4}y^{5/4}\Phi_{2,3}(\theta)$	...
$\varepsilon v_4(\zeta,y)$	$\eta^{-1}\Phi_{0,4}(\theta)$ ----- $\varepsilon y^{-1}\Phi_{0,4}(\theta)$	$\varepsilon\Phi_{1,4}(\theta)$ ----- $\varepsilon\Phi_{1,4}(\theta)$	$\varepsilon^2\eta\Phi_{2,4}(\theta)$ ----- $\varepsilon y\Phi_{2,4}(\theta)$	...
$\varepsilon^{3/2}\ln\varepsilon\, v_{6,1}(\zeta,y)$		$-\varepsilon^{3/2}y^{-1/2}\ln\varepsilon\,\Psi_{1,6}(\theta)$	$-\varepsilon^{3/2}y^{1/2}\ln\varepsilon\,\Psi_{2,6}(\theta)$	
$\varepsilon^{3/2}v_6(\zeta,y)$	$\eta^{-3/2}\Phi_{0,6}(\theta)$ ----- $\varepsilon^{3/2}y^{-3/2}\Phi_{0,6}(\theta)$	$\varepsilon\eta^{-1/2}(\Phi_{1,6}(\theta)+\ln\eta\Psi_{1,6}(\theta))$ ----- $\varepsilon^{3/2}y^{-1/2}(\Phi_{1,6}(\theta)+\ln y\Psi_{1,6}(\theta))$	$\varepsilon^2\eta^{1/2}(\Phi_{2,6}(\theta)+\ln\eta\Psi_{2,6}(\theta))$ ----- $\varepsilon^{3/2}y^{1/2}(\Phi_{2,6}(\theta)+\ln y\Psi_{2,6}(\theta))$	—
...	...	...	...	...

Table 9 is the matching table for the series (4.9) and (4.54). It is quite similar to Table 8 and requires no additional explanation. We only note that, according to the remark to Theorem 4.3, many of the functions $v_k(\xi,\eta)$ and $v_{k,1}(\xi,\eta)$ vanish identically. The corresponding rows in the table are naturally omitted.

The asymptotics of the solution $u(x,y,\varepsilon)$ is virtually constructed. In the vicinity of σ_2 (see Figure 23) one has to add the boundary layer series coinciding with the series S^* and $\widetilde{S}$ in §2. Their construction and purpose are perfectly clear, and in what follows all the statements are formulated for a domain detached from the upper boundary, namely for $\Omega_{\delta,\mu}$, where $\Omega_{\delta,\mu}=\Omega_\delta\cap\{x,y:y<\mu\}$. It is assumed that the number $\min_{|x|\le\delta}\varphi_2(x)-\mu$ is positive and sufficiently small.

The series W is matched to the series (4.6) in the vicinity of the lower branch of the parabola $x = y^2$ for $x \to 0$. It can also be proved that the series W is matched to the series (4.4) near the upper branch of the same parabola as $x \to 0$, and that the series V is matched to the series Z as $x \to 0$, $\zeta \to \infty$, and $y > \varepsilon^\nu$, $\nu < 1$. We will, however, take another route using Lemma 5.2 which makes it possible to estimate the solution of equation (4.1) inside a narrow strip $|x - x_0| < \rho$.

THEOREM 4.5. *Suppose that ν is a number such that $0 < \nu < 3/2$, and N is sufficiently large. Let $u(x, y, \varepsilon)$ be a solution of the problems* (4.1), (4.2), *U, W, and V the series* (4.3), (4.9) *and* (4.54) *constructed above, and*

$$T_N(x, y, \varepsilon) = A_{N,x,y}U + A_{N,\xi,\eta}W + A_{N,\zeta,y}V - A_{N,\xi,\eta}A_{N,\zeta,y}V - u(x, y, \varepsilon).$$

Then the following estimate is valid in the domain $\Omega_{\delta,\mu} \cap \{x, y : |x| < \varepsilon^\nu\}$:

$$|T_N(x, y, \varepsilon)| < M\varepsilon^{N_1},$$

where $N_1 \to \infty$ as $N \to \infty$, and the constant M depends only on μ, ν, and N.

PROOF. As in the proof of Theorem 1.4, it is easy to verify that $|T_N(x, y, \varepsilon)| < M\varepsilon^{N_2}$ on the curve σ_1, and on the parabola $x = y^2$ for $|x| < 2\varepsilon^\nu$ (see Figure 23). (Here and in what follows N_i are numbers tending to infinity as $N \to \infty$.) The functions $T_N(x, y, \varepsilon)$ are bounded in the domain $\overline{\Omega}_\delta$ uniformly with respect to ε: $|T_N(x, y, \varepsilon)| < M$. It remains to apply Lemma 5.2 setting $\rho = 2\varepsilon^\nu$, $\gamma = \varepsilon^\nu$, $\tau = \mu + \gamma$, and choosing the constant λ in such way that $\nu < \lambda < 3/2$. (Recall that since the coefficient of the second order operator in Lemma 5.2 is denoted by ε, while in this section it is denoted by ε^3, the condition $\lambda < 1/2$ is replaced with the condition $\lambda < 3/2$.) ∎

THEOREM 4.6. *Suppose that U, W, V and the constants ν, μ, and N are the same as in Theorem 4.5, and Z and S are the series* (4.4) *and* (4.6). *Then in the domain $\Omega_{\delta,\mu} \cap \{x, y : x < -\varepsilon^\nu\}$ the estimate*

$$|A_{N,x,y}U - u(x, y, \varepsilon)| < M\varepsilon^{N_1}$$

holds. In the domain $\Omega_{\delta,\mu} \cap \{x, y : x > \varepsilon^\nu, y \geq \sqrt{x}\}$ the estimate

$$|A_{N,x,y}U + A_{N,x,y}Z - u(x, y, \varepsilon)| < M\varepsilon^{N_1}$$

is valid, while in the domain $\Omega_{\delta,\mu} \cap \{x, y : x > \varepsilon^\nu, y \leq -\sqrt{x}\}$ the estimate

$$|A_{N,x,y}U + A_{N,x,\tau}S - u(x, y, \varepsilon)| < M\varepsilon^{N_1}.$$

The proof is completely analogous to that of the preceding theorem. One can easily verify that the values of each of the functions whose absolute values appear in the left-hand sides of the inequalities are small on the corresponding sections of the

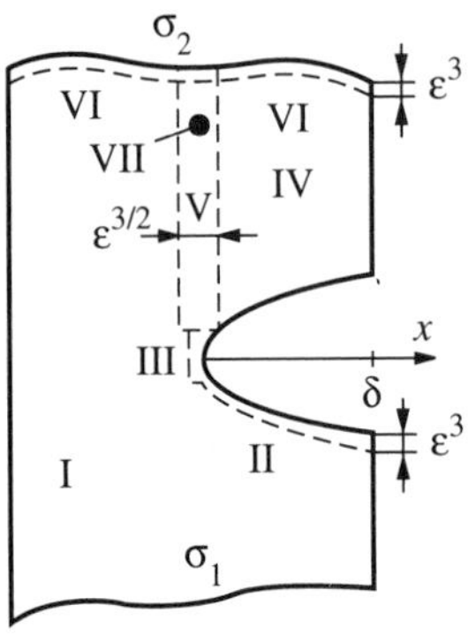

I	U	x, y	(4.3)
II	$U + S$	$x, (-y - \sqrt{x})\varepsilon^{-3}$	(4.6)
III	$U + W$	$\xi = \varepsilon^{-2}x, \eta = \varepsilon^{-1}y$	(4.9)
IV	$U + Z$	x, y	(4.4)
V	$U + V$	$\zeta = \varepsilon^{-3/2}x, y$	(4.54)
VI	$U + S^*$	$x, (-y + \varphi_2(x))\varepsilon^{-3}$	
VII	$U + V + \widetilde{S}$	$\zeta, (-y + \varphi_2(x))\varepsilon^{-3}$	

FIGURE 25

boundary $\partial\Omega$, as well as values of the operator applied to this function, while on the vertical sections of the boundary of the corresponding domain and for $y = \mu$ the function is bounded. Now it only remains to apply Lemma 5.2. The constants can be chosen in the following way: $\rho = \frac{1}{2}\varepsilon^\nu$, $\gamma = \frac{1}{4}\varepsilon^\nu$, $\tau = \mu + \gamma$, $\nu < \lambda < 3/2$. Now Ω_ρ denotes the intersection of the domain Ω with the strip $|x - x_0| < \rho$, where (x_0, y) is the point at which the estimate of the solution $u(x, y, \varepsilon)$ is to be obtained (and not with the strip $|x| < \rho$). Clearly, this modification of the domain Ω_ρ does not affect the validity of Lemma 5.2. ∎

Figure 25 shows various subdomains of Ω_δ. They are labeled by Roman numerals, and the legends to the right of the figure present the asymptotics of the solution in these subdomains, the corresponding inner variables, and the formulas defining the respective series.

§5. Remarks

1. First, a few words are in order concerning the asymptotics of the solution of the problem (0.3), (0.4) at "corner" points of the boundary of type c_5 in Figure 15. For equation (2.1) they are points on the upper part of the boundary, but their definition is, of course, invariant under the change of independent variables. Suppose that the operator $\mathcal{M}$ in equation (0.1) is locally negative definite (for an elliptic second order operator it means that the matrix of the coefficients of the second order derivatives is positive definite, but the notion can also be defined for operators of a more general type), and the first order operator is of the form $b(x, y)d/ds$, where d/ds is the derivative along the characteristic of the operator l. Choose the direction s and, therefore, the direction on the characteristic for which the coefficient $b(x, y)$ is negative (we assume that $b(x, y) \neq 0$ everywhere; in the examples of §§1–4 the direction s coincides with the direction of the y-axis, in §§1, 2, 4 $b(x, y) = -a(x, y)$, in §3 $b(x, y) \equiv -1$).

In general position the corner points of the boundary can be of one of the three types:

(1) the characteristic issues from the corner point into the domain Ω, and is, therefore, a singular one;

(2) the characteristic passing through the corner point behaves as in Figure 17; this means that a section of the characteristic lying in Ω enters the corner point, and its continuation also lies in Ω so that the characteristic is also a singular one;

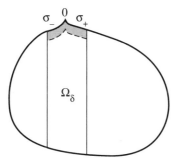

FIGURE 26

(3) the characteristic enters the corner point from inside Ω, but its continuation lies outside Ω.

The first and second cases were considered in §2 (see Figures 15 and 17) and required a rather complicated investigation. In the third case the investigation is much simpler. There is a usual exponential boundary layer on the sides of the angle (cf. the series S in formulas (1.5), (2.6), and (4.6)). Since the boundary at the corner point is not smooth, the boundary layer on one side of the angle gives rise to a residual in the boundary condition on the other side in precisely the same way as in Chapter I, §2, Example 4 (cf. Figures 4 and 26). Thus, it only remains to eliminate the resulting residual in the boundary condition.

Suppose, for example, that the corner point lies at the origin, and that the boundary in its neighborhood consists of two arcs: $\sigma_- = \{x, y: y = \varphi_-(x), -\delta \le x \le 0\}$ and $\sigma_+ = \{x, y: y = \varphi_+(x), 0 \le x \le \delta\}$, where $\varphi_\pm(0) = 0$, $\varphi'_-(0) > \varphi'_+(0)$. Suppose that the equation is of the form (2.1), and the points of the domain Ω_δ satisfy $y < \varphi_\pm(x)$. There is a boundary layer in the vicinity of the arcs $\sigma_\pm$, where the f.a.s. $U + S_\pm$ is already constructed (as in each of the preceding sections of Chapter IV). Here U is the outer expansion, and

$$S_\pm = \sum_{k=0}^{\infty} \varepsilon^{2k} s_{2k}^\pm(x, \tau^\pm), \qquad \tau^\pm = \varepsilon^{-2}(-y + \varphi_\pm(x)).$$

The functions $s_{2k}^\pm \in C^\infty$ decay exponentially as $\tau \to \infty$.

The series $U + S_\pm$ formally satisfy the condition $u(x, y, \varepsilon)|_{\partial\Omega} = 0$, each on its own arc $\sigma_\pm$. For simplicity, we will assume that the arcs $\sigma_\pm$ are straight-line segments, which involves no loss of generality because the equation can be reduced to this case by an appropriate change of the independent variables. The change of variables evidently modifies equation (2.1), replacing the Laplace operator with an elliptic operator with variable coefficients, but this does not affect the analysis in any noticeable way.

Thus, suppose that the equation of σ_- is of the form $y = \varphi_- \cdot x$, and that of σ_+ is $y = \varphi_+ \cdot x$, where $\varphi_- > \varphi_+$. We will assume that the coefficients of the series U are defined everywhere for $|x| \le \delta$, and the coefficients of the series $S_\pm$ are defined everywhere for $|x| \le \delta$, $y \le \varphi_\pm \cdot x$, so that the series $U + S_\pm$ formally vanish for $y = \varphi_\pm \cdot x$. Then the series $U + S_+ + S_-$ is an f.a.s. of the equation giving rise to the residual in the boundary condition equal to

$$\sum_{k=0}^{\infty} \varepsilon^{2k} s_{2k}^+(x, \varepsilon^{-2}(\varphi_+ - \varphi_-)x) \quad \text{for } -\delta \le x \le 0, \quad y = \varphi_- \cdot x,$$

$$\sum_{k=0}^{\infty} \varepsilon^{2k} s_{2k}^-(x, \varepsilon^{-2}(\varphi_- - \varphi_+)x) \quad \text{for } 0 \le x \le \delta, \quad y = \varphi_+ \cdot x. \tag{5.1}$$

Thus, it only remains to construct an f.a.s. of the equation satisfying conditions (5.1). This can easily be done along the same lines as in Chapter I, §2, Example 4, and in Chapter IV, §§1 and 3. After the change of variables $x = \varepsilon^2 \xi$, $y = \varepsilon^2 \eta$ the coefficients of the inner expansion

$$W = \sum_{k=0}^{\infty} \varepsilon^{2k} w_{2k}(\xi, \eta)$$

must satisfy the equations

$$\Delta w_{2k} - \frac{\partial}{\partial \eta} w_{2k} = F_k(\xi, \eta), \qquad k \ge 0, \tag{5.2}$$

at the boundary conditions

$$w_{2k}(\xi, \varphi_- \cdot \xi) = g_k^-(\xi) \quad \text{for } -\infty < \xi \le 0,$$

$$w_{2k}(\xi, \varphi_+ \cdot \xi) = g_k^+(\xi) \quad \text{for } 0 \le \xi < \infty. \tag{5.3}$$

Here $F_k(\xi, \eta)$ depend on the functions $w_{2j}(\xi, \eta)$ with $j < k$ and their derivatives, and the $g_k^\pm(\xi)$ are obtained by expanding the functions $s_k^\pm(\varepsilon^2 \xi, \pm(\varphi_+ - \varphi_-)\xi)$ in Taylor series with respect to the first argument. Consequently, the functions $g_k^\pm(\xi)$ decay exponentially at infinity.

One has to find the functions $w_{2k}(\xi, \eta)$ in the domain $\{\xi, \eta : \xi \in \mathbf{R}^1, \eta \le \varphi_+ \cdot \xi, \eta \le \varphi_- \cdot \xi\}$. In this domain, the problems (5.2), (5.3) can be shown to have solutions which also decay exponentially as $\xi^2 + \eta^2 \to \infty$. The proof is virtually the same as in §2 (cf. Lemmas 2.1, 2.2). Thus, in the case of corner points of the third type (such as the point c_5 in Figure 15) one has to add to the previous asymptotics only the corner boundary layer of exponential type affecting just a small neighborhood of the corner point.

2. In this subsection we prove lemmas on localization of asymptotic expansions for equation (0.3) repeatedly mentioned above and used in this chapter for the justification of the asymptotics. Let $\Omega \subset \mathbf{R}^2$ be a bounded domain with piecewise smooth boundary,

$$\mathscr{L}_\varepsilon = \varepsilon \mathscr{M} - a(x, y)\frac{\partial}{\partial y} + b(x, y), \tag{5.4}$$

where $\mathscr{M}$ is an elliptic second order operator of the form (0.2) with continuous coefficients, $a_{i,j}(x, y)$ is a positive definite matrix, $a(x, y) \in C(\overline{\Omega})$, $b(x, y) \in C(\Omega)$, $a(x, y) > 0$ in $\overline{\Omega}$, and $\varepsilon > 0$. Denote by Ω_ρ the domain $\Omega \cap \{x, y : |x| < \rho\}$. Suppose, to be specific, that the origin lies in the domain Ω.

LEMMA 5.1. *There exists* $\varepsilon_0 > 0$ *such that for all* $\varepsilon < \varepsilon_0$ *for any function* $u(x, y) \in C^2(\Omega)$ *the estimate*

$$\max_{\Omega_{\rho-\gamma}} |u(x, y)| \le M\beta_1 + M(\beta_1 + \beta_2)\exp(\gamma \varepsilon^{-\lambda}) \tag{5.5}$$

holds, where

$$\beta_1 = \max_{\partial\Omega\cap\Omega_\rho} |u(x,y)| + \max_{\overline{\Omega}_\rho} |\mathscr{L}_\varepsilon u(x,y)|, \qquad \beta_2 = \max_{|x|=\rho} |u(x,y)|,$$

γ *and* λ *are any numbers such that* $0 < \lambda < 1/2$, $0 < \gamma < \rho$, *and the constant* M *depends just on the coefficients of the operator* $\mathscr{M}$, *the coefficients* $a(x,y)$, $b(x,y)$, *and the domain* Ω.

PROOF. We begin with the construction of a function $v(x,y)$ such that $v(x,y) > 1$ and $\mathscr{L}_\varepsilon v(x,y) < -1$ in $\overline{\Omega}$. For $v(x,y)$ one can take the function $\mu_1 e^{\mu_2 y}$, where the constants μ_1 and μ_2 depend only on Ω and the above mentioned coefficients. Indeed,

$$\mathscr{L}_\varepsilon e^{\mu_2 y} = e^{\mu_2 y}\left[-a(x,y)\mu_2 + b(x,y) + O(\varepsilon(1+\mu_2^2)) \right] < -e^{-\mu_2 y}$$

for μ_2 sufficiently large, and ε sufficiently small. Now consider the functions

$$\Phi_\pm(x,y) = (\beta_1 + \beta_2)\frac{\cosh(x\varepsilon^{-\lambda})}{\cosh(\rho\varepsilon^{-\lambda})}v(x,y) + \beta_1 v(x,y) \pm u(x,y).$$

Evidently, one has $\Phi_\pm(x,y) \geq 0$ on the boundary of Ω_ρ, and in Ω_ρ itself

$$\mathscr{L}_\varepsilon \Phi_\pm(x,y) < \frac{\beta_1 + \beta_2}{\cosh(\rho\varepsilon^{-\lambda})}\cosh(x\varepsilon^{-\lambda}))\left[-1 + O\left(\varepsilon^{1-2\lambda}\right) \right] - \beta_1 + |\mathscr{L}_\varepsilon u| < 0$$

for all sufficiently small ε. Consequently, the estimate (5.5) follows from the maximum principle. ∎

Introduce another notation for a subdomain of Ω. Let $\Omega_{\rho,\tau} = \Omega_\rho \cap \{x, y: y < \tau\}$. It is assumed that the interval $\{x, y: y = \tau, |x| < \rho\}$ belongs to Ω_ρ.

LEMMA 5.2. *There exists* $\varepsilon_0 > 0$ *such that for all* $\varepsilon < \varepsilon_0$ *for any positive numbers* γ *and* λ, *where* $\lambda < 1/2$, *and* γ *is sufficiently small the estimate*

$$\max_{\Omega_{\rho-\gamma,\tau-\gamma}} |u(x,y)| \leq M\beta_1 + M(\beta_1 + \beta_2)\exp(-\gamma\varepsilon^{-\lambda})$$

holds for any function $u(x,y) \in C^2(\overline{\Omega})$. *Here*

$$\beta_1 = \max_{\partial\Omega\cap\Omega_{\rho-\gamma,\tau-\gamma}} |u(x,y)| + \max_{\overline{\Omega}_{\rho,\tau}} |\mathscr{L}_\varepsilon u|, \qquad \beta_2 = \max_{|x|=\rho} |u(x,y)|,$$

and the constant M *is the same as in Lemma* 5.1.

PROOF. It is sufficient to consider the auxiliary functions

$$(\beta_1 + \beta_2)\frac{\cosh(x\varepsilon^{-\lambda})}{\cosh(\rho\varepsilon^{-\lambda})}v(x,y) + \beta_1 v(x,y) + (\beta_1 + \beta_2)\exp((y-\tau)\varepsilon^{-\lambda}) \pm u(x,y),$$

where $v(x,y)$ is the same function as in Lemma 5.1 and apply the maximum principle. ∎

All the problems considered in this chapter belong to a rather narrow class of differential equations, viz., elliptic second order equations. Nevertheless, this approach to the construction of the asymptotics can be applied, without serious modifications, to elliptic equations of higher order, as well as to other types of equations containing a small parameter. However, the justification of the asymptotics was based on the maximum principle whose applicability is narrow. Other classes of problems call for different methods for estimating their solutions. In particular, energy estimates are widely used. We present, for illustration, the proof of a lemma which is similar to

Lemma 5.1 and also guarantees the localization principle for the asymptotics of the solutions of equation (0.3) near the characteristic of the limiting equation.

Thus, consider the operator (5.4) in a bounded domain Ω. The coefficients of the operator and the boundary $\partial\Omega$ are assumed to be smooth, $a(x, y) \geq a_0 > 0$. The domain Ω_ρ is defined as above. Introduce the additional notation

$$\|u\|^2_{0, \delta} = \iint_{\Omega_\delta} u^2(x, y) \, dx \, dy,$$

$$\|u\|^2_{1, \delta} = \iint_{\Omega_\delta} \left[\left(\frac{\partial u}{\partial x}\right) + \left(\frac{\partial u}{\partial y}\right) + u^2 \right] dx \, dy.$$

LEMMA 5.3. *Let ν, ρ, and δ be positive numbers such that $\nu < 1/2$, $\delta - \rho > \varepsilon^\nu$, and let n be a positive integer. Then there exists $\varepsilon_0 > 0$ depending just on ν and n such that for all $\varepsilon < \varepsilon_0$ and any function $u(x, y) \in C^2(\overline{\Omega})$ vanishing on $\partial\Omega$ the estimate*

$$\|u\|^2_{0, \rho} + \varepsilon\|u\|^2_{1, \rho} \leq \varepsilon^{(n+1)(1/2-\nu)}\|u\|^2_{1, \delta} + M\|u\|_{0, \delta} \|\mathscr{L}_\varepsilon u\|_{0, \delta} \qquad (5.6)$$

holds. Here the constant M depends just of the domain Ω, the coefficients, and the numbers ν and n.

PROOF. Denote by M positive constants independent of ε, $u(x, y)$, ρ, and δ. Denoting $\mathscr{L}_\varepsilon u = f$, write out the equation satisfied by the function $u(x, y)$:

$$\varepsilon\mathscr{M}u - a(x, y)\frac{\partial u}{\partial y} + b(x, y)u = f(x, y, \varepsilon). \qquad (5.7)$$

Let $\chi(z) \in C^\infty(\mathbf{R}^1)$ be a truncating function such that $\chi'(z) \geq 0$, $\chi(z) \equiv 0$ for $z \leq 0$, $\chi(z) \equiv 1$ for $z \geq 1$. Then the function

$$\psi(x, \mu, \gamma) = \chi\left(\frac{x+\mu}{\mu-\gamma}\right)\chi\left(\frac{\mu-x}{\mu-\gamma}\right),$$

where $0 < \gamma < \mu$ vanishes for $|x| \geq \mu$, and equals 1 for $|x| \leq \gamma$. Evidently, $|d^i\psi/dx^i| \leq M(\mu-\gamma)^{-i}$. Multiply equation (5.7) by $\psi^2(x, \mu, \gamma)\exp(\lambda y)\, u(x, y)$, where $0 < \gamma < \mu \leq \delta$ and the constant $\lambda > 0$ is to be chosen later, and integrate over the domain Ω_δ:

$$\varepsilon\iint_{\Omega_\delta} \psi^2 e^{\lambda y} u\mathscr{M}u \, dx \, dy - \iint_{\Omega_\delta} a(x, y)\psi^2 e^{\lambda y} u\frac{\partial u}{\partial y} \, dx \, dy$$

$$+ \iint_{\Omega_\delta} b(x, y)\psi^2 e^{\lambda y} u^2 \, dx \, dy = \iint_{\Omega_\delta} \psi^2 e^{\lambda y} uf \, dx \, dy. \quad (5.8)$$

Integration by parts transforms the second term in the left-hand side of this equality in the following way:

$$-\iint_{\Omega_\delta} a(x, y)\psi^2 e^{\lambda y} u\frac{\partial u}{\partial y} \, dx \, dy$$

$$= \frac{1}{2}\iint_{\Omega_\delta} u^2\psi^2 e^{\lambda y}\frac{\partial a}{\partial y} \, dx \, dy + \frac{1}{2}\iint_{\Omega_\delta} u^2\psi^2\lambda e^{\lambda y}a \, dx \, dy.$$

Choose and fix a constant λ such that

$$a(x, y)\lambda > \max_{\overline{\Omega}}\left|\frac{\partial a}{\partial y}(x, y)\right| + \max_{\overline{\Omega}}|b(x, y)| + 2.$$

Owing to this inequality, one obtains from (5.8) the following relation

$$\varepsilon \iint_{\Omega_\delta} \psi^2 e^{\lambda y} u \mathcal{M} u \, dx \, dy + \iint_{\Omega_\delta} \psi^2 u^2 e^{\lambda y} \, dx \, dy \le M \|\psi u\|_{0,\delta} \|f\|_{0,\delta}. \quad (5.9)$$

Integrating the first integral in the left-hand side of (5.9) by parts, one arrives at the inequality

$$\varepsilon \left(M^{-1} \|u\|_{1,\gamma}^2 - M \|\psi u\|_{0,\delta} \|u\|_{1,\mu} (\mu - \gamma)^{-1} \right) + \|\psi u\|_{0,\delta}^2 \le M \|\psi u\|_{0,\delta} \|f\|_{0,\delta},$$

which easily implies the estimate

$$\|\psi u\|_{0,\delta}^2 + 2\varepsilon \|u\|_{1,\gamma}^2$$
$$\le \varepsilon^{-3/2-\nu} \|u\|_{1,\mu}^2 + M(\varepsilon^{1/2+\nu}(\mu-\gamma)^{-2} \|\psi u\|_{0,\delta}^2 + \|\psi u\|_{0,\delta} \|f\|_{0,\delta}).$$

Since $\psi(x, \mu, \gamma) \equiv 1$ for $|x| \le \gamma$, one has

$$[1 - M\varepsilon^{1/2+\nu}(\mu-\gamma)^{-2}] \|u\|_{0,\gamma}^2 + 2\varepsilon \|u\|_{1,\gamma}^2 \le \varepsilon^{3/2-\nu} \|u\|_{1,\mu}^2 + M\|u\|_{0,\delta} \|f\|_{0,\delta}.$$

Writing these inequalities for $\gamma = \gamma_i = \delta - (\delta - \rho)2^{-i}$, $\mu = \mu_i = \gamma_{i+1}$, $i = 0, 1, 2, \ldots, n$, multiplying both parts of each inequality by $\varepsilon^{(1/2-\nu)i}$ and adding them up term-by-term one obtains the inequality

$$\sum_{i=0}^{n} \varepsilon^{(1/2-\nu)i} (1 - M\varepsilon^{1/2} 2^{2i+2} (\delta-\rho)^{-2}) \|u\|_{0,\gamma_i}^2 + \varepsilon \|u\|_{1,\rho}^2$$

$$\le \varepsilon^{(n+3)/2 - n\nu - \nu} \|u\|_{1,\gamma_{n+1}}^2 + M \sum_{i=0}^{n} \varepsilon^{(1/2-\nu)i} \|u\|_{0,\delta} \|f\|_{0,\delta}. \quad (5.10)$$

Since $\varepsilon^{1/2+\nu}(\delta-\rho)^{-2} < \varepsilon^{1/2-\nu}$, one can choose ε_0 in such a way that

$$1 - M\varepsilon^{1/2+\nu} 2^{2(n+1)} (\delta-\rho)^{-2} > 1/2.$$

Consequently, (5.10) implies (5.6). ∎

It follows from Lemma 5.3 that if $\|\mathcal{L}_\varepsilon u\|_{0,\delta} < M\varepsilon^N$, and $\|u\|_{1,\delta} < M$ then the estimate $\|u\|_{1,\rho} < M\varepsilon^{N_1}$ holds in the smaller domain Ω_ρ (it is sufficient to take an appropriate value of n).

CHAPTER V

Singular Perturbation
of a Hyperbolic System of Equations

In the preceding chapters the method of matched asymptotic expansions was applied either to ordinary differential equations, or to elliptic second order equations. However, the application range of the method is much wider. In this chapter, and in Chapter VI the method is demonstrated on two apparently quite dissimilar problems. Nevertheless, it turns out that in the situations of this kind the ideas developed above in combination with the special investigation of the auxiliary problems arising in each particular case make it possible to construct uniform asymptotic approximations of the desired solutions.

Here we consider the simplest problem of this kind for a system of hyperbolic equations—a system of two differential equations for the functions $u^+(x, t)$ and $u^-(x, t)$:

$$[\partial_t \pm \partial_x]u^\pm + \varepsilon[a^\pm(u^+, u^-)\partial_x u^\pm + b^\pm(u^+, u^-)\partial_x u^\mp + c^\pm(u^+, u^-)] = 0 \quad (0.1)$$

with the initial conditions

$$u^\pm(x, t, \varepsilon)|_{t=0} = \varphi^\pm(x), \qquad x \in \mathbf{R}^1, \ 0 < \varepsilon \ll 1. \tag{0.2}$$

The question is how to construct the asymptotics of the solution $u^\pm(x, t, \varepsilon)$ as $\varepsilon \to 0$.

The problem of constructing the asymptotic expansion of the functions $u^\pm(x, t, \varepsilon)$ for finite periods of time $(0 \le t \le M)$ is trivial. Such an asymptotic expansion is given by the direct series of the perturbation theory:

$$u^\pm(x, t, \varepsilon) = \sum_{n=0}^{\infty} \varepsilon^n u_n^\pm(x, t), \qquad \varepsilon \to 0,$$

where the functions $u_n^\pm(x, t)$ are determined from the linear equations $[\partial_t \pm \partial_x]u_n^\pm = f_n^\pm(x, t)$.

The singularity of perturbation in the problem (0.1), (0.2) shows up if time periods are long: for $t \approx \varepsilon^{-1}$ the direct expansion fails. For example, in the case of a single equation $u_t + u_x - \varepsilon u_x = 0$ the direct expansion of the exact solution $u = \varphi(x - t + \varepsilon t)$ yields

$$u(x, t) = \sum_{n=0}^{\infty} \varepsilon^n t^n \frac{1}{n!} \varphi^{(n)}(x - t). \tag{0.3}$$

Evidently, this series is an asymptotic one as $\varepsilon \to 0$ only for $t \ll \varepsilon^{-1}$. Such a situation is typical for systems of equations with a hyperbolic principal part (for $\varepsilon = 0$): owing to singular terms $\varepsilon^n t^n$ the direct expansion fails for $t \approx \varepsilon^{-1}$, despite the fact that there is an exact solution $u(x, t, \varepsilon)$ for $t \leq M\varepsilon^{-1}$.

Note that the presence of nonlinearities in perturbation plays no significant role in the emergence of singular terms. Nonlinearities can cause only the appearance of an upper time limit beyond which no exact solution of the original problem exists (to be more precise, at least no smooth solution exists). The presence of the small parameter ε in nonlinear terms guarantees the existence of a smooth solution for the times $t \leq M\varepsilon^{-1}$.

It is, therefore, natural to pose the problem which will be solved in this chapter: how to construct a uniform asymptotic expansion for the solution of the problem (0.1), (0.2) for large intervals of time: $\{x \in \mathbf{R}^1, 0 \leq t \leq O(\varepsilon^{-1})\}$.

In the above notation (x, t) are the inner variables. Writing the problem in the outer variables $\xi = \varepsilon x$, $\tau = \varepsilon t$ virtually does not affect the form of the equation, but the dependence on the small parameter ε appears in the initial data: $\varphi(\xi/\varepsilon)$. Here the singular character of the perturbation is due to the dependence of the initial data on the rapid "boundary layer" variable $x = \xi\varepsilon^{-1}$.

The form of the asymptotics of the solution essentially depends on the behavior of the initial functions at infinity. Here only initial data exhibiting slow stabilization at infinity are considered. Subject to this condition the problem is bisingular: as in the preceding chapters, the coefficients of the outer expansion have growing singularities at the origin, while the coefficients of the inner expansion, as shown in Example (0.3), have growing singularities as $t \to \infty$.

Thus, we suppose that the initial functions admit the following asymptotic expansions:

$$\varphi^{\pm}(x) = \begin{cases} \sum_{j=0}^{\infty} \varphi_{j,1}^{\pm} x^{-j}, & x \to +\infty, \\ \sum_{j=0}^{\infty} \varphi_{j,2}^{\pm} x^{-j}, & x \to -\infty, \end{cases} \tag{0.4}$$

where $\varphi_{j,1}^{\pm}$, $\varphi_{j,2}^{\pm}$ are constants. As everywhere before, all initial data are assumed to be infinitely differentiable. In particular, the coefficients $a^{\pm}$, $b^{\pm}$, $c^{\pm}(u^+, u^-)$ can be expanded in asymptotic Taylor series for $u^+ \to u$, $u^- \to v$ with the coefficients $a_{p,q}^{\pm}$, $b_{p,q}^{\pm}$, $c_{p,q}^{\pm}(u, v)$ being smooth functions of u and v.

§1. Construction of the inner expansion

First, we construct an f.a.s. in the form of the usual perturbation series:

$$u^{\pm}(x, t, \varepsilon) = \sum_{n=0}^{\infty} \varepsilon^n u_n^{\pm}(x, t). \tag{1.1}$$

For the coefficients $u_n^{\pm}$ one obtains the recurrence system of problems

$$(\partial_t \pm \partial_x)u_n^{\pm} = f_n^{\pm}(x, t), \tag{1.2}$$

$$u_n^{\pm}(x, 0) = \begin{cases} \varphi^{\pm}(x), & n = 0, \\ 0, & n \geq 1, \end{cases} \tag{1.3}$$

The right-hand sides $f_n^{\pm}(x, t)$ are multilinear forms of the preceding approximations u_k^+, u_k^- $(0 \leq k \leq n-1)$ and their derivatives with coefficients smoothly depending on u_0^+, $u_0^-(x, t)$:

$$f_0^{\pm} \equiv 0,$$

$$f_n^{\pm} = \sum_{|k|+|l|+m=n-1} \left[a_{k,l}^{\pm}(u_0^+, u_0^-)\partial_x u_m^{\pm} + b_{k,l}^{\pm}(u_0^+, u_0^-)\partial_x u_m^{\mp} \right] \prod_{i=1}^{p} u_{k_i}^+ u_{l_i}^-$$

$$+ \sum_{|k|+|l|=n-1} c_{k,l}^{\pm}(u_0^+, u_0^-) \prod_{i=1}^{p} u_{k_i}^+ u_{l_i}^-, \qquad n \geq 1, \tag{1.4}$$

$$k = (k_1, \ldots, k_p), \qquad l = (l_1, \ldots, l_p).$$

It is perfectly clear that the functions $u_n^{\pm}(x, t)$, $\forall n \geq 0$ are defined from equations (1.2) uniquely. The main purpose of this section is to examine the structure of asymptotic expansions of these functions as $|x| + t \to \infty$.

The asymptotics of the first approximation $u_0^{\pm}(x, t) = \varphi^{\pm}(x \mp t)$ follow from relations (0.4). It can evidently vary depending on the direction. For example,

$$u_0^+(x, t) = \sum_{j=0}^{\infty} \varphi_{j,1}^+(x - t)^{-j} \qquad \text{as } x - t \to \infty,$$

$$u_0^-(x, t) = \sum_{j=0}^{\infty} \varphi_{j,2}^-(t + x)^{-j} \qquad \text{as } t + x \to -\infty,$$

The asymptotics for the next approximations $u_n^{\pm}(x, t)$ for $|x| + t \to \infty$ may include terms growing with respect to x, t, and each component u_n^+, u_n^- may depend on both variables $x + t$ and $x - t$. In what follows, the variables $t + x$ and $t - x$, naturally, play an important role, and we denote

$$s = t + x, \qquad \sigma = t - x. \tag{1.5}$$

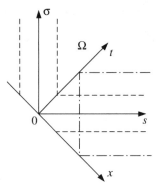

FIGURE 27

Denote by Ω the half-plane $\{x, t: x \in \mathbf{R}^1, t > 0\}$, or, which is the same, $\{s, \sigma: s \in \mathbf{R}^1, \sigma \in \mathbf{R}^1, s + \sigma > 0\}$ (see Figure 27). To describe the asymptotics of the coefficients of the series (1.1) the following definitions of the corresponding classes of functions are convenient. First, we define a class of functions of one variable. Denote by $\mathfrak{B}$ the set of functions $\varphi(s) \in C^\infty(\mathbf{R}^1)$ such that

$$\varphi(s) = \sum_{k=0}^\infty c_k^\pm s^{-k}, \qquad s \to \pm\infty, \tag{1.6}$$

and this series can be differentiated term-by-term to any order.

Denote by $\mathfrak{A}_0$ the class of functions $u(x, t) \in C^\infty(\overline{\Omega})$ possessing the following properties:

(a) the following asymptotic expansions are valid:

$$u(x, t) = \sum_{k=0}^\infty \varphi_k^\pm(s)\sigma^{-k}, \qquad \sigma \to \pm\infty, \tag{1.7}$$

$$u(x, t) = \sum_{i=0}^\infty \psi_i^\pm(\sigma)s^{-i}, \qquad s \to \pm\infty, \tag{1.8}$$

(b) the asymptotic expansion (1.7) is uniform with respect to $s \in \mathbf{R}^1$, the asymptotic expansion (1.8) is uniform with respect to $\sigma \in \mathbf{R}^1$;
(c) $\varphi_k^\pm, \psi_i^\pm \in \mathfrak{B}$;
(d) the series (1.7) and (1.8) can be differentiated term-by-term to any order.

Denote by $\mathfrak{A}_n$, $n > 0$, the class of functions $u(x, t) \in C^\infty(\overline{\Omega})$ having the following properties:

(a) in each of the domains $\{x, t: s > 1, \sigma > 1\}$, $\{x, t: \sigma > 1, s < -1, s + \sigma > 0\}$, $\{x, t: \sigma < -1, s > 1, \sigma + s > 0\}$ (see Figure 27) these functions are polynomials of degree n in the variables $\ln|s|$, $\ln|\sigma|$, s, and σ with coefficients belonging to $\mathfrak{A}_0$;

(b) in the domain $\{x, t : s > 2, |\sigma| < 2\}$ the function $u(x, t)$ is a polynomial of degree n in $\ln s$ and s with coefficients from $\mathfrak{A}_0$, in the domain $\{x, t : \sigma > 2, |s| < 2\}$ the function $u(x, t)$ is a polynomial of degree n in $\ln \sigma$ and σ with coefficients from $\mathfrak{A}_0$.

LEMMA 1.1. *Let $u(x, t) \in \mathfrak{A}_0$. Then the formal double series obtained from (1.7) by expanding the coefficients $\varphi_k^+(s)$ as $s \to \infty$ coincides with the similar series obtained from (1.8) by expanding $\psi_i^+(\sigma)$ as $\sigma \to \infty$. A similar coincidence of double series is also observed in other parts of the half-plane Ω.*

PROOF. The asymptotic series (1.7) and (1.8) should be replaced with their partial sums with remainders. The same procedure should be applied to the coefficients φ_k, ψ_i. After equating these two different expressions for the function $u(x, y)$, one concludes that the corresponding coefficients of the asymptotic expansions are equal. ∎

LEMMA 1.2. *The statement of the preceding lemma on the coincidence of the double series holds for any function $u(x, t) \in \mathfrak{A}_n$.*

The proof follows from the definition of the class $\mathfrak{A}_n$, and Lemma 1.1.   ∎

LEMMA 1.3. *If $u_k^\pm(x, t) \in \mathfrak{A}_k$ for $k \le n - 1$, then the functions $f_n^\pm(x, t)$ defined by equality (1.4) belong to the class $\mathfrak{A}_{n-1}$.*

The proof follows from the fact that a smooth function $a(u^+, u^-)$ belongs to $\mathfrak{A}_0$ if $u^\pm \in \mathfrak{A}_0$, and from the explicit form of the function $f_n^\pm$.   ∎

THEOREM 1.1. *The coefficients of the series (1.1), i.e., the functions $u_n^\pm(x, t)$, belong to the class $\mathfrak{A}_n$.*

The proof is achieved by induction. As mentioned above, the statement of the theorem holds for $n = 0$. According to Lemma 1.3, the right-hand sides $f_n^\pm(x, t) \in \mathfrak{A}_{n-1}$. The solutions of the problems (1.2), (1.3) for $n > 0$ are obtained by integrating along the corresponding characteristic. For example,

$$u_n^-(x, t) = \int_{-s}^{\sigma} F(s, \sigma_1) \, d\sigma_1, \tag{1.9}$$

where $F(s, \sigma) = \frac{1}{2} f_n \left(\frac{s-\sigma}{2}, \frac{s+\sigma}{2} \right)$. It is sufficient to consider just the function u_n^-, because u_n^+ is investigated in precisely the same way. Evidently, $u_n^-(x, t) \in C^\infty(\overline{\Omega})$, and one only has to examine the asymptotics of this function at infinity. We will consider only the most complicated case $s \to \infty$, $\sigma \to \infty$, but to make it less cumbersome let us assume that $F(s, \sigma) \in \mathfrak{A}_0$.

Split the integral (1.9) into two integrals: one from $-s$ to 1, and the other from 1 to σ. The first integral depends only on s and belongs to $\mathfrak{A}_1$ which is easily implied by the following representation:

$$\int_{-s}^{1} F(s, \sigma_1) \, d\sigma_1 = \int_{-s}^{1} \sum_{i=0}^{N} \psi_i(\sigma_1) s^{-i} \, d\sigma_1 + O\left(s^{-N}\right)$$

$$= \sum_{i=0}^{N} s^{-i} \left\{ \int_{-1}^{1} \psi_i(\sigma_1) \, d\sigma_1 \right.$$

$$\left. + \int_{-s}^{-1} \left[\sum_{j=0}^{N} c_{i,j} \sigma_1^{-j} + O\left(\sigma_1^{-N-1}\right) \right] d\sigma_1 \right\} + O\left(s^{-N}\right).$$

$$(1.10)$$

The second integral is first examined for $\sigma \to \infty$. Let

$$g_N(s, \sigma) = F(s, \sigma) - \sum_{k=0}^{N} \varphi_k^+(s) \sigma^{-k}.$$

By hypothesis, $g_N(s, \sigma) = O\left(\sigma^{-N-1}\right)$, whence

$$\int_{1}^{\sigma} F(s, \sigma_1) \, d\sigma_1 = \sum_{k=0}^{N} \varphi_k^+(s) \int_{1}^{\sigma} \sigma_1^{-k} \, d\sigma_1 + \int_{1}^{\infty} g_N(s, \sigma_1) \, d\sigma_1 + O\left(\sigma^{-N}\right).$$

The sum in the right-hand side of this equality evidently belongs to $\mathfrak{A}_1$, and one only has to investigate $\int_{1}^{\infty} g_N(s, \sigma) \, d\sigma$. Transforming this integral one gets

$$\int_{1}^{\infty} g_N(s, \sigma) \, d\sigma$$

$$= \int_{1}^{s} \left[\sum_{i=0}^{m} \psi_i^+(\sigma) s^{-i} + O\left(s^{-m-1}\right) - \sum_{k=0}^{m} \varphi_k^+(s) \sigma^{-k} \right] d\sigma + O\left(s^{-N}\right)$$

$$= \int_{1}^{s} \left\{ \sum_{i=0}^{m} \psi_i^+(\sigma) s^{-i} \right.$$

$$\left. - \sum_{k=0}^{m} \sigma^{-k} \left(\sum_{i=0}^{m} c_{k,i} s^{-i} + O\left(s^{-m-1}\right) \right) \right\} d\sigma + O\left(s^{-N}\right).$$

Denote $g_{i,m}(\sigma) = \psi_i^+(\sigma) - \sum_{k=0}^{m} c_{k,i} \sigma^{-k}$. According to Lemma 1.2, $g_{i,m}(\sigma) = O\left(\sigma^{-m-1}\right)$. Hence

$$\int_{1}^{\infty} g_N(s, \sigma) \, d\sigma = \sum_{i=0}^{m} s^{-i} \int_{1}^{\infty} g_{i,m}(\sigma) \, d\sigma + O\left(s^{-m} + s^{-N}\right).$$

It remains to consider $\int_{1}^{\sigma} F(s, \sigma_1) \, d\sigma_1$ as $s \to \infty$:

$$\int_{1}^{\sigma} F(s, \sigma_1) \, d\sigma_1 = \int_{1}^{\sigma} \sum_{k=0}^{N} \varphi_k^+(s) \sigma_1^{-k} \, d\sigma_1 + O\left(s^{-N}(\sigma + s)\right). \quad \blacksquare$$

REMARK. The asymptotic expansions of the functions $u_n^\pm(x, t)$ as $s \to \infty$, $\sigma \to -\infty$, and as $s \to -\infty$, $\sigma \to \infty$ (i.e. for $s < -1$, $s + \sigma > 0$, and for $\sigma < -1$, $s + \sigma > 0$; see Figure 27) include logarithmic terms only in the form $[\ln |s/\sigma|]^p$.

For the verification of this fact it is sufficient to analyze the above proof of Theorem 1.1. The explicit formulas for the integrals of partial sums of the asymptotic series imply that the logarithmic terms are precisely of the form mentioned above. ∎

§2. Construction of an f.a.s. in the outer domain (under discontinuity lines)

Since the series (1.1) forfeits its asymptotic character for large x and t of order ε^{-1}, we introduce the slow variables

$$\xi = \varepsilon x, \qquad \tau = \varepsilon t \tag{2.1}$$

and look for the outer expansion in the form

$$V^\pm = \sum_{n=0}^\infty \varepsilon^n v_n^\pm(\xi, \tau). \tag{2.2}$$

In the slow variables the equations (0.1) and initial conditions (0.2) for $v_n^\pm(\xi, \tau, \varepsilon) \equiv u_n^\pm(x, t, \varepsilon)$ take the form

$$\left(\frac{\partial}{\partial \tau} \pm \frac{\partial}{\partial \xi}\right) v^\pm + c^\pm(v^+, v^-) + \varepsilon \left[a^\pm(v^+, v^-)\frac{\partial v^\pm}{\partial \xi} + b^\pm(v^+, v^-)\frac{\partial v^\mp}{\partial \xi}\right] = 0, \tag{2.3}$$

$$v^\pm(\xi, 0, \varepsilon) = \varphi^\pm(\xi\varepsilon^{-1}), \qquad \xi \in \mathbf{R}^1, \tag{2.4}$$

while for the coefficients $v_n^\pm(\xi, \tau)$ one obtains the differential equations

$$\left(\frac{\partial}{\partial \tau} \pm \frac{\partial}{\partial \xi}\right) v_0^\pm + c^\pm(v_0^+, v_0^-) = 0, \tag{2.5}$$

$$\left(\frac{\partial}{\partial \tau} \pm \frac{\partial}{\partial \xi}\right) v_n^\pm + c_{1,0}^\pm(v_0^+, v_0^-)v_n^\pm + c_{0,1}^\pm(v_0^+, v_0^-)v_n^- = g_n^\pm(\xi, \tau), \quad n \geq 1. \tag{2.6}$$

As usual, the right-hand sides $g_n^\pm$ are defined through the preceding approximations v_k^+, v_k^-, $k \leq n - 1$, and their derivatives with respect to ξ, and are multilinear forms in $v_k^\pm$ with coefficients depending on $v_0^\pm$:

$$g_n^\pm(\xi, \tau) = \sum_{|k|+|l|+m=n-1} \left[a_{k,l}^\pm(v_0^+, v_0^-)\frac{\partial v_m^\pm}{\partial \xi} + b_{k,l}^\pm(v_0^+, v_0^-)\frac{\partial v_m^\mp}{\partial \xi}\right] \prod_{i=1}^p v_{k_i}^+ v_{l_i}^-$$

$$+ \sum_{2 \leq |k|+|l|=n} c_{k,l}^\pm(v_0^+, v_0^-) \prod_{i=1}^p v_{k_i}^+ v_{l_i}^-,$$

$$k = (k_1, \ldots, k_p), \qquad l = (l_1, \ldots, l_p), \qquad k_i \geq 1, \quad l_i \geq 1. \tag{2.7}$$

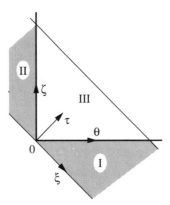

FIGURE 28

The initial conditions $v_n^\pm$ are obtained from (2.4) using the asymptotics (0.4) for $\varphi^\pm(\xi\varepsilon^{-1})$ as $|x| = |\xi|\varepsilon^{-1} \to \infty$:

$$v_0^\pm(\xi, 0) = \begin{cases} \varphi_{0,1}^\pm, & \xi > 0, \\ \varphi_{0,2}^\pm, & \xi < 0, \end{cases} \tag{2.8}$$

$$v_n^\pm(\xi, 0) = \begin{cases} \xi^{-n}\varphi_{n,1}^\pm, & \xi > 0, \\ \xi^{-n}\varphi_{n,2}^\pm, & \xi < 0. \end{cases} \tag{2.9}$$

Here $\varphi_{n,1}^\pm$, $\varphi_{n,2}^\pm$ are some constants.

As in the preceding section, it is convenient to introduce the characteristic variables in the outer domain:

$$\theta = \tau + \xi, \qquad \zeta = \tau - \xi.$$

At the initial step $n = 0$ the problem (2.5), (2.8) for $v_0^\pm$ causes no difficulty, having a solution at least in the domain $\{\xi, \tau : 0 \le \tau \le M\}$. The solutions $v_0^\pm(\xi, \tau)$ may have discontinuities of the first kind on the rays $\xi \pm \tau = 0$ (since the initial data are discontinuous at the point $\xi = 0$ if $\varphi_{0,1}^\pm \ne \varphi_{0,2}^\pm$). These rays are characteristics of the limiting system, and, as in Chapter IV, will be called *singular characteristics* (see Figure 28).

On the succeeding steps $n = 1, 2, \ldots$ the singularities of the initial functions (2.9) at the point $\xi = 0$ and the discontinuities of the right-hand sides $g_n^\pm(\xi, \tau)$ on the singular characteristics make the situation worse. Therefore, the problems (2.6), (2.9) can be guaranteed to have a solution under the discontinuity lines in the domains $\{\xi, \tau : \theta < 0, 0 \le \tau \le M\}$, $\{\xi, \tau : \zeta < 0, 0 \le \tau \le M\}$ shaded in Figure 28. In these domains all the functions $v_n^\pm(\xi, \tau)$ are defined uniquely and are smooth. However, these functions have singularities on the lines $\theta = 0$ and $\zeta = 0$. The main purpose of this section is to examine the structure of the asymptotics for $v_n^\pm(\xi, \tau)$ on these lines: as $\zeta \to -0$ and as $\theta \to -0$. We consider only the domain $\{\xi, \tau: \zeta < 0, \, 0 \le \tau \le M\}$ as the situation in the other one is quite similar.

Analyzing the solutions of the problems (2.4)–(2.9) one can directly find the form of the asymptotics of the functions $v_n^{\pm}$ as $\zeta \to 0$. It is, however, easier and more convenient to use the same approach as in Chapter IV (see, for example, Theorem 1.3, Chapter IV). Namely, the asymptotics of the solutions in the vicinity of the origin ($\xi = 0$, $\tau = 0$) can be obtained by matching the series (1.1) to (2.2). Then, using the asymptotics as a kind of initial data, one studies the functions $v_n^{\pm}$. The only difference is that in Chapter IV this approach was used in order to construct functions in the intermediate layer in the vicinity of a singular characteristic in order to choose a unique solution. Here the existence of the functions $v_n^{\pm}$ causes no doubt, and one only has to find the form of their asymptotics in the vicinity of the singular characteristic. The same approach is used in the next section to construct an f.a.s. in the intermediate layer, in a neighborhood of a singular characteristic.

First, we note that the series (1.1) provide an f.a.s. of the problem (0.1), (0.2) for $|x| + t < \varepsilon^{\alpha - 1}$, $0 < \alpha < 1$. This is easily verified in the same way as, for example, in Chapter II, §3. It is sufficient to substitute the partial sums $A_{N,x,t} U^{\pm}$ into the equations and initial conditions (0.1), (0.2) and use the asymptotics of the functions $u_n^{\pm}(x, t)$ obtained in Theorem 1.1.

Making the change of variables (2.1) in the sum $A_{N,x,t} U^{\pm}$ and applying the operator $A_{N,\xi,\tau}$, one gets the equality

$$A_{N,\xi,\tau} A_{N,x,t} U^{\pm} = \sum_{n=0}^{N} \varepsilon^n B_N \widetilde{V}_n^{\pm}(\xi, \tau), \qquad (2.10)$$

where $B_N \widetilde{V}_n^{\pm}$ are the partial sums of the formal series $\widetilde{V}_n^{\pm}(\xi, \tau)$ as $\xi \to 0$, $\tau \to 0$. (Owing to the remark in §1, equality (2.10) includes no powers of $\ln \varepsilon$.) Since the pair of series (1.1) are the f.a.s. of the problem (0.1), (0.2), the series $\sum_{n=0}^{\infty} \varepsilon^n \widetilde{V}_n^{\pm}(\xi, \tau)$ are f.a.s. of the same problem for $|\xi| + \tau < \varepsilon^{\alpha}$ ($0 < \alpha < 1$), while the series $\widetilde{V}_n^{\pm}(\xi, \tau)$ are f.a.s. of the problem (2.5)–(2.9) as $|\xi| + \tau \to 0$. Clearly, for the matching conditions for the series (1.1) and (2.2) to be satisfied, it is necessary that the series $\widetilde{V}_n^{\pm}$ be asymptotic expansions of the functions $v_n^{\pm}(\xi, \tau)$—solutions of the problem (2.5)–(2.9). The uniqueness of the solutions for these problems implies that

$$v_n^{\pm}(\xi, \tau) = \widetilde{V}_n^{\pm}(\xi, \tau), \qquad |\xi| + \tau \to 0. \qquad (2.11)$$

Consider the structure of the asymptotic series $\widetilde{V}_n^{\pm}$ defined by equality (2.10) in more detail. Theorem 1.1 and the remark thereto implies

$$\widetilde{V}_n^{\pm} = \sum \zeta^i \theta^j \ln^p |\zeta/\theta| \, c_{n,i,j,p}. \qquad (2.12)$$

Here the sum is taken over all i, j, and p such that $0 \le p \le n$, $p - i - j \le n$, $-i \le n$, $-j \le n$. In other words, the series (2.12) are polynomials of degree n in ζ^{-1}, θ^{-1}, $\ln|\zeta/\theta|$ whose coefficients are asymptotic series in nonnegative powers of ζ and θ (the form of the series (2.12) can be appropriately compared with the asymptotic series for the functions from the class $\mathfrak{A}_n$).

For the investigation of the asymptotics of the solutions $v_n^{\pm}(\xi, \tau)$ near the singular characteristic $\zeta = 0$ the following classes of functions are convenient. Denote by $\mathfrak{B}$ the set of functions $\varphi(\theta) \in C^{\infty}$ for $\theta \geq 0$. Denote by $\mathfrak{U}_n$ $(n \geq 0)$ the set of functions $\varphi(\theta) \in C^{\infty}$ for $\theta > 0$ which are polynomials of degree n in θ^{-1} and $\ln\theta$ with coefficients from $\overline{\mathfrak{B}}$.

Denote by $\mathfrak{N}_n$ the set of functions $v(\xi, \tau) \in C^{\infty}$ in the domain $I = \{\xi, \tau: \zeta < 0, 0 \leq \tau \leq M\}$ (see Figure 28), having the asymptotic expansions

$$v(\xi, \tau) = \sum \zeta^i \ln^p |\zeta| \varphi_{i,p,k}(\theta), \qquad \zeta \to 0 \qquad (2.13)$$

where $\varphi_{i,p,k}(\theta) \in \mathfrak{U}_k$, and the sum is taken over all i, p, k for which $0 \leq p \leq n$, $p + k - i \leq n$. It follows from this definition that any function $v(\xi, \tau) \in \mathfrak{N}_n$ yields a double series of the form (2.13) after expanding the coefficients $\varphi_{i,p,k}(\theta)$ as $\theta \to 0$.

THEOREM 2.1. *The functions* $v_n^{\pm}(\xi, \tau)$ *solving the recurrence systems* (2.5)–(2.9) *belong to the classes* $\mathfrak{N}_n$, *i.e., admit asymptotic expansions of the form* (2.13). *On expanding the coefficients* $\varphi_{i,p,k}(\theta)$ *into the series as* $\theta \to 0$, *the resulting double series for* $v_n^{\pm}(\xi, \tau)$ *coincide with the series* (2.12).

The proof consists of two steps. First, an f.a.s. as $\zeta \to 0$ is constructed for the sequence of problems (2.5)–(2.9) in the form of the series (2.12):

$$v_n^{\pm}(\xi, \tau) = \sum \zeta^i \ln^p |\zeta| v_{n,i,p}^{\pm}(\theta) \qquad (2.14)$$

$$(v_{n,i,p}^{\pm}(\theta) \in \mathfrak{N}_k, \qquad 0 \leq p \leq n, \qquad p + k - i \leq n).$$

In general, the coefficients $v_{n,i,p}^{\pm}$ in this expansion are not defined uniquely. At the second step, one proves that there exists a unique choice of the functions $v_{n,i,p}^{\pm}$ for which the series (2.14) provide the asymptotic expansions of the solutions $v_{n,i,p}^{\pm}$.

Looking for an f.a.s. of the problems (2.5)–(2.9) in the form of the series (2.14) results in the recurrence system of problems for the coefficients:

$$2v_{n,i,p}^{\pm\,'} + c_{1,0}(\theta)v_{n,i,p}^{+} + c_{0,1}(\theta)v_{n,i,p}^{-} = g_{n,i,p}^{+}(\theta), \qquad (2.15)$$

$$2iv_{n,i,p}^{-} = g_{n,i,0}^{-}(\theta) \qquad \forall i \neq 0,$$
$$2pv_{n,i,p}^{-} = g_{n,i,p}^{-}(\theta) \qquad \forall p > 0, \forall i. \qquad (2.16)$$

Here

$$c_{1,0}(\theta) = c_{1,0}^{+}(v_0^{+}, v_0^{-})|_{\zeta=-0},$$
$$c_{0,1}(\theta) = c_{0,1}^{+}(v_0^{+}, v_0^{-})|_{\zeta=-0},$$

the right-hand sides $g_{n,i,p}^{\pm}(\theta)$ are determined through the functions $v_{m,j,l}^{\pm}$ and their derivatives from the preceding steps: either $m < n$, or $m = n$ and $l < p$, or $m = n$, $l = p$ and $j < i$.

For the desired functions $v^\pm(\xi, \tau)$ the asymptotics (2.11), (2.12) is known. Comparing this asymptotics with the desired series (2.14) immediately yields the asymptotic expansion for the coefficients $v^\pm_{n,i,p}(\theta)$ as $\theta \to +0$:

$$v^\pm_{n,i,p}(\theta) = \sum_{j,l} \theta^j \ln^l \theta\, v^\pm_{n,i,p,j,l}. \tag{2.17}$$

These series evidently are f.a.s. for equations (2.15), (2.16) as $\theta \to +0$. The functions $v^\pm_{n,i,p}(\theta)$ are easily reconstructed from them as exact solutions of equations (2.15), (2.16) with the asymptotics (2.17). We look for $v^\pm_{n,i,p}$ in the form of the sum of segments of the series (2.17) of arbitrary length N and remainders $\tilde{v}^\pm_{n,i,p}(\theta)$. One can easily see that the problems for the remainders obtained from (2.15), (2.16) with the initial conditions $\tilde{v}^\pm_{n,i,p}(0) = 0$ and given functions $\tilde{v}^\pm_{n,0,0}(\theta) = O(\theta^{N-n})$ have the solutions $\tilde{v}^\pm_{n,i,p}(\theta) = O(\theta^{N-n})$ $\forall n \le N/2$.

This approach yields all functions $v^\pm_{n,i,p}(\theta)$, and hereby the problem of constructing the f.a.s. (2.14) is solved.

Note that at each step $n = 1, 2, \ldots$ the f.a.s. (2.14) are not defined uniquely due to the choice of $v^\pm_{n,0,0}(\theta)$. The reason is that (2.16) includes no equations for $v^-_{n,0,0}$, while the series (2.17) is not sufficient to construct $v^-_{n,0,0}$ for all $\theta \in (0, M)$.

Nevertheless, the series (2.14) provide f.a.s. for the problem (2.5)–(2.9) for any choice of smooth functions $v^-_{n,0,0}(\theta)$ satisfying (2.17) making it possible to reconstruct the asymptotics of the exact solutions $v^\pm_n(\xi, \tau)$ in the usual way. The functions $v^\pm_n$ are constructed in the form of the sum of segments of the series (2.14) of arbitrary length N and remainders $\tilde{v}^\pm_n(\xi, \tau)$. For the remainders $\tilde{v}^\pm_n$ one obtains equations of the type (2.5) with the right-hand sides $\tilde{g}^\pm_n = O\left(|\zeta|^{N-n}\right)$ $\forall \theta \in [0, M]$ and homogeneous initial conditions $(n \le N-1)$. Solutions of these problems can be differentiated $N - n$ times up to the boundary $\zeta = -0$, and can, therefore, be expanded in Taylor series

$$\tilde{v}^\pm_n(\xi, \tau) = \sum_{j=0}^{N-n-1} \zeta^j \tilde{v}^\pm_{n,j}(\theta) + O\left(\zeta^{N-n}\right). \tag{2.18}$$

The coefficients $\tilde{v}^\pm_{n,j}(\theta)$ can be differentiated to the order $N - n - j$ up to the point $\theta = 0$, and, therefore, also admit a Taylor expansion to the order $N - n - j - 1$. Thus, the asymptotics of the exact solutions $v^\pm_n(\xi, \tau)$ of the problems (2.5)–(2.9) can be represented as a sum of segments of the series (2.14) and (2.18). Since N is an arbitrary number, the structure of the complete asymptotic expansion for $v^\pm_n(\xi, \tau)$ is of the form given in (2.13). ■

§3. Construction of f.a.s. in the vicinity of singular characteristics

The series (2.2) constructed and studied in the preceding section is not asymptotic in the vicinity of singular characteristics. Its coefficients $v^\pm_n(\xi, \tau)$ exhibit growing singularities as $\zeta \to -0$ (and, similarly, as $\theta \to -0$ in the domain II; see Figure 28). The series (2.2) has an asymptotic character, and is an f.a.s. of the problem (0.1), (0.2) in the domain I only for $|\zeta| > \varepsilon^\alpha$, $\theta > \varepsilon^\alpha$, $\forall \alpha: 0 < \alpha < 1$.

Consider the situation in the vicinity of one singular characteristic, e.g., $\zeta = 0$, and construct in its neighborhood an f.a.s. of the problem. The natural independent variables in such a neighborhood are

$$\sigma = t - x = (\tau - \xi)\varepsilon^{-1} \quad \text{and} \quad \theta = \varepsilon(t + x) = \tau + \xi.$$

The original equations (0.1) for $w^\pm(\sigma, \theta, \varepsilon) \equiv u^\pm(x, t, \varepsilon)$ take the form

$$2\frac{\partial w^+}{\partial \theta} + a^+(w^+, w^-)\left[-\frac{\partial}{\partial \sigma} + \varepsilon\frac{\partial}{\partial \theta}\right]w^+$$

$$- b^+(w^+, w^-)\left[\frac{\partial}{\partial \sigma} - \varepsilon\frac{\partial}{\partial \theta}\right]w^- + c^+(w^+, w^-) = 0,$$

$$2\frac{\partial w^-}{\partial \sigma} + \varepsilon\left\{a^-(w^+, w^-)\left[-\frac{\partial}{\partial \sigma} + \varepsilon\frac{\partial}{\partial \theta}\right]w^-\right.$$

$$\left. + b^-(w^+, w^-)\left[-\frac{\partial}{\partial \sigma} + \varepsilon\frac{\partial}{\partial \theta}\right]w^+ + c^-(w^+, w^-)\right\} = 0.$$

$$(3.1)$$

These differential equations should be supplemented by the matching conditions for the asymptotics of $w^\pm(\sigma, \theta, \varepsilon)$ as $\varepsilon \to 0$ and the f.a.s. (1.1) and (2.2) constructed above. It follows from the form of the asymptotics of the functions $u_n^\pm(x, t)$ as $s \to \infty$ (Theorem 1.1) and the functions $v_n^\pm(\xi, \tau)$ as $\zeta \to -0$ (Theorem 2.1) that each of the expressions $A_{N,\sigma,\theta}A_{N,x,t}U^\pm$ and $A_{N,\sigma,\theta}A_{N,\xi,\tau}V^\pm$ is a sum of the terms of the form $\varepsilon^n \ln^m \varepsilon Z_{n,m}(\sigma, \theta)$, where $m \le n$. It is, therefore, natural to look for f.a.s. of equations (3.1) in the form

$$W^\pm = \sum_{0 \le m \le n}^{\infty} \varepsilon^n \ln^m \varepsilon\, w_{n,m}^\pm(\sigma, \theta) \tag{3.2}$$

in order to satisfy the equations

$$A_{N,x,t}A_{N,\sigma,\theta}W^\pm = A_{N,\sigma,\theta}A_{N,x,t}U^\pm \quad \forall N, \tag{3.3}$$

$$A_{N,\xi,\tau}A_{N,\sigma,\theta}W^\pm = A_{N,\sigma,\theta}A_{N,\xi,\tau}V^\pm \quad \forall N. \tag{3.4}$$

The equations for $w_{n,m}^\pm(\sigma, \theta)$ are obtained from (3.1) in the usual way:

$$\frac{\partial w_{n,m}^-}{\partial \sigma} = h_{n,m}^-(\sigma, \theta) \quad \forall n, m \ge 0\ (h_{0,0} \equiv 0), \tag{3.5}$$

$$2\frac{\partial w_0^+}{\partial \theta} - a(w_0^+, w_0^-)\frac{\partial w_0^+}{\partial \sigma} + c^+(w_0^+, w_0^-) - b^+(w_0^+, w_0^-)\frac{\partial w_0^-}{\partial \sigma} = 0 \tag{3.6}$$

$$(w_0^\pm \equiv w_{0,0}^\pm),$$

$$2\frac{\partial w_{n,m}^+}{\partial \theta} - a_0^+(\sigma, \theta)\frac{\partial w_{n,m}^+}{\partial \sigma} + c_0^+(\sigma, \theta)w_{n,m}^+ = h_{n,m}^+(\sigma, \theta), \tag{3.7}$$

$$m + n \ge 1, \quad \sigma \in \mathbf{R}^1, \quad \theta > 0.$$

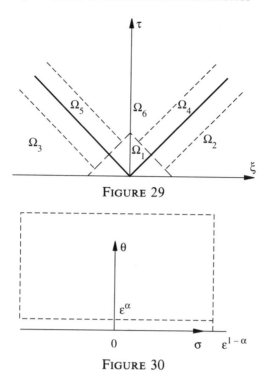

FIGURE 29

FIGURE 30

Here $a_0^+ = a_{1,0}^+(w_0^+, w_0^-)$, $c_0^+ = c_{1,0}^+(w_0^+, w_0^-) - b_{1,0}^+(w_0^+, w_0^-)\partial w_0^- / \partial \sigma$; the right-hand sides $h_{n,m}^\pm$ are defined through the preceding approximation $w_{q,r}^\pm$ and their derivatives, with the functions $h_{n,m}^+$ depending also on $w_{n,m}^-$.

The series (3.2), whose coefficients will be constructed below, is an f.a.s. of the original problem in the domain $\{\sigma, \theta : |\sigma| < \varepsilon^{1-\alpha}, \varepsilon^\alpha < \theta < M\}$ (Ω_4 in Figure 29, and the domain bounded by broken lines in Figure 30). It is, however, more convenient to construct the solutions $w_{n,m}^\pm(\sigma, \theta)$ in the entire strip on the plane σ, θ, as it is done everywhere in other chapters (see Figure 30). Here the functions $w_{n,m}^\pm(\sigma, \theta)$ have given asymptotics as $\theta \to 0$ and as $\sigma \to -\infty$. The form of these asymptotics follow from the matching conditions (3.3), (3.4).

Theorem 1.1 easily yields the form of each term in the right-hand side of (3.3), and, therefore, relations for $w_{n,m}^\pm(\sigma, \theta)$:

$$w_{n,m}^\pm(\sigma, \theta) = \sum_{\substack{0 \le p \le n-m \\ p-j \le n-m}} r_{n,m,j,p}^\pm(\sigma)\theta^j \ln^p \theta, \qquad \theta \to 0, \qquad (3.8)$$

where the coefficients $r_{n,m,j,p}^\pm(\sigma) \in C^\infty(\mathbf{R}^1)$ can be expanded in asymptotic series as $\sigma \to -\infty$ in powers of σ^{-1} and $\ln|\sigma|$; the restriction on the exponents of these powers, as in the definition of the classes $\mathfrak{A}_n$ in §1, are conveniently formulated in the following way: inserting the asymptotic series for the functions $r_{n,m,j,p}^\pm(\sigma)$ into (3.8) yields polynomials of degree $n-m$ in

$\ln\theta$, $\ln|\sigma|$, θ^{-1}, and σ with coefficients of the form $\sum_{i\geq 0,\, q\geq 0} c_{i,q}\theta^i\sigma^{-q}$.

Similarly, Theorem 2.1 and condition (3.4) yield the asymptotic expansions

$$w_{n,m}^{\pm}(\sigma,\theta) = \sum_{\substack{0\leq p\leq n-m \\ p-j\leq n-m}} z_{n,m,p,j}^{\pm}(\theta)\sigma^{-j}\ln^p|\sigma|, \qquad \sigma\to-\infty, \tag{3.9}$$

where the coefficients $z_{n,m,p,j}^{\pm}(\theta)\in C^\infty(0,M)$ can be expanded in asymptotic series as $\theta\to 0$ in powers of θ and $\ln\theta$. The double series obtained from (3.9) on expanding the coefficients as $\theta\to 0$, coincide, by virtue of the construction procedure for the functions $v_n^{\pm}(\xi,\tau)$ in §2, with the double series obtained in the same manner from (3.8).

THEOREM 3.1. *There exist solutions of equations* (3.5)–(3.7) *satisfying conditions* (3.8), (3.9).

PROOF. Consider the problem on the initial step $n=m=0$. Equation (3.5) is a homogeneous one. Therefore, the solution w_0^- does not depend on σ, and is determined through the values of the corresponding approximation in the outer domain I (see Figure 28) evaluated on the discontinuity curve: $w_0^-(\theta)\equiv v_0^-(\theta,\theta)$. Recall that $v_0^-(\xi,\tau)$ is a smooth function on the characteristic $\xi=\tau$. Hence $w_0^-(\theta)\in C^\infty$ for $\theta\geq 0$.

As w_0^- is already determined, equation (3.6) is a single quasilinear equation of the first order with respect to w_0^+. The initial function for it coincides with the original one:

$$w_0^+(\sigma,\theta) = \varphi^+(\sigma), \qquad \sigma\in\mathbf{R}^1.$$

A solution of equation (3.6) for $w_0^+(\sigma,\tau)$ subject to this condition exists in a strip $\{\sigma,\theta\colon\sigma\in\mathbf{R}^1,\ 0\leq\theta\leq M\}$ (quasilinear equations of the first order are discussed in more detail at the beginning of Chapter VI). One easily verifies that, owing to relation (0.4) for $\varphi^+(\sigma)$, the function $w_0^+(\sigma,\theta)$ admits asymptotic expansions (3.8), (3.9) with $m=n=0$.

The succeeding steps involve the solution of the linear equations (3.5), (3.7). The proof that the problems for these equations are solvable is based on the fact that the series (1.1) and (2.2) are f.a.s. of equations (3.1) in the overlapping domains as $\varepsilon\to 0$. This property means that the substitution of segments of the series (3.8), (3.9) of arbitrary length N into equations (3.5)–(3.7) yields residuals of the order of magnitude $O\big(\theta^N(1+|\sigma|)^{n+1}\big)$, $\theta\to+0$, $\forall\sigma\in\mathbf{R}^1$, or $O\big(|\sigma|^{-N}\theta^{-n-1}\big)$, $|\sigma|\to\infty$, $\forall\theta\in(0,M)$. However, this is not enough; one has to construct functions $w_{N,n,m}^{\pm}$ whose substitution into equations (3.5)–(3.7) gives rise to residuals that are small for both $\theta\to+0$ and $\sigma\to+\infty$ simultaneously. For this purpose the composite asymptotic expansion can be used. An appropriate segment of the f.a.s. for equations (3.5)–(3.7) can be taken in the form

$$w_{N,n,m}^+(\sigma,\theta) = \sum_{j=-n}^{N}\theta^j\sum_{\substack{m+p\leq n \\ m+p\leq n+j}} r_{n,m,j,p}^{\pm}(\sigma)\ln^p|\theta|$$

$$+ \sum_{j=-n}^{N}\sigma^{-j}\sum_{\substack{m+p\leq n \\ m+p\leq n+j}} z_{n,m,j,p}^{\pm}(\theta)\ln^p|\sigma| - \Lambda_{N,n,m}^{\pm}(\sigma,\theta). \tag{3.10}$$

Here $\Lambda^{\pm}_{N,n,m}(\sigma,\theta)$ means the partial sum of the double series obtained from (3.8) by expanding the coefficients $r^{\pm}_{n,m,j,p}(\sigma)$ as $\sigma \to -\infty$, which coincides with the analogous sum obtained from the series (3.9).

The substitution of the functions (3.10) into equations (3.5)–(3.7) for $\forall n \le N-1$, $m \le n$, gives rise to residuals of the order of magnitude $O\left(\theta^{N-n-2}\right)$ for $\theta \le |\sigma|^{-1}$ and $O\left(\sigma^{-N+n+2}\right)$ for $\theta \ge |\sigma|^{-1}$. Looking for the exact solutions in the form $w^{\pm}_{n,m} = w^{\pm}_{N,n,m} + \tilde{w}^{\pm}_{N,n,m}$ results in recurrence equations for the remainders $\tilde{w}^{\pm}_{N,n,m}(\sigma,\theta)$ of the type (3.5)–(3.7) with small inhomogeneities of the order of magnitude $O\left(\theta^{N-n-2}\right)$, $\theta \le |\sigma|^{-1}$, $O\left(\sigma^{-N+n+2}\right)$, $\theta > |\sigma|^{-1}$ and with homogeneous boundary conditions

$$\tilde{w}^{+}_{N,n,m}(\sigma,0) = 0, \qquad \tilde{w}^{-}_{N,n,m}(\sigma,\theta) \underset{\sigma \to -\infty}{\longrightarrow} 0.$$

Since for $\theta \to 0$, $\sigma \to -\infty$ the right-hand sides are small on each step, one can easily see, by induction, that the solutions for the remainders are small:

$$\tilde{w}^{\pm}_{N,n,m}(\sigma,\theta) = \begin{cases} O\left(\theta^{N-n-2}\right), & \theta \to 0, \ \theta \le |\sigma|^{-1}, \\ O\left(|\sigma|^{-N+n-2}\right), & \sigma \to -\infty, \ \theta \ge |\sigma|^{-1}, \ n \le N-2. \end{cases}$$

The arbitrariness of N and the structure of the functions $w^{\pm}_{N,n,m}(\sigma,\theta)$ imply that the exact solutions $w^{\pm}_{n,m}(\sigma,\theta)$ admit expansions of the form (3.8), (3.9).

The asymptotics of the functions $w^{\pm}_{n,m}(\sigma,\theta)$ as $\sigma \to \infty$ can be established by examining the solutions of the system (3.5)–(3.7). They have the same asymptotics as (3.9) (but, of course, with different coefficients $z^{\pm}_{n,m,j,p}(\theta)$). ∎

A similar situation is observed in the vicinity of the second singular characteristic $\theta = 0$ (see Figure 28). Here one introduces the variables $s = t + x = (\tau+\xi)\varepsilon^{-1}$ and $\zeta = \tau - \xi = \varepsilon(t-x)$. An f.a.s. of equation (1.1) is constructed in the form of the series

$$Y^{\pm} = \sum_{0 \le m \le n} \varepsilon^n \ln^m \varepsilon \cdot y^{\pm}_{n,m}(s,\zeta). \tag{3.11}$$

The problems obtained for the coefficients $y^{\pm}_{n,m}$ are similar to (3.5)–(3.7) with the only difference that the superscripts $+$ and $-$ change places. Relations similar to (3.3), (3.4) and meaning that the series (3.11) is matched to the series (1.1), and the outer expansion in the domain II (see Figure 28) are also valid.

§4. Construction of an f.a.s. in the outer domain (above discontinuity curves)

Consider our original problem in the domain $\{\xi, \tau: \tau+\xi > 0, \ \tau-\xi > 0\}$ (domain III in Figure 28 containing the domain Ω_6 in Figure 29). Here the variables to be considered are $\zeta = \tau - \xi$ and $\theta = \tau + \xi$. The equations for

$v^{\pm}(\xi, \tau, \varepsilon) \equiv u^{\pm}(x, t, \varepsilon)$ are the same as in §2:

$$
2\frac{\partial v^+}{\partial \theta} + c^+(v^+, v^-)
$$
$$
+ \varepsilon\left\{ a^+(v^+, v^-)\left[\frac{\partial v^+}{\partial \theta} - \frac{\partial v^+}{\partial \zeta}\right] + b^+(v^+, v^-)\left[\frac{\partial v^-}{\partial \theta} - \frac{\partial v^-}{\partial \zeta}\right] \right\} = 0,
$$

$$
2\frac{\partial v^-}{\partial \zeta} + c^-(v^+, v^-)
$$
$$
+ \varepsilon\left\{ a^-(v^+, v^-)\left[\frac{\partial v^-}{\partial \theta} - \frac{\partial v^-}{\partial \zeta}\right] + b^-(v^+, v^-)\left[\frac{\partial v^+}{\partial \theta} - \frac{\partial v^+}{\partial \zeta}\right] \right\} = 0.
$$
$$(4.1)$$

Additional conditions for $v^{\pm}$ are due to the requirements that the asymptotics of $v^{\pm}$ as $\varepsilon \to 0$ should be matched to the asymptotics (3.2) and (3.11) in the vicinity of singular characteristics.

An f.a.s. in domain III (see Figure 28) is looked for (as in §2, but for different values ζ and θ) in the form

$$
\overline{V}^{\pm} = \sum_{0 \le m \le n} v^{\pm}_{n, m}(\zeta, \theta)\varepsilon^n \ln^m \varepsilon, \qquad \zeta > 0, \quad \theta > 0, \qquad (4.2)
$$

with the matching conditions written in the standard way:

$$
A_{N,\sigma,\theta}A_{N,\zeta,\theta}V^{\pm} = A_{N,\zeta,\theta}A_{N,\sigma,\theta}W^{\pm},
$$
$$
A_{N,s,\zeta}A_{N,\zeta,\theta}\overline{V}^{\pm} = A_{N,\zeta,\theta}A_{N,s,\zeta}Y^{\pm} \quad \forall N.
$$
$$(4.3)$$

The presence of logarithmic terms in (4.2) is due to the fact that such terms appear in the right-hand sides of equalities (4.3). Indeed, the asymptotics of the functions $w^{\pm}_{n,m}$ as $\sigma \to \infty$ include terms of the form $\ln^p \sigma$ which, after the change of variable $\sigma = \varepsilon^{-1}\zeta$, yield $\ln^m \varepsilon$.

Substituting the series (4.2) into equation (4.1) results in the recurrence system of equations for the coefficients:

$$
2\frac{\partial v^+_0}{\partial \theta} + c^+(v^+_0, v^-_0) = 0,
$$
$$
2\frac{\partial v^-_0}{\partial \zeta} + c^-(v^+_0, v^-_0) = 0, \qquad v^{\pm}_0 \equiv v^{\pm}_{0,0},
$$
$$(4.4)$$

$$
2\frac{\partial v^+_{n,m}}{\partial \theta} + c^+_{1,0}(\zeta, \theta)v^+_{n,m} + c^+_{0,1}(\zeta, \theta)v^-_{n,m} = g^+_{n,m}(\zeta, \theta),
$$
$$
2\frac{\partial v^-_{n,m}}{\partial \zeta} + c^-_{1,0}(\zeta, \theta)v^+_{n,m} + c^-_{0,1}(\zeta, \theta)v^-_{n,m} = g^-_{n,m}(\zeta, \theta).
$$
$$(4.5)$$

The right-hand sides $g^{\pm}_{n,m}$ are determined as in §2; $c^{\pm}_{1,0}$, $c^{\pm}_{0,1}$ are defined through v^+_0, v^-_0.

Conditions (4.3) generate the given asymptotic expansions for the functions $v_{n,m}^{\pm}(\zeta, \theta)$ as $\zeta \to +0$ and as $\theta \to +0$:

$$v_{n,m}^{\pm}(\zeta, \theta) = \sum_{\substack{0 \leq p \leq n-m \\ p-j \leq n-m}} \mu_{n,m,p,j}^{\pm}(\theta)\zeta^j \ln^p \zeta, \qquad \zeta \to +0, \qquad (4.6)$$

$$v_{n,m}^{\pm}(\zeta, \theta) = \sum_{\substack{0 \leq p \leq n-m \\ p-j \leq n-m}} \nu_{n,m,p,j}^{\pm}(\zeta)\theta^j \ln^p \theta, \qquad \theta \to +0. \qquad (4.7)$$

Here the coefficients $\mu^{\pm}$ are easily expressed via the functions $z_{n,m,p,j}^{\pm}(\theta)$ in formula (3.9), while the coefficients $\nu^{\pm}$ are as easily expressed through the asymptotics of the functions $y_{n,m}^{\pm}(s, \zeta)$ in formula (3.11).

THEOREM 4.1. *The recurrence system of equations* (4.4), (4.5) *admits solutions satisfying conditions* (4.6), (4.7).

PROOF. At the initial step $n = m = 0$ equations (2.5), (4.4) and the matching conditions imply that the principal terms of the f.a.s. (4.2) coincide with those obtained in §2: $v_{0,0}^{\pm}(\zeta, \theta) = v_0^{\pm}(\frac{\theta-\zeta}{2}, \frac{\theta+\zeta}{2})$. These functions are smooth for $\zeta \geq 0$, $\theta \geq 0$ up to the singular characteristics.

At the next steps the functions $v_{n,m}^{\pm}(\zeta, \theta)$ have singularities as $\zeta \to +0$, $\theta \to +0$. The proof now follows along the same lines as in Theorem 3.1, and is based on the fact that the series (4.6), (4.7) are f.a.s. of the recurrence system of problems (4.4)–(4.7) as $\zeta \to 0$, or as $\theta \to 0$, or as $\zeta \to 0$, $\theta \to 0$. This property makes it possible to construct the functions $v_{N,n,m}^{+}$ from segments of the series (4.6), (4.7) of length N. Inserting these functions into equations (4.4), (4.5) results in residuals of the order of magnitude $O(\zeta^{N-n})$, $\zeta \to 0$ $(\zeta \leq \theta)$, or $O(\theta^{N-n})$ $\theta \to 0$ $(\theta \leq \zeta)$, or $O((\zeta + \theta)^{N-n})$, $\zeta \to 0$, $\theta \to 0$, $\zeta \sim \theta$. The equations for the remainders $\tilde{v}_{N,n,m}^{\pm} = v_{n,m}^{\pm} - v_{N,n,m}^{\pm}$ are similar to (4.4), (4.5), and, are supplemented with the homogeneous boundary conditions. They have solutions which in the order of magnitude do not exceed these residuals:

$$\tilde{v}_{n,m}^{\pm}(\zeta, \theta) = \begin{cases} O\left(\zeta^{N-n}\right), & \zeta \leq \theta, \\ O\left(\theta^{N-n}\right), & \theta \leq \zeta. \end{cases}$$

Thus, since N is an arbitrary number, it follows that the solutions of the equations (4.4), (4.5) have the asymptotics (4.6), (4.7). ∎

§5. Justification of the asymptotic expansion

In the preceding sections, the series (1.1), (2.2), (3.2), (3.11), (4.2) have been constructed providing an f.a.s. of the problem (0.1), (0.2) as $\varepsilon \to 0$ in the subdomains

$$\Omega_1 = \{\tau + |\xi| \leq 4\varepsilon^{\alpha}\}, \qquad \Omega_2 = \{\xi - \tau \geq \varepsilon^{\beta}\},$$
$$\Omega_3 = \{\xi + \tau \leq -\varepsilon^{\beta}\}, \qquad \Omega_4 = \{|\xi - \tau| \leq 2\varepsilon^{\gamma}\}, \qquad (5.1)$$
$$\Omega_5 = \{|\xi + \tau| \leq 2\varepsilon^{\gamma}\}, \qquad \Omega_6 = \{\tau - \xi \geq \varepsilon^{\delta}, \ \tau + \xi \geq \varepsilon^{\delta}\}$$

in the strip $\Pi = \{\xi \in \mathbf{R}^1, \ 0 \leq \tau \leq M\}$ (see Figure 29); here $\alpha, \beta, \gamma, \delta \in (0, 1)$ are arbitrary numbers. In the intersection of these domains the corresponding series asymptotically coincide. The aim of the present section is to prove that these series provide asymptotic expansions for the exact solution $u^{\pm}(x, t, \varepsilon)$ of the problem (0.1), (0.2) in the respective domains.

Starting from the f.a.s., the construction of $u^{\pm}(x, t, \varepsilon)$ proceeds as follows. Let in (5.1) $\alpha = \beta = \gamma = \delta = 1/2$, and introduce a partition of unity in the strip Π:

$$1 \equiv \sum_{j=1}^{6} \chi_j(\xi, \tau, \varepsilon)$$

such that each function $\chi_j(\xi, \tau, \varepsilon)$ vanishes outside the jth subdomain (5.1), and is smooth in ξ, τ. Since the width of the overlap domain is of the order of magnitude $O(\varepsilon^{1/2})$, one can choose the functions in such a way that $|\partial \chi_j / \partial \tau| + |\partial \chi^j / \partial \xi| \leq M_1 \varepsilon^{-1/2} \; \forall \xi, \tau, \varepsilon$. Then the f.a.s. in the strip Π is easily assembled as a sum of the series multiplied by the corresponding segments:

$$\begin{aligned}
U_N^{\pm}(x, t, \varepsilon) &= \chi_1(\xi, \tau, \varepsilon) A_{N, x, t} U^{\pm} + [\chi_2(\xi, \tau, \varepsilon) + \chi_3(\xi, \tau, \varepsilon)] A_{N, \xi, \tau} V^{\pm} \\
&\quad + \chi_4(\xi, \tau, \varepsilon) A_{N, \sigma, \theta} W^{\pm} + \chi_5(\xi, \tau, \varepsilon) A_{N, s, \zeta} Y^{\pm} \\
&\quad + \chi_6(\xi, \tau, \varepsilon) A_{N, \zeta, \theta} V^{\pm}.
\end{aligned} \tag{5.2}$$

Note that the specific choice of α, β, γ, δ plays no role in forming the f.a.s. (5.2) provided that the domains (5.1) overlap.

THEOREM 5.1. *The expression* (5.2) *is an asymptotic representation as* $\varepsilon \to 0$ *of the exact solution* $u^{\pm}(x, t, \varepsilon)$ *of the problem* (0.1), (0.2) *in the strip* $\Pi = \{\xi \in \mathbf{R}^1, \; 0 \leq \tau \leq M\}$ *so that the series* (1.1), (2.2), (3.2), (3.11), (4.2) *are asymptotic expansions for* $u(x, t, \varepsilon)$ *as* $\varepsilon \to 0$ *uniformly in the respective domains* Ω_j.

PROOF. The exact solution of the problem (0.1), (0.2) is sought in the form of the sum:

$$u^{\pm}(x, t, \varepsilon) = U_N^{\pm}(x, t, \varepsilon) + \varepsilon^{N-1} r^{\pm}(\xi, \tau, \varepsilon), \qquad \xi = \varepsilon x, \qquad \tau = \varepsilon t. \tag{5.3}$$

For the functions $r^{\pm}(\xi, \tau, \varepsilon)$ one obtains the nonlinear system equations

$$\begin{aligned}
\left(\frac{\partial}{\partial \tau} \pm \frac{\partial}{\partial \xi} \right) r^{\pm} &+ c_{1,0}^{\pm}(\xi, \tau, \varepsilon) r^+ + c_{0,1}^{\pm}(\xi, \tau, \varepsilon) r^- \\
&= \varepsilon \left\{ f^{\pm}(r^+, r^-, \xi, \tau, \varepsilon) \frac{\partial r^+}{\partial \xi} + g^{\pm}(r^+, r^-, \xi, \tau, \varepsilon) \frac{\partial r^-}{\partial \xi} \right. \\
&\qquad \left. + h^{\pm}(r^+, r^-, \xi, \tau, \varepsilon) \right\},
\end{aligned} \tag{5.4}$$

$$r^{\pm}(\xi, 0, \varepsilon) = \varphi^{\pm}(\xi, \varepsilon). \tag{5.5}$$

The properties of f.a.s. imply that the initial data in this problem are functions that are smooth in r^+, r^-, ξ, τ, uniformly bounded with respect to ε, while their derivatives with respect to ξ, τ are estimated in terms of $O(\varepsilon^{-1})$. In view of the fact that nonlinear terms in (5.4) are preceded by a small factor ε, the problem (5.4), (5.5) is solvable in the strip of finite width $\{\xi \in \mathbf{R}^1, 0 \leq \tau \leq M\}$ for all $\varepsilon \in (0, \varepsilon_0]$, $\varepsilon_0 > 0$ [108, Chapter 2]. The solution $r^{\pm}(\xi, \tau, \varepsilon)$ is bounded uniformly with respect to ε, and its derivatives with respect to ξ, τ are estimated in terms $O(\varepsilon^{-1})$. This proves the existence of a solution of the original problem (0.1), (0.2) in the form (5.3). Since N is an arbitrary number, this provides the justification of the asymptotics for $u^{\pm}(x, t, \varepsilon)$ in the form (5.2) uniformly in the strip Π. The properties of the functions $\chi_j(\xi, \tau, \varepsilon)$, and the matching conditions for different series in the overlap domains imply that each of the series (1.1), (2.2), (3.2), (3.11), (4.2) is an asymptotic expansion of the exact solution $u^{\pm}(x, t, \varepsilon)$ as $\varepsilon \to 0$ in the corresponding domains Ω_j.

REMARK 5.1. The terms in the expansion of $u^{\pm}(x, t, \varepsilon)$ containing $\varepsilon^p \ln^q \varepsilon$ are present only in the domains in the vicinity of singular characteristics and above them $\{\tau + \xi > 0, \tau - \xi > 0\}$, and arise only if the initial functions decay slowly: $\exists p \quad \varphi^{\pm}(x) = O\left(|x|^{-p}\right)$, $|x| \to \infty$ $(0 < p < \infty)$. If the initial data decay rapidly $(\forall p < \infty)$, the series contain no $\ln \varepsilon$ [51]. If $\varphi^{\pm}(x) = O\left(|x|^{-1}\right)$, the presence of a leading logarithmic term with $\varepsilon \ln \varepsilon$ depends on the structure of the initial equations (0.1). In the general case there always are terms with $\varepsilon \ln \varepsilon$ in the layers near the lines $\tau \pm \xi = 0$. In the domain $\{\tau + \xi > 0, \tau - \xi > 0\}$ similar terms can arise only if the first terms (i.e., the coefficients of x^{-1}) in the asymptotics $\varphi^{\pm}(x)$ at $+\infty$ and $-\infty$ do not coincide, i.e., if $\varphi^+_{1,1} \neq \varphi^+_{1,2}$, $\varphi^-_{1,1} \neq \varphi^-_{1,2}$ (see (0.4)).

REMARK 5.2. A similar approach solves the problem in the case where the asymptotics of the initial functions for $\varphi^{\pm}(x)$ is of a more complicated form, e.g., involving fractional powers of $|x| \to \infty$, or powers and logarithms of $|x|$. Naturally, this complicates the structure of the gauge functions causing the appearance of fractional powers of ε and additional powers of $\ln \varepsilon$.

CHAPTER VI

Cauchy Problem for Quasilinear Parabolic Equation with a Small Parameter

Here we consider the following initial value problem([1]):

$$\frac{\partial u}{\partial t} + \frac{\partial \varphi(u)}{\partial x} = \varepsilon^4 \frac{\partial^2 u}{\partial x^2}, \tag{0.1}$$

$$u(x, t_0, \varepsilon) = \psi(x), \qquad x \in \mathbf{R}^1. \tag{0.2}$$

In what follows we assume that $\varphi(u) \in C^\infty$, $\varphi''(u) > 0$, $\psi(x)$ is a bounded and piecewise smooth function. For $\varepsilon > 0$ it is known (see, for example, [60, Chapter 5, §8]) that the problem (0.1), (0.2) has a solution $u(x, t, \varepsilon)$ for $t \geq t_0$. This solution is bounded and infinitely differentiable everywhere for $t \geq t_0$ with the exception of the discontinuity points of the initial function. Our aim is to examine the behavior of the solution $u(x, t, \varepsilon)$ as $\varepsilon \to 0$. A natural first step is to attempt solving the limit problem, i.e., the problem (0.1), (0.2) for $\varepsilon = 0$. At first sight its solution is very simple. Indeed, write the equation in the form

$$\frac{\partial u}{\partial t} + \varphi'(u)\frac{\partial u}{\partial x} = 0. \tag{0.3}$$

Fix some value u^* and consider the line $x - \varphi'(u^*) = \text{const}$ on the plane (x, t). The derivative of the smooth function $u(x, t)$ in the direction of this line equals, up to a scalar multiple,

$$\frac{d}{dt}(u(x(t), t)) = \frac{d}{dt}(u(\varphi'(u^*)t + \text{const}, t)) = \left[\frac{\partial u}{\partial t} + \varphi'(u^*)\frac{\partial u}{\partial x}\right]_{x=\varphi'(u^*)t+\text{const}}$$

Thus, if the function $u(x, t)$ equals the value u^* on the line $x = \varphi'(u^*) + \text{const}$, then $\frac{d}{dt}(u(x(t), t)) = 0$, and the function $u(x, t)$ satisfies equation (0.3) on this line. It is now clear how a solution of the problem (0.3), (0.2) can be constructed. Through each point (y, t_0) on the initial line draw the line

$$x - y = \varphi'(\psi(y))(t - t_0) \tag{0.4}$$

([1])The reason for denoting the coefficient by ε^4 instead of ε will become clear much later in §4. Before that it would be more natural to denote the coefficient by ε, and the only motivation for using ε^4 is to facilitate references from §4 to preceding sections.

on the plane (x, t) and define the function $u(x, t)$ on this line to be equal to its value at the initial moment, i.e., $\psi(y)$.

The lines (0.4) will be called the characteristics of equation (0.3) (although it should be noted that this name is more often given to the lines $x = y + \varphi'(u^*)(t - t_0)$, $u = u^*$ in the three-dimensional space (x, t, u), while their projections on the plane (x, t), i.e., the lines (0.4), are called rays). Then one can simply say that each smooth solution of equation (0.3) is constant on its characteristics. Thus, the solution of the problem (0.3), (0.2) is a function satisfying the initial condition (0.2) and constant on the characteristics.

The simplest interpretation of equation (0.3) is as follows. Let u be the density (depending on a single coordinate x) of a substance moving along the x-axis with local velocity v. If the substance is neither created, nor destroyed in the process, then the variation in its quantity on the interval $[x_1, x_2]$ during the time (t_1, t_2) is generated only by the flow of substance through the endpoints of the interval. Hence

$$\int_{x_1}^{x_2} u\,dx\Big|_{t=t_2} - \int_{x_1}^{x_2} u\,dx\Big|_{t=t_1} = \int_{t_1}^{t_2} vu\,dt\Big|_{x=x_1} - \int_{t_1}^{t_2} vu\,dt\Big|_{x=x_2}$$

$$\Longrightarrow \int_{t_1}^{t_2}\int_{x_1}^{x_2}\left[\frac{\partial u}{\partial t} + \frac{\partial}{\partial x}(vu)\right]dx\,dt = 0,$$

if u and v are smooth functions. Owing to the arbitrariness in the choice of x_1, x_2, t_1, t_2 this implies that $\partial u/\partial t + (\partial/\partial x)(vu) = 0$. This equation bears the name of the continuity equation, and is satisfied, subject to the above assumptions, for the flow of gas or fluid having neither sources, nor sinks. Equation (0.3) is obtained from the continuity equation under the additional assumption that the velocity v depends on the density u: $\varphi(u) = uv(u)$. Such an assumption is reasonable in the so-called traffic problem (where u is the density of cars on the highway), in the problem about flood waves (where u is the area of the river cross-section), in the problem of a glacier motion, and a number of others.

For the problem (0.3), (0.2) the situation is simple and easy in the strip $t_0 \le t \le T$ if the characteristics do not intersect and cover the entire strip. But if some of the characteristics intersect (as in Figures 31 or 32), then, evidently, no smooth solution exists. Note that in Figure 31 the initial function $\psi(x)$ is discontinuous, and, consequently, no smooth solutions exist anywhere for $t > t_0$. In Figure 32 the initial function $\psi(x)$ is smooth so that a smooth solution $u(x, t)$ exists for the values of t close to t_0 and such that $t > t_0$ but beyond some t_1 the smoothness property is lost, and on approaching the point O the gradient of the function $u(x, t)$ increases becoming infinite at the point itself. For that reason the situation depicted in Figure 32 is sometimes called the *gradient catastrophe*. (The point O in

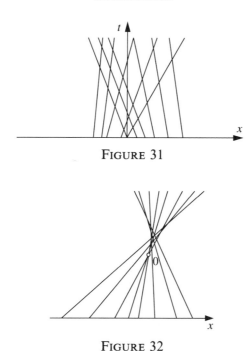

FIGURE 31

FIGURE 32

Figure 32 is also called the *wave breaking point*. On the origin of this name, and more details on the interpretation of equation (0.3) and the structure of its solution the reader is advised to consult [128, Chapter 2].)

Since the problem (0.3), (0.2) admits no smooth solution, it is natural to try to construct a piecewise smooth solution of this problem satisfying equation (0.3) outside the discontinuity curves. However, it is not an easy task because there are many such solutions. For example, in the case shown in Figure 31, one can construct a piecewise smooth solution with a single smooth discontinuity curve $l = \{x, t: x = s(t)\}$ shown in Figure 33 (next page). The discontinuity curve is chosen in such a way that for each point lying to the left or to the right of the curve l there is only one characteristic connecting this point with the initial line $t = t_0$. Clearly, there are infinitely many such curves, and, consequently, infinitely many piecewise smooth solutions. How is the only one between them that is correct to be chosen, and on what principles? The problem can be approached in different ways, but one of them is connected with equation (0.1). It turns out that, subject to the above restrictions on the functions φ and ψ, there exists the limit of the solution $u(x, t, \varepsilon)$ as $\varepsilon \to 0$. This limit is called the generalized solution $u_0(x, t)$ of the problem (0.3), (0.2). In [98] the existence of a generalized solution $u_0(x, t)$ is proved even under more general assumptions (the function $\psi(x)$ is bounded and measurable, and the equation is of a more general form than (0.3)), but in what follows no use is made of the theorem on the existence of a generalized solution.

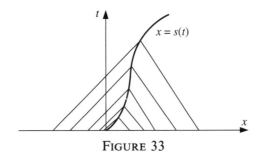

FIGURE 33

Even for a piecewise smooth initial function the structure of the generalized solution may be very complicated. Naturally, the asymptotics of the solution of the problem (0.1), (0.2) as $\varepsilon \to 0$ is in that case even more complicated. Thus, in order to obtain any clear and constructive results on the asymptotics of $u(x, t, \varepsilon)$, we restrict our attention to those cases where the structure of the function $u_0(x, t)$, i.e., the generalized solution of the problem (0.3), (0.2), is simple enough.

Everywhere in this chapter we assume that the generalized solution of the problem (0.3), (0.2) in the strip $\{x, t: t_0 \leq t \leq T, x \in \mathbf{R}^1\}$, i.e., the function $u_0(x, t)$, is smooth everywhere except a single discontinuity curve $l = \{x, t: t_1 \leq t \leq T, x = s(t)\}$. A generalized solution $u_0(x, t)$ is defined as the limit of the solution $u(x, t, \varepsilon)$ as $\varepsilon \to 0$. In what follows, another independent definition for the particular case considered here will be given. §§2 and 3 include conditions on $\varphi(u)$ and $\psi(x)$ sufficient for the generalized solution to have a single smooth discontinuity curve l.

§1. Outer expansion. Asymptotics of the solution near the discontinuity curve

In this section we consider an inner section l^* of the discontinuity curve l, and construct an f.a.s. of equation (0.1) in a neighborhood of l^*. As a by-product, it is clarified what properties the function $s(t)$ defining the discontinuity curve $s(t)$ must have. It turns out that $s(t)$ cannot be a totally arbitrary smooth function—it must satisfy a rather rigid condition. Although at this stage of the investigation the function $s(t)$ and the coefficients of the asymptotic expansion are constructed with a certain degree of arbitrariness, the choice is not especially wide: the function $s(t)$ depends on a single arbitrary constant, and so does each term of the asymptotic expansion. The constants are to be determined in the following sections.

Thus, suppose that $l^* = \{x, t: x = s(t), t_2 \leq t \leq t_3\}$, and for some fixed $p > 0$ there exists a function $u_0(x, t)$ in the domain $\Omega_p^- = \{x, t: s(t) - p < x < s(t), t_2 \leq t \leq t_3\}$ and in the domain $\Omega_p^+ = \{x, t: s(t) < x < s(t) + p, t_2 \leq t \leq t_3\}$ satisfying equation (0.3) outside l^*, has a discontinuity of the first kind on l^*, so that $u_0(s(t) - 0, t) \neq u_0(s(t) + 0, t)$, and becomes infinitely differentiable in the closure of Ω_p^+ and Ω_p^- after being

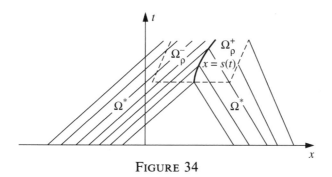

FIGURE 34

extended in a continuous manner to l^* from the left and from the right. We will assume that for each point $(x^*, t^*) \in \Omega_\rho^+ \cup \Omega_\rho^-$ there is a characteristic $\{x, t: x - x^* = (t - t^*)\varphi'(u_0(x^*, t^*))\}$ passing through it and intersecting the initial line at the point (y, t_0). Thus, $u_0(x^*, t^*) = \psi(y)$. Denote the portion of the plane spanned by these characteristics by Ω^* (see Figure 34). Let us also assume that the characteristics depend on the parameter y smoothly everywhere in Ω^*. More precisely, suppose that for each fixed t the mapping $y \mapsto x$ defined by the equation of the characteristics in Ω^* ($x = y + \varphi'(\psi(y))(t - t_0)$) is a diffeomorphism, i.e., $\partial x/\partial y > 0$. In what follows we denote $g(y) = \varphi'(\psi(y))$,

$$\omega(y, t) = 1 + g'(y)(t - t_0), \tag{1.1}$$

and always assume that x, y, and t are connected by the relation

$$x = y + g(y)(t - t_0). \tag{1.2}$$

Thus, under the assumption that

$$1 + g'(y)(t - t_0) > 0 \quad \text{in } \Omega^*, \tag{1.3}$$

we construct an outer expansion in Ω^* in the form

$$U = \sum_{k=0}^{\infty} \varepsilon^{4k} u_{4k}(x, t). \tag{1.4}$$

Here $u_0(x, t)$ is the function defined above ($u_0(x, t) = \psi(y)$), while for $k > 0$ the functions $u_{4k}(x, t)$ must be defined from equation (0.1) and the initial condition (0.2). On inserting the series (1.4) into (0.1) one gets the recurrence system of equations

$$\frac{\partial u_4}{\partial t} + \frac{\partial}{\partial x}(\varphi'(u_0)u_4) = \frac{\partial^2 u_0}{\partial x^2},$$

$$\frac{\partial u_{4k}}{\partial t} + \frac{\partial}{\partial x}(\varphi'(u_0)u_{4k}) \tag{1.5}$$

$$= \frac{\partial^2 u_{4k-4}}{\partial x^2} - \frac{\partial}{\partial x} \sum_{j=2}^{k} \frac{1}{j!} \varphi^{(j)}(u_0) \sum_{\sum i_p = k} \prod_{p=1}^{j} u_{4i_p}, \quad k \geq 2.$$

Condition (0.2) implies that $u_{4k}(x, t_0) = 0$ for $k > 0$. Since the left-hand side of equations (1.5) includes, up to a scalar multiple, the derivative along the characteristic (1.2), it is natural to pass in these equations (sometimes called the transfer equations) to the variables y and t. This change of variables turns equation (1.5) into

$$\frac{\partial}{\partial t}(\omega(y, t)u_{4k}^*(y, t)) = F_k(y, t),$$

where $\omega(y, t)$ is defined by formula (1.1)

$$u_{4k}^*(y, t) \equiv u_{4k}(x, t), \qquad F_1(y, t) = \frac{\partial}{\partial y}\left[\frac{1}{\omega}\frac{\partial u_0^*}{\partial y}\right],$$

$$F_k(k, t) = \frac{\partial}{\partial y}\left[\frac{1}{\omega}\frac{\partial u_{4k-4}^*}{\partial y}\right] \tag{1.6}$$

$$-\frac{\partial}{\partial y}\left[\sum_{j=2}^{k}\frac{1}{j!}\varphi^{(j)}(\psi(y))\sum_{\sum i_p=k}\prod_{p=1}^{j}u_{4i_p}^*\right], \qquad k \geq 2.$$

Hence

$$u_{4k}^*(y, t) = \frac{1}{\omega(y, t)}\int_{t_0}^{t} F_k(y, \theta)\,d\theta. \tag{1.7}$$

It is now clear that all the functions $u_{4k}(x, t)$ are smooth everywhere in $\overline{\Omega^*}$ with the exception of the curve l^* on which all of them together with their derivatives have discontinuities of the first kind. Because of these discontinuities the series (1.4) is not an f.a.s. of equation (0.1) everywhere in $\overline{\Omega^*}$. Evidently, one has to introduce new independent variables in the neighborhood of l^* and look for a new asymptotic expansion.

The natural choice of independent variables is $\zeta = \varepsilon^{-\alpha}(x - s(t))$, t, where $\alpha > 0$. Denoting $u(x, t, \varepsilon) \equiv v(\zeta, t, \varepsilon)$, one obtains the following equation for v:

$$\frac{\partial v}{\partial t} - \varepsilon^{-\alpha}s'(t)\frac{\partial v}{\partial \zeta} + \varepsilon^{-\alpha}\frac{\partial \varphi(v)}{\partial \zeta} = \varepsilon^{4-2\alpha}\frac{\partial^2 v}{\partial \zeta^2}. \tag{1.8}$$

The principal part of equation (1.8) expressed in the inner variable must clearly include the term appearing in the right-hand sides of equation (1.8). However, it is not the only term of the principal part. Hence, $-\alpha = 4 - 2\alpha$, i.e., $\alpha = 4$.

Thus the change of variables is of the form

$$\zeta = \varepsilon^{-4}(x - s(t)), \tag{1.9}$$

and one looks for the inner expansion in the form

$$V = \sum_{i=0}^{\infty}\varepsilon^{4i}v_{4i}(\zeta, t). \tag{1.10}$$

Insertion of this series into the equation

$$\frac{\partial v}{\partial t} = \varepsilon^{-4}\left(\frac{\partial^2 v}{\partial \zeta^2} + s'(t)\frac{\partial v}{\partial \zeta} - \frac{\partial \varphi(v)}{\partial \zeta}\right) \tag{1.11}$$

results in the recurrence system of ordinary differential equations

$$\frac{\partial^2 v_0}{\partial \zeta^2} + s'(t)\frac{\partial v_0}{\partial \zeta} - \frac{\partial \varphi(v_0)}{\partial \zeta} = 0, \tag{1.12}$$

$$L_1 v_{4i} \equiv \frac{\partial^2 v_{4i}}{\partial \zeta^2} + \frac{\partial}{\partial \zeta}\{[s'(t) - \varphi'(v_0)]v_{4i}\}$$
$$= \frac{\partial v_{4i-4}}{\partial t} + \frac{\partial}{\partial \zeta}G_i(v_0, v_4, \dots, v_{4i-4}), \qquad i > 0, \tag{1.13}$$

where

$$G_i(v_0, v_4, \dots, v_{4i-4}) = \sum_{q \geq 2}\frac{\varphi^{(q)}(v_0)}{q!}\sum_{\sum j_p = i}\prod_{p=1}^{q}v_{4j_p}. \tag{1.14}$$

Here the variable t plays the role of a parameter.

For $\zeta \to \infty$ the series (1.10) has to be matched to the series (1.4) for $x \to s(t) + 0$, and for $\zeta \to -\infty$ the series (1.10) has to be matched to the series (1.4) for $x \to s(t) - 0$. The matching condition involving all terms of the series (1.4) and (1.10) will be dealt with later; we turn our attention now just to the conditions $v_0(\zeta, t)$ has to satisfy at infinity. Clearly, the following relations must be satisfied:

$$v_0(-\infty, t) = u_0(s(t) - 0, t),$$
$$v_0(+\infty, t) = u_0(s(t) + 0, t). \tag{1.15}$$

Integrating equation (1.12) with respect to ζ, one has

$$\frac{\partial v_0}{\partial \zeta} = \varphi(v_0) - s'(t)v_0 - C(t). \tag{1.16}$$

Since the function $v_0(\zeta, t)$ must have limits as $\zeta \to \pm\infty$, equation (1.16) implies the existence of the limits of $\partial v_0/\partial \zeta$ for $\zeta \to \pm\infty$. Hence, $(\partial v_0/\partial \zeta)$ $\cdot (\zeta, t) \to 0$ as $\zeta \to \pm\infty$. Thus, a necessary condition for the solvability of the problem (1.12), (1.15) is given by the equalities

$$C(t) = \varphi(u_0(s(t) - 0, t)) - s'(t)u_0(s(t) - 0, t)$$
$$= \varphi(u_0(s(t) + 0, t)) - s'(t)u_0(s(t) + 0, t) \tag{1.17}$$

or, what is the same,

$$\frac{ds}{dt} = \frac{\varphi(u_0(s(t) + 0, t)) - \varphi(u_0(s(t) - 0, t))}{u_0(s(t) + 0, t) - u_0(s(t) - 0, t)}. \tag{1.18}$$

Thus, one can now see that the discontinuity curve $l = \{x, t: x = s(t)\}$ for the limit generalized solution $u_0(x, t)$ cannot be defined by an arbitrary smooth function, if in a neighborhood of l the asymptotics of the solution of the perturbed problem (0.1), (0.2) is to be described by the series (1.10). Equality (1.18) bears the name of the *Hugoniot condition*. Since the function $\varphi(u)$ is convex downward, and, by virtue of (1.17), the right-hand side of equality (1.16) vanishes for both $v_0 = u_0(s(t)-0, t)$ and $v_0 = u_0(s(t)+0, t)$, this right-hand side is negative in the interval between these two values of v_0. Thus, $\partial v_0 / \partial \zeta < 0$ for a solution of the problem (1.12), (1.15) and

$$u_0(s(t) - 0, t) > u_0(s(t) + 0, t). \tag{1.19}$$

We now show that (1.18), (1.19) provide not only necessary, but also sufficient conditions for the existence of the series (1.10) providing an f.a.s. of equation (0.1) in a neighborhood of l^* that is matched to the series (1.4). First, we construct a solution of the problem (1.12), (1.15). Define $C(t)$ by formulas (1.17), which, in view of (1.18), lead to no contradiction, and then write out the explicit solution of equation (1.16):

$$\int\limits_{a}^{v_0} [\varphi(z) - s'(t)z - C(t)]^{-1} \, dz = \zeta + c_0(t). \tag{1.20}$$

Here a is a number lying between $v_0(-\infty, t)$ and $v_0(+\infty, t)$, and $c_0(t)$ is an arbitrary function. We assume that a is a constant independent of t. The convexity of $\varphi(u)$, equalities (1.17), and inequality (1.19) imply that the function $v_0(\zeta, t)$ defined by relation (1.20) is a smooth function of ζ for $\zeta \in \mathbf{R}^1$ which tends to its limits at $\pm\infty$ exponentially. Denote by $\tilde{v}(\zeta, t)$ the function defined by equality (1.20) with $c_0(t) \equiv 0$. Thus, if the Hugoniot condition (1.18) and inequality (1.19) are satisfied, there exists a solution of the problem (1.12), (1.15) determined to within a shift of the independent variable:

$$v_0(\zeta, t) = \tilde{v}(\zeta + c_0(t), t). \tag{1.21}$$

We now proceed with the construction of $v_{4i}(\zeta, t)$ for $i \geq 1$. First, let us examine the conditions $v_{4i}(\zeta, t)$ should satisfy at infinity if the matching conditions for the series (1.4) and (1.10) are satisfied. Since, after being continuously extended to l^*, the functions $u_k(x, t)$ are smooth everywhere in $\overline{\Omega_\rho^-}$ and $\overline{\Omega_\rho^+}$, one has

$$A_{4m, \zeta, t} A_{4n, x, t} U = \sum_{i=0}^{m} \varepsilon^{4i} P_i^{\pm}(\zeta, t)$$

as $x \to s(t) \pm 0$. Here $P_i^{\pm}(\zeta, t)$ are polynomials in ζ with coefficients smoothly depending on t:

$$P_i^{\pm}(\zeta, t) = \sum_{j=0}^{i} \zeta^j \frac{1}{j!} \frac{\partial^j u_{i-j}}{\partial x^j}(s(t) \pm 0, t). \tag{1.22}$$

The matching condition

$$A_{4n,x,t}A_{4m,\zeta,t}V = A_{4m,\zeta,t}A_{4n,x,t}U$$

implies the relations

$$v_{4i}(\zeta,t) - P_i^{\pm}(\zeta,t) \underset{\zeta\to\pm\infty}{\longrightarrow} 0. \tag{1.23}$$

Here the function must decay faster than any power of ζ. Since the functions $u_{4k}(x,t)$ are smooth, the problem is not bisingular in a neighborhood of the curve l^*. In fact, $v_{4i}(\zeta,t)$ are boundary layer functions of the type considered in §1, and can be represented as sums of polynomials and functions decaying exponentially at infinity. The only difference is that now the functions v_{4i} do not serve to eliminate the residual in the boundary condition, but rather to smooth the discontinuity on the curve l^*. They are, therefore, defined for all $\zeta \in \mathbf{R}^1$, and have different asymptotic representations at $\pm\infty$. This difference is, however, inessential.

Thus, we have to examine the boundary value problems (1.13), (1.23). Note that the polynomials $P_i^{\pm}(\zeta,t)$ satisfy exactly the same recurrence system as the system (1.12), (1.14):

$$\frac{\partial^2 P^i}{\partial\zeta^2} + \frac{\partial}{\partial\zeta}\{[s'(t) - \varphi'(P_0(t))]P_i\}$$

$$= \frac{\partial P_{i-1}}{\partial t} + \frac{\partial}{\partial\zeta}G_i(P_0, P_1, \ldots, P_{i-1}), \qquad i > 0, \tag{1.24}$$

where G_i are the same functions as in the right-hand sides of equalities (1.14). The system (1.24) can be easily obtained by taking into account the fact that the series is an f.a.s. of equation (0.1); it suffices to pass to the variables ζ, t in the corresponding equality for $A_{n,x,t}U$. Let

$$Z(\zeta,t) = \frac{\partial\tilde{v}}{\partial\zeta}(\zeta,t), \tag{1.25}$$

where $\tilde{v}(\zeta,t)$ is the solution of equation (1.12) defined above. Differentiating this equation with respect to ζ, one concludes that $Z(\zeta,t)$ is a solution of the homogeneous equation $L_1Z = 0$.

For $i = 1$ equations (1.13) and (1.24) yield the equality

$$L_1 Q^{\pm} = f^{\pm}(\zeta,t), \tag{1.26}$$

where

$$Q^{\pm}(\zeta,t) = v_4(\zeta,t) - P_1^{\pm}(\zeta,t), \tag{1.27}$$

and

$$f^{\pm}(\zeta,t) = \frac{\partial v_0}{\partial t} - \frac{\partial P_0^{\pm}}{\partial t} + \frac{\partial}{\partial\zeta}\left\{\left[\varphi'(v_0) - \varphi'(P_0^{\pm}(t))\right]\sqrt{P_1^{\pm}}\right\}.$$

We seek solutions of equations (1.26) in the following form:

$$Q^{\pm}(\zeta, t) = Z(\zeta, t) \int_0^{\zeta} [Z(\xi, t)]^{-1} \int_{\pm\infty}^{\xi} f^{\pm}(\eta, \, , t) \, d\eta \, d\xi + K^{\pm}(t) Z(\zeta, t),$$

(1.28)

where $Z(\zeta, t)$ is the function (1.25). The functions $Q^{\pm}(\zeta, t)$ obviously satisfy equations (1.26) and decay exponentially as $\zeta \to \pm\infty$. Therefore, the function $v_4(\zeta, t)$ defined, according to (1.27), as $Q^+(\zeta, t) + P_1^+(\zeta, t)$ for $\zeta > 0$ and as $Q^-(\zeta, t) + P_1^-(\zeta, t)$ for $\zeta \leq 0$, is the solution of equation (1.13) for $\zeta \neq 0$ satisfying conditions (1.23). Thus, a sufficient condition for the function $v_4(\zeta, t)$ constructed to be the desired smooth solution everywhere for $\zeta \in \mathbf{R}^1$, is that it be continuous together with its first derivative for $\zeta = 0$:

$$Q^+(0, t) - Q^-(0, t) = P_1^-(0, t) - P_1^+(0, t),$$
$$\frac{\partial Q^+}{\partial \zeta}(0, t) - \frac{\partial Q^-}{\partial \zeta}(0, t) = \frac{\partial P_1^-}{\partial \zeta}(0, t) - \frac{\partial P_1^+}{\partial \zeta}(0, t).$$

(1.29)

One can easily check that these conditions are not only necessary, but also sufficient for the solvability of the problem (1.13), (1.23) for $i = 1$.

Substituting expressions (1.28) for $Q^{\pm}$ into equalities (1.29), one has

$$\left[K^+(t) - K^-(t) \right] Z(0, t) = P_1^-(0, t) - P_1^+(0, t),$$

$$\int_{+\infty}^0 f^+(\eta, t) \, d\eta - \int_{-\infty}^0 f^-(\eta, t) \, d\eta + \left[K^+(t) - K^-(t) \right] \frac{dZ}{d\zeta}(0, t)$$

$$= \frac{\partial P_1^-}{\partial \zeta}(0, t) - \frac{\partial P_1^+}{\partial \zeta}(0, t).$$

Taking into account the form of $f^{\pm}(\zeta, t)$ and the equation $L_1 Z = 0$, one obtains the equality

$$\int_{-\infty}^0 \left[\frac{dP_0^-(t)}{dt} - \frac{\partial v_0(\zeta, t)}{\partial t} \right] d\zeta + \int_0^{\infty} \left[\frac{dP_0^+(t)}{dt} - \frac{\partial v_0(\zeta, t)}{\partial t} \right] d\zeta$$

$$+ \left[\varphi'(v_0(0, t)) - \varphi'(P_0^+(t)) \right] P_1^+(0, t)$$

$$- \left[\varphi'(v_0(0, t)) - \varphi'(P_0^-(t)) \right] P_1^-(0, t)$$

$$- \left[P_1^-(0, t) - P_1^+(0, t) \right] [s'(t) - \varphi'(v_0(0, t))]$$

$$= \frac{\partial P_1^-}{\partial \zeta}(0, t) - \frac{\partial P_1^+}{\partial \zeta}(0, t).$$

(1.30)

Since $v_0(\zeta, t) = \tilde{v}(\zeta + c_0(t), t)$ (see (1.21)), relation (1.30) yields the following equation for $c_0(t)$:

$$\frac{d}{dt}\left\{c_0(t)\left[P_0^+(t) - P_0^-(t)\right]\right\}$$

$$= \frac{d}{dt}\int_{-\infty}^{0}\left[P_0^-(t) - \tilde{v}(\zeta, t)\right] d\zeta$$

$$+ \frac{d}{dt}\int_{0}^{\infty}\left[P_0^+(t) - \tilde{v}(\zeta, t)\right] d\zeta + \frac{\partial}{\partial\zeta}\left[P_1^+(\zeta, t) - P_1^-(\zeta, t)\right]$$

$$- \left[\varphi'(P_0^+(t)) - s'(t)\right]P_1^+(0, t) + \left[\varphi'(P_0^-(t)) - s'(t)\right]P_1^-(0, t). \quad (1.31)$$

Thus, the function $c_0(t)$ in formula (1.21) is not arbitrary, and has to satisfy equation (1.31). Both the function $s(t)$, and the function $c_0(t)$ are, therefore, determined, up to a single arbitrary constant, as solutions of the first order equations (1.18) and (1.31). We have seen that for such $s(t)$ and $c_0(t)$ there exists a solution of the problem (1.13), (1.23) for $i = 1$. The function $v_4(\zeta, t)$ is determined up to the term $c_1(t)Z(\zeta, t)$. The existence of solutions of the problems (1.13), (1.23) for other i can then be established by induction, but first we prove the following auxiliary lemma.

Denote by $\mathfrak{M}^+$ the set of smooth functions $v(\zeta, t)$ defined for $\zeta \in \mathbf{R}^1$, $t_2 \le t \le t_3$, such that for some $\gamma > 0$

$$\left|\frac{\partial^{i+j}}{\partial\zeta^i\partial t^j}v(\zeta, t)\right| \le M_{i,j}\exp(-\gamma\zeta) \quad \forall i, j.$$

Denote by $\mathfrak{M}^-$ the analogous set of functions $v(\zeta, t)$ for which

$$\left|\frac{\partial^{i+j}}{\partial\zeta^i\partial t^j}v(\zeta, t)\right| \le M_{i,j}\exp(\gamma\zeta) \quad \forall i, j.$$

LEMMA 1.1. *Suppose that the functions $P^+(\zeta, t)$, $P^-(\zeta, t)$, $F(\zeta, t)$ are smooth for $\zeta \in \mathbf{R}^1$, $t_2 \le t \le t_3$, and $F(\zeta, t) - L_1P^+(\zeta, t) \in \mathfrak{m}^+$ and $F(\zeta, t) - L_1P^-(\zeta, t) \in \mathfrak{m}^-$, where L_1 is the operator (1.13). The problem*

$$L_1v = F(\zeta, t),$$
$$v(\zeta, t) - P^+(\zeta, t) \in \mathfrak{m}^+, \qquad v(\zeta, t) - P^-(\zeta, t) \in \mathfrak{m}^- \quad (1.32)$$

has a solution if and only if

$$\left\{\frac{\partial}{\partial\zeta}(P^+ - P^-) + [s'(t) - \varphi'(v_0)](P^+ - P^-)\right\}_{\zeta=0}$$

$$= \int_{-\infty}^{0}\left[F(\zeta, t) - L_1P^-(\zeta, t)\right] d\zeta + \int_{0}^{\infty}\left[F(\zeta, t) - L_1P^+(\zeta, t)\right] d\zeta. \quad (1.33)$$

PROOF. The necessity of condition (1.33) is easily obtained by integrating the equation for $v(\zeta, t) - P^+(\zeta, t)$ termwise from zero to infinity, integrating the equation for $v(\zeta, t) - P^-(\zeta, t)$ from $-\infty$ to zero, and then subtracting the resulting equalities term-by-term.

The sufficiency of condition (1.33) follows, for example, from the explicit formula for the solution $v(\zeta, t)$. Denote

$$F^-(\zeta, t) = \int_{-\infty}^{\zeta} \left[F(\zeta_1, t) - L_1 P^-(\zeta_1, t) \right] d\zeta_1,$$

$$F^+(\zeta, t) = -\int_{\zeta}^{\infty} \left[F(\zeta_1, t) - L_1 P^+(\zeta_1, t) \right] d\zeta_1$$

and let

$$v(\zeta, t) = \begin{cases} P^-(\zeta, t) + Z(\zeta, t) \left\{ \int_0^{\zeta} \dfrac{F^-(\zeta_1, t)}{Z(\zeta_1, t)} d\zeta_1 + \dfrac{P^+(0, t)}{Z(0, t)} \right\} & \text{for } \zeta < 0, \\[2em] P^+(\zeta, t) + Z(\zeta, t) \left\{ \int_0^{\zeta} \dfrac{F^+(\zeta_1, t)}{Z(\zeta_1, t)} d\zeta_1 + \dfrac{P^-(0, t)}{Z(0, t)} \right\} & \text{for } \zeta \geq 0. \end{cases}$$

$$(1.34)$$

By construction, the function $v(\zeta, t)$ satisfies the equation and conditions (1.32) for $\zeta \neq 0$. For $\zeta = 0$ this function is evidently continuous. A direct verification easily shows that, in view of condition (1.33) and the equation $L_1 Z = 0$, the derivative $\partial v / \partial \zeta$ is also continuous at zero. ∎

THEOREM 1.1. *There exist solutions of the problems* (1.13), (1.23) *for* $\zeta \in \mathbf{R}^1$, $t_2 \leq t \leq t_3$ *such that* $v_{4i}(\zeta, t) - P^\pm(\zeta, t) \in \mathfrak{M}^\pm$. *Assume that all* $v_{4j}(\zeta, t)$ *are already determined for* $j < i$, $i > 0$; *then* $v_{4i}(\zeta, t)$ *is determined up to the additive term* $c_i [\varkappa(t)]^{-1} Z(\zeta, t)$, *where* c_i *is an arbitrary constant,* $\varkappa(t) = P_0^+(t) - P_0^-$, *and* $Z(\zeta, t)$ *is defined by formula* (1.25).

PROOF. We have already constructed the function $v_0(\zeta, t) = \tilde{v}(\zeta + c_0(t), t)$ such that the problem (1.13), (1.23) for $i = 1$ has a solution. Now the proof proceeds by induction. Suppose that solutions of the problems (1.13), (1.23) for $i < n$ are already constructed in such a way that this problem has a solution for $i = n$. Denote such a fixed solution by $v_{4n}^*(\zeta, t)$. Then any solution can be presented in the form $v_{4n}^*(\zeta, x) + c_n Z(\zeta, t)$. Let us find the function $c_n(t)$ such that the problem (1.13), (1.23) has a solution for $i = n+1$. Apply Lemma 1.1 to the function v_{4n+4} taking for $P^\pm(\zeta, t)$ the polynomials $P_{n+1}^\pm(\zeta, t)$ in formula (1.22). Equations (1.13) and (1.24) for $i = n + 1$ and the inductive hypothesis imply that the conditions of Lemma 1.1 are satisfied, where $F(\zeta, t) = \partial v_{4n} / \partial t + (\partial / \partial \zeta) G_{n+1}(v_0, v_4, \ldots, v_{4n})$, $P^\pm(\zeta, t) = P_{n+1}^\pm(\zeta, t)$. According to Lemma 1.1 the problem (1.13), (1.23) for $i = n + 1$ has a solution if and only if condition (1.33) is satisfied. Substituting the functions $v_{4i}(\zeta, t)$ for $i < n$ and $v_{4n}(\zeta, t) = v_{4n}^*(\zeta, t) + c_n(t) Z(\zeta, t)$ into this equality we see that the function $c_n(t)$ enters the resulting relation only in the following form

$$\int_{-\infty}^{0} \frac{\partial}{\partial t} (c_n(t)Z(\zeta, t)) \, d\zeta + \int_{0}^{\infty} \frac{\partial}{\partial t} (c_n(t)Z(\zeta, t)) \, d\zeta$$

$$= \frac{d}{dt} \left[c_n(t) \int_{-\infty}^{\infty} \frac{\partial \tilde{v}}{\partial \zeta} (\zeta, t) \, d\zeta \right] = \frac{d}{dt} \left\{ c_n(t) \left[P_0^+(t) - P_0^-(t) \right] \right\}.$$

Thus, condition (1.33) is of the form $\frac{d}{dt}(c_n(t)\varkappa(t)) = f_n(t)$, where $f_n(t)$ is a known smooth function. Therefore, the solution $v_n(\zeta, t)$ is defined up to the term $c_n[\varkappa(t)]^{-1}Z(\zeta, t)$, and the problem (1.13), (1.23) has a solution for $i = n + 1$. ∎

§2. Shock wave caused by discontinuity of the initial function

Consider the problem (0.1), (0.2) in the case where the function $\psi(x)$ has a single point of discontinuity of the first kind x_0 such that $\psi(x_0 + 0) < \psi(x_0 - 0)$. Without any loss of generality, we can assume that $x_0 = 0$ and $t_0 = 0$. Suppose that the characteristics are positioned on the plane (x, t) as in Figure 33. The solution $u(x, t, \varepsilon)$ will be considered in the strip $\Omega = \{x, t : x \in \mathbf{R}^1, 0 \le t \le T\}$. Choose the number $T > 0$ such that there exists a generalized solution $u_0(x, t)$ of the problem (0.3), (0.2) in Ω with a single discontinuity curve $l = \{x, t : x = s(t), 0 \le t \le T\}$. Such a value of T evidently exists provided the function $\psi(x)$ is bounded for $x \ne 0$ together with its derivatives of sufficiently high order. Indeed, since the function $\psi(x)$ is smooth for $x \ne 0$, the characteristics passing through the points $(y, 0)$, $y > 0$, do not intersect for $0 \le t \le T$. The same applies also for characteristics passing through the negative half of the x-axis. Thereby, a smooth solution $u^+(x, t)$ of equation (0.3) satisfying condition (0.2) for $x > 0$ is defined for $0 \le t \le T$, $\varphi'(\psi(+0))t \le x < \infty$. Similarly, a smooth solution $u^-(x, t)$ is defined for $-\infty < x \le \varphi'(\psi(-0))t$. The continuous function

$$\frac{\varphi(u^+(x, t)) - \varphi(u^-(x, t))}{u^+(x, t) - u^-(x, t)}$$

is defined inside the angle $\varphi'(\psi(+0))t \le x \le \varphi'(\psi(-0))t$ so that the smooth function $s(t)$ is defined for $0 \le t \le T$ as a solution of equation (1.18) in a neighborhood of zero ($s(0) = 0$). For $x < s(t)$, the function $u_0(x, t)$ is set to be equal to $u^-(x, t)$, and for $x > s(t)$ to $u^+(x, t)$.

The function constructed in this way is a generalized solution of the problem (0.2), (0.3). Next, according to the analysis of §1, one constructs the outer expansion (1.4) which is an f.a.s. of the problem (0.3), (0.2) everywhere for $x \ne s(t)$. In a neighborhood of the curve l one can, as in §1, define the inner asymptotic expansion (1.10). The composite asymptotic expansion

$$Y_{4n}(x, t, \varepsilon) = A_{4N, x, t} U + A_{4N, \zeta, t} V - A_{4N, x, t} A_{4N, \zeta, t} V$$

is, by construction, an approximate solution of equation (0.1) everywhere in $\overline{\Omega}$:

$$\left| \frac{\partial Y_{4N}}{\partial t} + \frac{\partial}{\partial x} \varphi(Y_{4N}) - \varepsilon^4 \frac{\partial^2 Y_{4N}}{\partial x^2} \right| \leq M\varepsilon^{4N}. \tag{2.1}$$

Inequality (2.1) follows directly from the properties of the functions $u_{4k}(x, t)$ and $v_{4i}(\zeta, t)$.

Two related factors now stand in the way of finally constructing the asymptotics of the solution $u(x, t, \varepsilon)$ of the problem (0.1), (0.2). First, as we have seen in §1, the coefficients $v_{4i}(\zeta, t)$ are not defined uniquely. The function $v_0(\zeta, t)$ is defined up to the shift of the first argument by $c_0(t)$, where $c_0(t)$ is a solution of equation (1.31), while $v_{4i}(\zeta, t)$ for $i > 0$ are defined up to the term $c_i[\varkappa(t)]^{-1} Z(\zeta, t)$. Second, the composite asymptotic expansion $Y_{4N}(x, t, \varepsilon)$ does not approximate the initial function $\psi(x)$ in a neighborhood of zero. The approximation is satisfactory for $x \gg \varepsilon^4$, but fails for $x = O(\varepsilon^4)$: the function $\psi(x)$ is discontinuous while $Y_{4N}(x, 0, \varepsilon)$ is continuous to within $O(\varepsilon^{4N})$. For that reason we consider a new asymptotic expansion in the neighborhood of the origin. It is intuitively clear that after a change of variables all terms in equation (0.1) must become principal. Therefore, the stretching coefficient for both x and t must be the same. It is convenient to introduce the variables

$$\zeta = \varepsilon^{-4}(x - s(t)), \qquad \tau = \varepsilon^{-4}t. \tag{2.2}$$

The equation for the function $w(\zeta, \tau, \varepsilon) \equiv u(x, t, \varepsilon)$ is of the following form:

$$\frac{\partial^2 w}{\partial \zeta^2} + s'(\varepsilon^4 \tau)\frac{\partial w}{\partial \zeta} - \frac{\partial \varphi(w)}{\partial \zeta} - \frac{\partial w}{\partial \tau} = 0.$$

The inner asymptotic expansion is looked for in the form

$$W = \sum_{i=0}^{\infty} \varepsilon^{4i} w_{4i}(\zeta, \tau). \tag{2.3}$$

The equations and initial conditions for the functions w_{4i} are obtained in the usual way:

$$\frac{\partial^2 w_0}{\partial \zeta^2} + s'(0)\frac{\partial w_0}{\partial \zeta} - \frac{\partial \varphi(w_0)}{\partial \zeta} - \frac{\partial w_0}{\partial \tau} = 0, \tag{2.4}$$

$$L_2 w_{4i} \equiv \frac{\partial^2 w_{4i}}{\partial \zeta^2} + \frac{\partial}{\partial \zeta}\{[s'(0) - \varphi'(w_0)]w_{4i}\} - \frac{\partial w_{4i}}{\partial \tau}$$

$$= \frac{\partial}{\partial \zeta}G_i(w_0, w_4, \dots, w_{4i-4}) - \sum_{i=1}^{i} \frac{\tau^j}{j!}\frac{d^{j+1}s}{dt^{j+1}}(0)\frac{\partial w_{4(i-j)}}{\partial \zeta}, \quad i \geq 1, \tag{2.5}$$

where G_i are the same functions as in (1.14),

$$w_{4i}(\zeta, 0) = \begin{cases} \dfrac{1}{i!}\dfrac{d^i\psi}{dx^i}(+0)\zeta^i & \text{for } \zeta > 0, \\[3mm] \dfrac{1}{i!}\dfrac{d^i\psi}{dx^i}(-0)\zeta^i & \text{for } \zeta < 0. \end{cases} \tag{2.6}$$

The asymptotics of solutions of the problems (2.4)–(2.6) as $\tau \to \infty$ is quite easily examined by the same methods as those used in [44]. The following theorem holds.

THEOREM 2.1. *There exists a solution* $w_0(\zeta, \tau)$ *of the problem* (2.4), (2.6) *bounded for* $\tau \geq 0$ *and infinitely differentiable for* $|\zeta| + \tau > 0$. *The following estimate holds*:

$$|w_0(\zeta, \tau) - \tilde{v}(\zeta + h, 0)| < M \exp(\gamma(|\zeta| + \tau)), \tag{2.7}$$

where $\gamma > 0$, $\tilde{v}(\zeta, t)$ *is the function defined in* §1 *(see* (1.21)*), and the constant* h *is such that*

$$\int_{-\infty}^{\infty} [\tilde{v}(\zeta + h, 0) - w_0(\zeta, 0)]\, d\zeta = 0.$$

It is, therefore, clear that the matching conditions for the series (1.10) and (2.3) require $c_0(0) = h$ which defines the function $c_0(t)$ from equation (1.31), and thereby the function $v_0(\zeta, t)$, uniquely.

THEOREM 2.2. *There exist solutions of the problems* (2.5), (2.6) *for* $i > 0$ *which grow at infinity not faster than some power of* $|\zeta| + \tau$, *and are infinitely differentiable for* $|\zeta| + \tau > 0$.

In order to examine the asymptotics of the solutions $w_{4i}(\zeta, \tau)$ as $\tau \to \infty$ and thereby find the values $c_i(0)$ (see Theorem 1.1), consider the partial sum $A_{4N,\zeta,t}V$ of the series (1.10). Assuming that the functions $c_k(t)$ are smooth, we obtain

$$A_{4N,\zeta,\tau}A_{4N,\zeta,t}V = \sum_{j=0}^{N} \varepsilon^{4j} R_j(\zeta, \tau), \tag{2.8}$$

$$|A_{4N,\zeta,t}V - A_{4N,\zeta,\tau}A_{4N,\zeta,t}V| < M\varepsilon^{4N+1}(1 + |\zeta|^N)(1 + \tau^{N+1}), \tag{2.9}$$

where $R_j(\zeta, \tau)$ are polynomials with respect to τ. One can easily see that $R_j(\zeta, \tau)$ satisfies the system (2.4), (2.5), and $R_j(\zeta, 0) = v_{4j}(\zeta, 0)$. Since $v_{4j}(\zeta, t) = P_j^{\pm}(\zeta, t) + O(\exp(-\gamma|\zeta|))$,

$$R_j(\zeta, 0) = w_{4j}(\zeta, 0) + O(\exp(-\gamma|\zeta|)). \tag{2.10}$$

THEOREM 2.3. *There exist initial data* $c_j(0)$ *such that the functions* $c_j(t)$ *and* $v_{4j}(\zeta, t)$ *constructed from them satisfy the estimates*

$$\left| \frac{\partial^{l+m}}{\partial\zeta^l \partial\tau^m}(w_{4j}(\zeta, \tau) - R_j(\zeta, \tau)) \right| < M \exp\{-\gamma(|\zeta| + \tau)\} \quad \forall l, m \tag{2.11}$$

for $|\tau| + |\zeta| \geq 1$. Here $R_j(\zeta, \tau)$ are the functions defined by equality (2.8), and M and γ are positive constants depending on j, l, and m.

For smooth initial data the existence of solutions for parabolic equations (2.4), (2.5) is, for example, established in the article [99]. The initial data (2.6) are not smooth at zero which makes the direct application of the theorem from [99] impossible. There is, however, no difficulty in writing out a formal asymptotic expansion of the solution in a neighborhood of zero in the form of the series $\sum_{l=0}^{\infty} \tau^{l/2} q_l(\zeta \tau^{-1/2})$. Then, multiplying the partial sum of this series by a truncating function and subtracting $w_i(\zeta, \tau)$, we arrive at a problem for which the existence of a solution for small τ can be proved, for example, by the method of successive approximations. The smoothness of the solution for $|\xi| + \tau > 0$ is proved in a similar way. The estimate (2.7) is proved as in [44 (Theorem 1)].

The estimates (2.11) are proved in the same way. It suffices to introduce the functions $\int_{-\infty}^{\zeta} [w_{4j}(\eta, \tau) - R_j(\eta, \tau)] d\eta$, and consider the equations and initial conditions arising for them. ∎

Once the asymptotic series (1.4), (1.10), and (2.3) are constructed, it is, as usual, easy to form the function providing a uniform approximation to the initial solution $u(x, t, \varepsilon)$ everywhere in the strip Ω. Indeed, consider $X_N(x, t, \varepsilon) = A_{4N,x,t}U + A_{4N,\zeta,t}V - A_{4N,x,t}A_{4N,\zeta,t}V + A_{4N,\zeta,\tau}W - A_{4N,\zeta,t}A_{4N,\zeta,\tau}W$, where U, V, and W are the series (1.4), (1.10), and (2.3), while ζ and τ are defined by (2.2). It follows from (2.1), (2.9)–(2.11) that

$$\frac{\partial X_N}{\partial t} + \frac{\partial}{\partial x}\varphi(X_N) - \varepsilon^4 \frac{\partial^2 X_N}{\partial x^2} = O\left(\varepsilon^{\alpha N}\right),$$

$$X_N(x, 0) = \psi(x) + O\left(\varepsilon^{\alpha N}\right),$$

where $\alpha > 0$. Then the maximum principle, for example, implies that $|X_N(x, t, \varepsilon) - u(x, t, \varepsilon)| < M\varepsilon^{\beta N}$, $\beta > 0$, everywhere in $\overline{\Omega}$.

It has already been mentioned that the problem considered in this section is not a bisingular one. The difference between individual asymptotic expansions exhibits the character of exponentially decaying boundary layer functions. Nevertheless, the matching ideology is helpful in this case as well.

The problem becomes essentially bisingular if the discontinuity curve (the shock wave) does not arise at the initial moment, but only after a certain period of time. This is a much more interesting and complicated case to be treated in the following sections of the present chapter.

§3. Breaking of waves. Smoothness of the discontinuity curve. Asymptotics of the outer expansion coefficients

We begin by examining the asymptotics of the solution of the problem (0.1), (0.2) in the case where the initial function $\psi(x)$ is smooth, and the characteristics for equation (0.3) are positioned as in Figure 32. We assume, without any loss of generality, that $t_0 = -1$, and the discontinuity curve l issues from the point $(0, 0)$: $l = \{x, t: x = s(t), 0 \leq t \leq T\}$. Suppose that the Hugoniot condition (1.18) is satisfied on the line l.

Let Ω be the strip $\{x, t: -1 \leq t \leq T, x \in \mathbf{R}^1\}$. For what follows it is convenient to introduce the set Ω' obtained from $\Omega \setminus \bar{l}$ by adding the edges of the cut $(s(t) \pm 0, t)$ for $t > 0$. Denote by $C^\infty(\Omega')$ the set of functions continuous in $\Omega \setminus \bar{l}$ together with all their derivatives, and having limits for $t \to t^* > 0$, $x \to s(t^*) \pm 0$. In general, these functions may have strong singularities at the origin.

Suppose that the following condition is satisfied:

(a) $u_0(x, t) \in C^\infty(\Omega')$, $\omega(y, t) > 0$ in Ω', where the function $\omega(y, t)$ is defined by formula (1.1). A more constructive condition on the functions φ and ψ sufficient for condition (a) to be satisfied will be given below.

Without any loss of generality, we assume that $y_0 = 0$, $\psi(0) = 0$, $\varphi(0) = \varphi'(0) = 0$. These equalities can be easily obtained by a linear change of the independent variables, and a linear change of the functions u and $\varphi(u)$. Thus $g'(0) = -1$, where $g(y) = \varphi'(\psi(y))$. Furthermore, since, by condition (a), the discontinuity curve is unique, this implies that the function $g'(y)$ attains its minimum at zero. Hence, $g''(0) = 0$, $g'''(0) \geq 0$. Another essential assumption is that $g'''(0) > 0$. The rate with which the function $g''(y)$ tends to zero makes an essential impact on the form of the asymptotic expansion of the solution $u(x, t, \varepsilon)$. Consider the case of the general position corresponding to the inequality $g'''(0) > 0$. By stretching the independent variable x one can, in this case, make $g'''(0)$ equal to any positive number. Suppose that $g'''(0) = 6$.

Thus, we assume that

(b) $\varphi(0) = \varphi'(0) = 0$, $\psi(0) = 0$, $g(0) = g''(0) = 0$, $g'(0) = -1$, $g'''(0) = 6$, where $g(y) = \varphi'(\psi(y))$.

The purpose of our investigation is to find an asymptotic expansion of the solution of equation (0.1) under the assumption that

$$u(x, -1, \varepsilon) = \psi(x) \tag{3.1}$$

as $\varepsilon \to 0$, and conditions (a) and (b) are satisfied.

Denote by $y_+(t)$ and $y_-(t)$ the values of the parameter on the characteristics passing through the discontinuity curve $l = \{x, t, : x = s(t)\}$ from the right and left, respectively. Thus, for $t > 0$, according to (1.2), the equalities

$$s(t) = y_\pm(t) + (1+t)g(y_\pm(t)), \qquad y_+(t) > 0, \quad y_-(t) < 0 \tag{3.2}$$

hold. We now examine the asymptotics of the functions $s(t)$, $y_\pm(t)$ for $t \to 0$.

Condition (b) implies that

$$g(y) = -y + y^3 + \sum_{k=4}^{\infty} g_k y^k, \qquad y \to 0,$$

$$\varphi(u) = b^2 u^2 + \sum_{k=3}^{\infty} \varphi_k u^k, \qquad u \to 0, \ b > 0, \tag{3.3}$$

$$\psi(y) = -\frac{y}{2b^2} + \sum_{k=2}^{\infty} \psi_k y^3, \qquad y \to 0.$$

THEOREM 3.1. *For some $T > 0$, there exist functions $s(t)$ and $y_\pm(t)$ satisfying equations* (1.18), (3.2) *and the relations*

$$s(t) = t^2 \sigma(t), \qquad y_\pm(t) = \pm \sqrt{t}(1 + t\alpha(t)) + t\beta(t), \tag{3.4}$$

where $\sigma(t), \alpha(t), \beta(t) \in C^\infty[0, T]$.

PROOF. The change of variables (3.4) yields a system equivalent to the original one for $t > 0$. In what follows we denote by Φ_i smooth functions omitting the subscripts wherever this does not cause any misunderstanding.

It follows from (3.4) that

$$\begin{aligned}
y_\pm^3(t) = {} & \pm t^{3/2}(1 + 3t\alpha(t)) + 3t^2\beta(t) \pm 3t^{5/2}\beta^2(t) \\
& \pm t^{7/2}\Phi(t, \alpha(t), \beta(t)) + t^3\Phi(t, \alpha(t), \beta(t)), \\
y_\pm^4(t) = {} & t^2 \pm 4t^{5/2}\beta(t) \pm t^{7/2}\Phi(t, \alpha(t), \beta(t)) + t^3\Phi(t, \alpha(t), \beta(t)), \\
y_\pm^5(t) = {} & \pm t^{5/2} \pm t^{7/2}\Phi(t, \alpha(t), \beta(t)) + t^3\Phi(t, \alpha(t), \beta(t)).
\end{aligned}$$

Substituting these expressions into equation (3.2) and using equalities (3.3), (3.4) results in the two equations:

$$\begin{aligned}
\sigma &= 2\beta + g_4 + t\Phi(t, \alpha, \beta), \\
2\alpha + 3\beta^2 + 4g_4\beta + g_5 + 1 &= t\Phi(t, \alpha, \beta).
\end{aligned} \tag{3.5}$$

Now we have to substitute expressions (3.4) into equation (1.18) taking into account equalities (3.3) and formula (3.1). The resulting equation is

$$(t^2\sigma)' = -t\beta + 2b^2\psi_2 t + (2b^2)^{-2}\varphi_3 t - t^2\Phi(t\alpha, \beta). \tag{3.6}$$

It follows from equations (3.5) that

$$\alpha = \Phi_1(t, \beta), \qquad \sigma = 2\beta + g_4 + t\Phi_2(t, \beta).$$

Substituting this expression for $\sigma(t)$ into equation (3.6) yields the following equation for $\beta(t)$:

$$(2t^2\beta)' = -t\beta + Ct + t^2\Phi_3(t, \beta),$$

where $C = 2\beta^2\psi_2 + (2b^2)^{-2}\varphi_3 - 2g_4$. This equation can be conveniently rewritten in the form

$$2(t^{5/2}\beta)' = Ct^{3/2} + t^{5/2}\Phi_3(t, \beta). \tag{3.7}$$

It remains to prove that equation (3.7) has a unique smooth solution in a neighborhood of zero. With that in mind, consider the integral equation equivalent to (3.7):

$$\beta(t) = \frac{C}{5} + \frac{t^{-5/2}}{2} \int_0^t \tau^{5/2} \Phi_3(\tau, \beta(\tau)) \, d\tau,$$

and apply the standard method of successive approximations. ∎

In conclusion, we present conditions sufficient for relations (a) to be satisfied. We demonstrate that relations (a) follow from (b) and the inequalities

$$g'(y) < 0 \quad \text{for } y \in \mathbf{R}^1,$$
$$yg''(y) > 0 \quad \text{for } y \neq 0. \tag{3.8}$$

Indeed, the function $s(t)$ is constructed above, and it suffices to show that the inequality $\partial x/\partial y = \omega(y, t) > 0$ holds outside the curve l. For small positive t relations (3.8) imply that there exist exactly two values $z_\pm(t)$ such that $\omega(z_\pm(t), t) = 0$. It follows from (1.1) that $z_\pm(t) = \pm\sqrt{\frac{t}{3}} + O(t)$. Let $v_\pm(t) = x(z_\pm(t))$. According to (1.2), one has $v_\pm(t) = \mp 2t^{3/2}/3\sqrt{3} + O(t^2)$, while (3.4) implies that $y_-(t) < z_-(t) < 0$, $y_+(t) > z_+(t) > 0$ for small $t > 0$. Thus, for these values of t,

$$v_+(t) < s(t) < v_-(t) \tag{3.9}$$

with the equality $\omega(y, t) = 0$ being attained only at those points where the function $y(x, t)$ is multivalued, while for the values of $y(x, t)$ corresponding to the generalized solution $u_0(x, t)$ one has $\omega(y, t) > 0$ for t small. It remains to show that relation (3.9) holds for all t.

Suppose, for example, that at some moment of time the function $s(t)$ first becomes equal to $v_-(t)$. Then one has $z_-(t) = y_-(t)$, $ds/dt \geq dv_-/dt$ at this point. Since $v_-(t) = z_-(t) + g(z_-(t))(t + 1)$, then $dv_-/dt = g(z_-(t))$. Therefore, $ds/dt \geq \varphi'(y_-(t))$ at this point, which contradicts the Hugoniot condition (1.18).

Thus, conditions (a) and (b) can be replaced by the more constructive, but stronger conditions (b) and (3.8).

We now proceed with the examination of the outer asymptotic expansion (1.4). The coefficients of this series, i.e., the functions $u_{4k}(x, t)$, were already constructed in §1. Indeed, one can take for the domain Ω^* considered there the entire strip Ω with the exception of the triangle

$$\{x, t: y_-(\delta) + g(y_-(\delta))(t + 1) < x < y_+(\delta) + g(y_+(\delta))(t + 1), \ -1 \leq t < \delta\},$$

where $\delta > 0$ is arbitrarily small. Thus, $u_{4k}(x, t) \in C^\infty(\Omega')$ for all k, and it only remains to examine the behavior of these functions in the vicinity of the origin. This is achieved using the formulas (1.6), (1.7) obtained above for $u_{4k}^*(y, t) \equiv u_{4k}(x, t)$. Since $\partial \omega/\partial t = g'(y) \neq 0$ for y small, and by virtue of condition (b) one can pass in the integral (1.7) to the new independent variable ω:

$$u_{4k}^*(y, t) = \frac{1}{\omega(y, t)g'(y)} \int_1^\omega F_k(y, \theta(\omega_1)) \, d\omega_1. \tag{3.10}$$

For $k = 1$, it follows from representation (3.10) and formulas (1.6) that

$$u_4(x, t) = \frac{\omega(y, t) - 1}{\omega^2(y, t)} \frac{g''(y)\psi'(y)}{[g'(y)]^2} + \frac{\ln \omega(y, t)}{\omega(y, t)} \left[\frac{1}{\varphi''(\psi(y))}\right]'. \tag{3.11}$$

Note that in the vicinity of zero the coefficient of $\ln \omega$ may vanish only at those points where $\varphi'''(\psi(y)) = 0$. Therefore, this coefficient vanishes identically only for the Burgers equation ($\varphi(u) = \text{const} \cdot u^2$). In this case the solution of equation (0.1) is written out explicitly, and the limit of the solution for $\varepsilon \to 0$ has been analyzed in [34]. In the remaining cases the asymptotics of the functions u_{4k} at the origin is rather complicated. In order to describe these functions the following notation is convenient:

$$\Psi_{k,p}(y, t) = \sum y^i [\omega(y, t)]^{-i} \Phi_{i,j}(y),$$

where the sum is finite, $1 \le i \le p$, $j \ge 0$, $2i - j \le k$, and $\Phi_{i,j} \in C^\infty$; if in this sum $2 \le i \le p$ then the sum is denoted by $\Psi^*_{k,p}(y, t)$.

The form of the functions $\Psi_{k,p}$, equality (1.1), and condition (b) easily imply the following relations:

$$\Psi_{k,p} \cdot \Psi_{m,n} = \Psi^*_{k+m,p+n}, \qquad \frac{\partial}{\partial y}\Psi_{k,p} = \Psi_{k+1,p+1}, \tag{3.12}$$

$$\int_{-1}^{t} \frac{[\ln \omega(y, \xi)]^m}{[\omega(y, \xi)]^k} \, d\xi = [\omega(y, t)]^{-k+1} \sum_{m=0}^{k} \Phi_i(y)[\ln \omega(y, t)]^i. \tag{3.13}$$

THEOREM 3.2. *For $k \ge 1$ and sufficiently small y*

$$u_{4k}(x, t) \equiv u^*_{4k}(y, t) = \sum_{m=0}^{k} \Psi_{4k-m-1, 3k-m-1}(y, t)[\ln \omega(y, t)]^m. \tag{3.14}$$

PROOF. For $k = 1$, formula (3.14) follows from (3.11) because $g''(0)$. The proof then proceeds by induction. Suppose that (3.14) is satisfied for $k < n$, and transform the function $F_n(y, t)$ according to formula (1.6) using (3.12) and equality $(\partial \omega / \partial y)(0, t) = 0$:

$$\frac{\partial}{\partial y}\left[\frac{1}{\omega}\frac{\partial u^*_{4n-4}(y, t)}{\partial y}\right] = \sum_{m=0}^{n-1} \Psi_{4n-m-1, 3n-m-1}(y, t)[\ln \omega(y, t)]^m,$$

$$\prod_{\substack{p=1 \\ \sum i_p = n}}^{q} u^*_{4i_p} = \sum_{m=0}^{q} \Psi^*_{4q-m-2, 3q-m-2}(y, t)[\ln \omega(y, t)]^m.$$

Hence

$$F_n(y, t) = \sum_{m=0}^{n} \Psi^*_{4n-m-1, 3n-m-1}(y, t)[\ln \omega(y, t)]^m.$$

This equality and relations (1.7), (3.13) imply (3.14). ∎

§4. Asymptotics of solutions near the origin

1. The f.a.s. (1.4) constructed above evidently fails in the vicinity of the origin because the coefficients $u_k(x, t)$ of this expansion have strong singularities as $x \to 0$, $t \to 0$. Therefore, a correct asymptotic expansion in the vicinity of the origin must involve functions of other, stretched variables. Since equation (0.1) is nonlinear, and $u(0, 0) = 0$, the scale of the function

u itself can also vary. The magnitudes of these variations can be found along the following lines. Let $x = \varepsilon^{\alpha}\xi$, $t = \varepsilon^{\beta}\tau$, and let $u(x, t)$ have the order of magnitude ε^{γ}. A necessary condition for all terms of equation (0.1) to be of the same order of magnitude with respect to ε (taking into account that $\varphi(u) \sim b^2 u^2$) is that $-\beta = \gamma - \alpha = 4 - 2\alpha$. Another equality is obtained from the equation for the characteristics (1.2) and the relation $u_0(x, t) = \psi(y)$. Since $\psi'(0) \neq 0$, one has $y \asymp \varepsilon^{\gamma}$. In view of condition (b) in §3, the principal terms in equation (1.2) are x, y^3, and yt. Hence $\alpha = 3\gamma = \beta + \gamma$ and, consequently, $\alpha = 3$, $\beta = 2$, $\gamma = 1$.

These nonrigorous considerations suggest the following change of variables:

$$\xi = \varepsilon^{-3}x, \qquad \tau = \varepsilon^{-2}t; \tag{4.1}$$

denote $u(\varepsilon^3\xi, \varepsilon^2\tau, \varepsilon) \equiv w(\xi, \tau, \varepsilon)$.

The equation for w is of the form

$$\frac{\partial w}{\partial \tau} + \varepsilon^{-1}\frac{\partial \varphi(w)}{\partial \xi} = \frac{\partial^2 w}{\partial \xi^2}. \tag{4.2}$$

In view of the above, the asymptotic expansion for w must begin with the term $\varepsilon w_1(\xi, \tau)$. The detailed form of this series follows from matching it to the series (1.4). Making the formal change of variables (4.1) in the terms of the series (1.4), and taking into account formulas (3.14), one obtains a series including terms with the coefficients $\varepsilon^k \ln^i \varepsilon$, $i \leq k - 1$, $k \geq 1$. It is, therefore, natural to seek an asymptotic expansion for w in the form

$$W = \sum_{k=1}^{\infty} \varepsilon^k \sum_{j=0}^{k-1} w_{k,j}(\xi, \tau) \ln^j \varepsilon. \tag{4.3}$$

Inserting this series into equation (4.2) and expanding the functions $\varphi(w)$ in Taylor series yield the recurrence system

$$\frac{\partial w_{1,0}}{\partial \tau} + 2b^2 w_{1,0}\frac{\partial w_{1,0}}{\partial \xi} - \frac{\partial^2 w_{1,0}}{\partial \xi^2} = 0, \tag{4.4}$$

$$L_3 w_{k,j} \equiv \frac{\partial w_{k,j}}{\partial \tau} + 2b^2\frac{\partial}{\partial \xi}(w_{1,0}w_{k,j}) - \frac{\partial^2 w_{k,j}}{\partial \xi^2} = E_{k,j}, \qquad k \geq 2, \tag{4.5}$$

where

$$E_{k,j} = -\frac{\partial}{\partial \xi}\left[b^2 \sum_{\substack{2\leq i\leq k-1 \\ 0\leq m\leq j}} w_{i,m}w_{k+1-i,j-m} + \sum_{q=3}^{k+1}\frac{\varphi^{(q)}(0)}{q!}\sum_{\substack{\sum i_p=k+1 \\ \sum m_p=j}}\prod_{p=1}^{q} w_{i_p,m_p}\right], \tag{4.6}$$

$$2b^2 = \varphi''(0). \tag{4.7}$$

The central and most laborious part of the problem is the investigation of solutions of the system (4.4), (4.5). The most important for what follows is the asymptotics of the functions $w_{k,j}(\xi, \tau)$ as $\tau \to \infty$. In the subsequent sections it will enable us to uniquely determine the coefficients of the asymptotic expansion in a neighborhood of the discontinuity curve of the limit function $u_0(x, t)$, i.e., near the curve $l = \{x, t, : x = s(t), 0 < t \le T\}$.

At first sight, the analysis of the solutions for system (4.4), (4.5) looks very simple. Indeed, (4.4) is the Burgers equation which, as is known, can be reduced to the heat equation. Each of equations (4.5) also reduces to a nonhomogeneous heat equation. However, a considerable difficulty is due to the fact that the solutions $w_{k,j}(\xi, \tau)$ are defined on the entire plane (ξ, τ), and have rather complicated asymptotics at infinity (in particular, the corresponding solutions of the heat equation for a fixed τ, in general, grow as $\exp(\mu^2|\xi|^{4/3})$). In order to determine the solutions of the system (4.4), (4.5) uniquely, one has to impose additional conditions on $w_{k,j}(\xi, \tau)$ as $\tau \to -\infty$. These conditions are naturally obtained from the requirement of matching the series (4.3) to (1.4). The procedure of deriving such conditions is standard and can be described as follows.

The purpose of the matching procedure is to make the series (1.4) and (4.3) asymptotically coincide in the intermediate domain where the value $|x| + |t|$ is small, and $|\xi| + |\tau|$ large. With that in mind, each term of the series (1.4) is considered for $x \to 0$, $t \to 0$, and represented in the form (3.14). Next, after the change of variables (4.1), each function $u_k(x, t)$ can be written as a formal series in powers of ε and $\ln \varepsilon$ with coefficients depending on ξ and τ. On substituting these expressions into the series (1.4), one obtains the series

$$\sum_{m=1}^{\infty} \varepsilon^m \sum_{j=0}^{m-1} W_{m,j}(\xi, \tau) \ln^j \varepsilon. \tag{4.8}$$

Here $W_{m,j}$ is the sum of all coefficients of $\varepsilon^m \ln^j \varepsilon$ appearing in the expansion of $\varepsilon^k u_k(x, t)$ for all k. One can easily see that each expression $W_{m,j}$ contains infinitely many terms, and is a formal asymptotic series for $\tau < 0$, $|\xi| + |\tau| \to \infty$. Thus, the equations (4.4), (4.5) must be supplemented by the conditions

$$w_{k,j}(\xi, \tau) = W_{k,j}, \qquad \tau \to -\infty. \tag{4.9}$$

Equality (4.9) means that the solutions $w_{k,j}(\xi, \tau)$ can be expanded in the asymptotic series constructed above as $\tau \to -\infty$ uniformly with respect to ξ.

2. Now we have to make the form of these asymptotic series $W_{k,j}(\xi, \tau)$ more precise by finding out the exact form of the series corresponding to each function $u_k(x, t)$. By Theorem 3.2, they can be represented in the form of series in y, t, and $\ln \omega(y, t)$ as $t \to 0$, $y \to 0$. The variable t is replaced by $\varepsilon\tau$, while the "Lagrangian variable" y depends on ξ and τ in a more

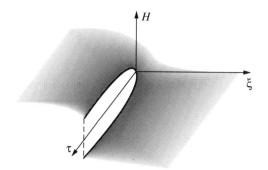

FIGURE 35

complex way. This dependence can be found from formula (1.2). Taking into account formula (3.3) for $g(y)$, making the change of variables (4.1) in formula (1.2), and representing y in the form of the series

$$y(\xi, \tau, \varepsilon) = \sum_{k=1}^{\infty} \varepsilon^k y_k(\xi, \tau), \qquad (4.10)$$

one obtains the recurrence system for y_k:

$$\xi = -\tau y_1 + y_1^3, \qquad (4.11)$$

$$\tau y_2 = 3y_1^2 y_2 + g_4 y_1^4,$$
$$(\tau - 3y_1^2) y_k = f_k(y_1, y_2, \dots, y_{k-1}, \tau). \qquad (4.12)$$

Equation (4.11) defines the so-called *Whitney fold function*. Since this function is not single-valued for $\tau > 0$ and some ξ, let us introduce a single-valued but discontinuous function $H(\xi, \tau)$.

In what follows by $H(\xi, \tau)$ we mean the function continuous everywhere in $\mathbf{R}^2$ with the exception of the nonnegative part of the τ-axis and satisfying the equation $H^3 - \tau H + \xi = 0$. Thus, $H \in C^\infty$ everywhere in this domain, and $H(\pm 0, \tau) = \mp\sqrt{\tau}$ for $\tau \geq 0$ (see Figure 35). It follows from equation (4.11) that $y_1(\xi, \tau) = -H(\xi, \tau)$, and equations (4.12) define the functions $y_k(\xi, \tau)$ uniquely. Everywhere below we denote

$$R(\xi, \tau) = 3H^2(\xi, \tau) - \tau. \qquad (4.13)$$

The function $R(\xi, \tau) \in C^\infty$ everywhere in $\mathbf{R}^2$ with the exception of the half-axis $\{\xi, \tau : \xi = 0, \tau \geq 0\}$. It is continuous and positive for $|\xi| + |\tau| > 0$. It follows from equations (4.12) and the explicit form of the functions $f_k(\xi, \tau)$ that all $y_k(\xi, \tau)$ are polynomials in τ, $[R(\xi, \tau)]^{-1}$, H, and homogeneous functions of degree k with respect to H, $|\tau|^{1/2}$, and $R^{1/2}$. It is easy to see that the asymptotic expansion (4.10) is valid for bounded ξ and τ. Moreover, the series (4.10) preserves its asymptotic character for $R(\xi, \tau) < \varepsilon^{-\nu}$ for any $\nu < 1/2$.

Now, after having established the form of the coefficients of the series (4.10), one can use formulas (3.14) to obtain the asymptotics of the functions $u_k(x, t)$ in the variables ξ, τ. From now on we denote functions that are polynomials in H, τ, R^{-1} homogeneous of degree k with respect to H, $|\tau|^{1/2}$, $R^{1/2}$ by $h_{k,\cdot}(\xi, \tau)$, or simple $h_{k,\cdot}$. Then condition (b) and equalities (4.7), (4.10) imply that

$$u_0(x, t) = \psi(y) = -(2b^2)^{-1}y + \sum_{l=2}^{\infty} c_l y_l = \varepsilon(2b^2)^{-1}H(\xi, \tau) + \sum_{l=2}^{\infty} \varepsilon^l h_{l,0}(\xi, \tau).$$

It follows from formula (1.1) that

$$\omega(y, t) = \varepsilon^2 R(\xi, \tau) + \sum_{l=3}^{\infty} \varepsilon^l h_l(\xi, \tau),$$

$$\ln \omega(y, t) = \frac{1}{2}\ln \varepsilon^4 R^2 + \sum_{l=1}^{\infty} \varepsilon^l h_l(\xi, \tau).$$

Substituting these expressions into (3.14), we have

$$u_k(x, t) = \sum_{m=0}^{k}(\ln \varepsilon^4 R^2)^m \sum_{l=m+1-4k} \varepsilon^l h_{l,k,m}(\xi, \tau).$$

These equalities enable one to make the form of the asymptotic series $W_{k,j}(\xi, \tau)$ in relations (4.9) more precise:

$$w_{k,j}(\xi, \tau) = \sum_{p=j}^{k-1}(\ln R)^{p-j} \sum_{l=p}^{\infty} h_{k-4l,p,k,j}(\xi, \tau), \qquad \tau \to -\infty. \qquad (4.14)$$

3. We now write out the asymptotic series $W_{1,0}(\xi, \tau)$, preliminarily denoting for convenience $\Gamma(\xi, \tau) = 2b^2 w_{1,0}(\xi, \tau)$. Equation (4.4) then looks as follows:

$$\Gamma_\tau + \Gamma\Gamma_\xi - \Gamma_{\xi,\xi} = 0, \qquad (4.15)$$

and relation (4.9) for $k = 1$, $j = 0$ is of the form

$$\Gamma(\xi, \tau) = H(\xi, \tau) + \sum_{l=1}^{\infty} h_{1-4l,0}(\xi, \tau), \qquad \tau \to -\infty. \qquad (4.16)$$

Omitting the heuristic considerations leading to the explicit form of $\Gamma(\xi, \tau)$, we immediately write out the formula

$$\Gamma(\xi, \tau) = -2[\Lambda(\xi, \tau)]^{-1}\frac{\partial \Lambda(\xi, \tau)}{\partial \xi}, \qquad (4.17)$$

where

$$\Lambda(\xi, \tau) = \int_{-\infty}^{\infty} \exp\left(-\frac{1}{8}(z^4 - 2z^2\tau + 4z\xi)\right) dz. \qquad (4.18)$$

The function $\Lambda(\xi, \tau)$ bears the name of the Pearcey function. It is immediately evident that $\Lambda(\xi, \tau)$ satisfies a homogeneous heat equation, and, consequently, $\Gamma(\xi, \tau)$ is a solution of equation (4.15).

In order to investigate the asymptotics of $\Gamma(\xi, \tau)$ one has to examine the asymptotics of the function $\Lambda(\xi, \tau)$ first. For $|\xi| + |\tau| \to \infty$, the stationary points in the integral (4.18) are the solutions of the equation

$$z^3 - z\tau + \xi = 0. \tag{4.19}$$

For negative τ and also for $0 \leq \tau \leq 3|\xi|^{2/3} 2^{-2/3}$ the only point of maximum of the integrand is $H(\xi, \tau)$. For $\tau > 3|\xi|^{2/3} 2^{-2/3}$ there is one more local maximum which we denote $H_1(\xi, \tau)$. It is most convenient to change the variable in the integral (4.18) by

$$z = H(\xi, \tau) + \frac{2}{\sqrt{R}} \zeta, \tag{4.20}$$

where R is defined by formula (4.13). This change of variables yields

$$\Lambda(\xi, \tau) = \frac{2}{\sqrt{R}} \exp\left(\frac{3}{8} H^4(\xi, \tau) - \frac{\tau}{4} H^2(\xi, \tau)\right)$$

$$\times \int_{-\infty}^{\infty} \exp\left(-\zeta^2 - \frac{4H}{R^{3/2}} \zeta^3 - \frac{2}{R^2} \zeta^4\right) d\zeta.$$

Therefore, in the domain where $H(\xi, \tau)$ is the only solution of equation (4.19), the function $\Lambda(\xi, \tau)$ as $R \to \infty$ can be expanded in the asymptotic series

$$\Lambda(\xi, \tau) = \frac{2}{\sqrt{R}} \exp\left(\frac{3}{8} H^4 - \frac{\tau}{4} H^2\right) \left\{\sqrt{\pi} + \sum_{l=1}^{\infty} h_{-4l}(\xi, \tau)\right\}. \tag{4.21}$$

Differentiating equality (4.18) with respect to ξ, and making the change of variable (4.20) in the resulting integral, one obtains the equality

$$\frac{\partial \Lambda}{\partial \xi} = \frac{2}{\sqrt{R}} \exp\left(\frac{3}{8} H^4 - \frac{\tau}{4} H^2\right) \left\{-\frac{\sqrt{\pi}}{2} H + \sum_{l=1}^{\infty} h_{1-4l}(\xi, \tau)\right\}. \tag{4.22}$$

Hence it follows that

$$\Gamma(\xi, \tau) = H(\xi, \tau) + \sum_{l=1}^{\infty} h_{1-4l}(\xi, \tau), \qquad R \to \infty. \tag{4.23}$$

For the values of ξ, τ for which there exists a second local maximum of the integrand in (4.18), the asymptotic expansions of the functions $\Lambda(\xi, \tau)$ and $\partial \Lambda(\xi, \tau)/\partial \xi$ are sums of two series: one has to add to the right-hand sides in (4.21) and (4.22) the same series with $H(\xi, \tau)$ replaced by $H_1(\xi, \tau)$, and $R(\xi, \tau)$ by $R_1(\xi, \tau) = 3H_1^2(\xi, \tau) - \tau$. One can see without difficulty that the values of the function $3H^4(\xi, \tau) - 2\tau H^2(\xi, \tau)$ are substantially larger than those of $3H_1^4(\xi, \tau) - 2\tau H_1^2(\xi, \tau)$ everywhere for R sufficiently large with the exception of a narrow strip along the positive half of the τ-axis. More precisely, for $|\xi|\tau^{1/2} > R^\nu$, $\nu > 0$, the additional series depending on H_1 in formulas (4.21), (4.22) are exponentially smaller than those already written out. Thus, formulas (4.21)–(4.23) hold for $\tau < 0$, as well as for $|\xi|\tau^{1/2} > R^\nu$ for $\nu > 0$. In particular, the principal term of the asymptotics $\Gamma(\xi, \tau)$ as $\tau \to -\infty$ coincides with the principal term of the asymptotic expansion (4.16).

THEOREM 4.1. *The function $\Gamma(\xi, \tau)$ defined by formulas (4.17), (4.18) is a solution of the problem (4.15), (4.16).*

THEOREM 4.2. *There exist solutions of the problems (4.5), (4.14) for $k \geq 2$, $0 \leq j \leq k - 1$ infinitely differentiable for all ξ and τ.*

PROOF OF THEOREM 4.1. As shown above, $\Gamma(\xi, \tau)$ satisfies equation (4.15), and the principal term of the asymptotics of $\Gamma(\xi, \tau)$ as $\tau \to -\infty$ coincides with the principal term of the series (4.16). It is, therefore, sufficient to check that all functions h_{1-4l} in formula (4.23) coincide with the functions $h_{1-4l,0}$ in the asymptotic expansion (4.16). The easiest way to do that is apparently as follows.

It follows from the explicit form of $H(\xi, \tau)$ and $R(\xi, \tau)$ that for any function $h_l(\xi, \tau)$ for $\tau < 0$ the following representation holds:

$$h_l(\xi, \tau) = |\tau|^{l/2} Z_l(\theta),$$

where $\theta = \xi|\tau|^{-3/2}$, and $Z_l(\theta) \in C^\infty(\mathbf{R}^1)$. Thus, the asymptotic series for the function $\Gamma(\xi, \tau)$ as $\tau \to -\infty$ is of the form

$$\Gamma(\xi, \tau) = |\tau|^{1/2} \left(Z_0(\theta) + \sum_{j=1}^\infty \tau^{-2j} Z_j(\theta) \right), \tag{4.24}$$

where

$$Z_0^3 + Z_0 + \theta = 0. \tag{4.25}$$

The series (4.16) is of the same form with the same function $Z_0(\theta)$. Equation (4.15) yields the recurrence system of ordinary differential equations for $Z_j(\theta)$:

$$-Z_0 + 3\theta Z_0' + 2Z_0 Z_0' = 0,$$
$$(4j - 1)Z_j + 3\theta Z_j' + 2(Z_0 Z_j)' = f_j(Z_0, Z_1, \ldots, Z_{j-1}), \qquad j \geq 1.$$

The first of these equations holds by virtue of (4.25), and each of the remaining ones has a unique smooth solution. This is a consequence of the fact that the corresponding homogeneous equations have solutions $(Z_0(\theta))^{3-4j}(1 + 3Z_0^2(\theta))^{-1}$ with singularities at $\theta = 0$. Thus, all coefficients $Z_j(\theta)$ in the asymptotic expansion (4.24) are uniquely defined from equation (4.15). The fact that the series (4.16) formally satisfies equation (4.15) follows from the way the functions $h_{1-4l,0}(\xi, \tau)$ were constructed. Consequently, the function $\Gamma(\xi, \tau)$ constructed by formulas (4.17), (4.18) is a solution of the problem (4.15), (4.16). Moreover, as shown above, formulas (4.21), (4.22) imply that the asymptotic expansion (4.16) is valid not just as $\tau \to -\infty$, but also as $R \to \infty$ everywhere except the strip $\{\xi, \tau : \tau > 1, |\xi|\tau^{1/2} < R^\nu\} \forall \nu > 0$. ∎

REMARK. It is of interest to note that replacing condition (4.16) with the condition $\Gamma(\xi, \tau) - H(\xi, \tau) \to 0$ as $\tau \to -\infty$ results in loss of the uniqueness of solution. There is a continuum of functions satisfying this condition and equation (4.15). They can be constructed by formula (4.17) where, instead of the function (4.18), one takes for Λ the integral $\int\limits_{-\infty}^{\infty} \rho(z)\exp\left(-\frac{1}{8}(z^4 - 2z^2\tau + 4z\xi)\right)dz$ with a smooth positive function $\rho(z)$.

4. This subsection proves Theorem 4.2. We begin with a number of auxiliary lemmas. An appropriate change of the unknown function reduces the linear equations (4.5) to nonhomogeneous heat equations. However, these equations are not easy to analyze because the desired solutions and the right-hand sides grow rapidly both as $|\xi| \to \infty$, and $\tau \to -\infty$. It is, therefore, more convenient to investigate the equations for differences between the functions $w_{k,j}$ and sufficiently long partial sums of their asymptotic series. Thus, the central problem of this subsection is the construction and investigation of a solution of the equation

$$\frac{\partial w}{\partial \tau} + \frac{\partial(\Gamma w)}{\partial \xi} - \frac{\partial^2 w}{\partial \xi^2} = F(\xi, \tau), \qquad (4.26)$$

where both the right-hand side F and solution w decay rapidly enough for $\tau <$ const, $R \to \infty$. The change of the unknown function

$$w(\xi, \tau) = \frac{\partial}{\partial \xi}(v(\xi, \tau)[\Lambda(\xi, \tau)]^{-1}) \qquad (4.27)$$

yields the equation

$$\frac{\partial}{\partial \xi}\left\{[\Lambda(\xi, \tau)]^{-1}\left[\frac{\partial v}{\partial \tau} - \frac{\partial^2 v}{\partial \xi^2}\right]\right\} = F(\xi, \tau), \qquad (4.28)$$

and for what follows we will require rather precise estimates of the function $\Lambda(\xi, \tau)$ and its derivatives.

LEMMA 4.1. *There exist constants $M_l > 0$ $(l \geq 0)$ such that for all ξ, τ such that $|\xi| + |\tau| > 1$ the estimates*

$$M_0^{-1} < \sqrt{R(\xi, \tau)}\Lambda(\xi, \tau)\exp(-S(\xi, \tau)) < M_0, \qquad (4.29)$$

$$\left|\frac{\partial^l \Lambda(\xi, \tau)}{\partial \xi^{l_1}\partial \tau^{l_2}}\right| \leq M_l[R(\xi, \tau)]^{l_2 + (l_1 + 1)/2}\exp(S(\xi, \tau)) \qquad (4.30)$$

hold. Here the function $R(\xi, \tau)$ is defined by formula (4.13), the function $\Lambda(\xi, \tau)$ by formula (4.18), $S(\xi, \tau) = \frac{3}{8}H^4(\xi, \tau) - \frac{\tau}{4}H^2(\xi, \tau)$, and the function $H(\xi, \tau)$ was defined at the beginning of subsection 2.

PROOF. The change of variables (4.20) yields

$$\Lambda(\xi, \tau) = \frac{2}{\sqrt{R}}J(\xi, \tau)\exp(S(\xi, \tau)),$$

where

$$J(\xi, \tau) = \int\limits_{-\infty}^{\infty}\exp\left(-\zeta^2 - \frac{4H}{R^{3/2}}\zeta^3 - \frac{2}{R^2}\zeta^4\right)d\zeta.$$

This immediately implies the left-hand side of inequality (4.29). To obtain an upper bound, let us assume, to be definite, that $\xi \leq 0$, $H(\xi, \tau) \geq 0$. Then

$$\Lambda(\xi, \tau) \leq 2 \int_0^\infty \exp\left\{-\frac{1}{8}(z^4 - 2z^2\tau + 4z\xi)\right\} dz$$

$$= 4R^{-1/2} \exp S(\xi, \tau) \int_{-2^{-1}HR^{1/2}}^\infty \exp\left(-\zeta^2 - \frac{4H}{R^{3/2}}\zeta^3 - 2\frac{\zeta^4}{R^2}\right) d\zeta.$$

Splitting the domain of integration in the last integral in two: from $-2^{-1}HR^{1/2}$ to $-\beta HR^{1/2}$, and from $-\beta HR^{1/2}$ to ∞ for sufficiently small $\beta > 0$, and estimating each of the two integrals separately, one arrives at the right-hand inequality (4.29).

To estimate the derivatives one has to differentiate equality (4.18) under the integral sign, and then make the change of variables (4.20). The resulting integral differs from $J(\xi, \tau)$ only by the presence of the factor $c(H + 2\zeta R^{-1/2})^{l_1+2l_2}$. This implies inequality (4.30) if one takes into account the relation $R \geq 2H^2$ which is easily verified. ■

Everywhere below we use the notation

$$G(\xi, \tau) = \frac{1}{2\sqrt{\pi\tau}} \exp\left(-\frac{\xi^2}{4\tau}\right).$$

LEMMA 4.2. *Suppose that for sufficiently large N we have $F(\xi, \tau) \in C^N$ for $\tau \leq -1$, and that the estimates*

$$|D^k F(\xi, \tau)| \leq M(\tau^2 + \xi^2)^{-N}$$

hold for $k \leq N$. Then the function

$$v(\xi, \tau) = \int_{-\infty}^\tau \int_{-\infty}^\infty G(\xi - y, \tau - \theta)\Lambda(y, \theta)F_1(y, \theta)\,dy\,d\theta, \qquad (4.31)$$

where $F_1(\xi, \tau) = \int_{-\infty}^\xi F(y, \tau)\,dy$, belongs to C^m for $\tau \leq -1$. The function $w(\xi, \tau)$ defined by formula (4.27) satisfies equation (4.28), and the estimates

$$|D^k w(\xi, \tau)| \leq M_1(\xi^2 + \tau^2)^{-m} \quad for \ k \leq m. \qquad (4.32)$$

Here $m \to \infty$ as $N \to \infty$, and D^k denotes the derivative of order k.

PROOF. First, we have to establish the convergence of the integral (4.31) and find an estimate for it. Lemma 4.1 implies the inequality

$$|v(\xi, \tau)| \leq M \int_{-\infty}^\tau \frac{1}{\sqrt{\tau - \theta}} \int_{-\infty}^\infty \frac{|F_1(y, \theta)|}{\sqrt{R(y, \theta)}} \exp Q(\xi, y, \tau, \theta)\,dy\,d\theta, \qquad (4.33)$$

where $Q(\xi, y, \tau, \theta) = 4^{-1}(\tau - \theta)^{-1}(\xi - y)^2 + S(y, \theta)$.

Since

$$\frac{\partial^2 Q}{\partial y^2} = -\frac{1}{2(\tau - \theta)} + \frac{1}{2R(y, \theta)} \leq -\left|\frac{t}{\theta}\right|\frac{1}{2(\tau - \theta)},$$

the function $Q(\xi, y, \tau, \theta)$ has a single point of maximum with respect to the variable y. Denote this point by $Y(\xi, \tau, \theta)$. Thus,

$$Q(\xi, y, \tau, \theta) \leq Q(\xi, Y(\xi, \tau, \theta), \tau, \theta) - \frac{1}{4(\tau - \theta)}\left|\frac{\tau}{\theta}\right|(y - Y(\xi, \tau, \theta))^2. \quad (4.34)$$

We now show that

$$H(Y(\xi, \tau, \theta), \theta) \equiv H(\xi, \theta). \quad (4.35)$$

Indeed, the function $Y(\xi, \tau, \theta)$ satisfies the equation $(\partial Q/\partial y)(\xi, Y, \tau, \theta) = 0$ by definition or, what is the same, $Y(\xi, \tau, \theta) - \xi - (\tau - \theta)H(Y(\xi, \tau, \theta), \theta) = 0$. Hence, using the equalities $\partial H/\partial y = -R^{-1}$, $\partial H/\partial \theta = HR^{-1}$, one easily gets that $\partial H(Y(\xi, \tau, \theta), \theta)/\partial \theta \equiv 0$. Therefore, equality (4.35) holds because $Y(\xi, \tau, \tau) = \xi$. It follows here this that the maximum value of the function Q with respect to y is also independent of θ:

$$Q(\xi, Y(\xi, \tau, \theta), \tau, \theta) = -\frac{1}{4}(\tau - \theta)H^2(\xi, \tau) + \frac{3}{8}H^4(\xi, \tau) - \frac{\theta}{4}H^2(\xi, \tau) = S(\xi, \tau).$$

Inequalities (4.33) and (4.34) imply the estimate

$$v(\xi, \tau) \leq Me^{S(\xi, \tau)} \int\limits_{-\infty}^{\tau} \frac{1}{\sqrt{\tau - \theta}} \int\limits_{-\infty}^{\infty} \frac{|F_1(y, \theta)|}{\sqrt{R(y, \theta)}}$$

$$\times \exp\left(-\left|\frac{\tau}{\theta}\right| \frac{(y - Y(\xi, \tau, \theta))^2}{4(\tau - \theta)}\right) dy\, d\theta. \quad (4.36)$$

By the hypothesis, the function $F_1(y, \theta)$ satisfies the inequalities

$$|D^k F_1(y, \theta)| \leq M(y^2 + \theta^2)^{-N+1} \quad \text{for } y \leq 0,$$
$$|D^k F_1(y, \theta)| \leq M\theta^{-2N+2} \quad \text{for } y \geq 0 \text{ for } k \leq N.$$

By considering the cases $|\xi| < |\tau|$ and $|\xi| \geq |\tau|$ separately, one can see without difficulty, owing to (4.36), that for $\xi \leq 0$ one has $|v(\xi, \tau)| \leq M(\xi^2 + \tau^2)^{-m} \exp S(\xi, \tau)$ where $m \to \infty$ as $N \to \infty$.

Since

$$D_\xi^k v(\xi, \tau) = \int\limits_{-\infty}^{\tau} \int\limits_{-\infty}^{\infty} G(\xi - y, \tau - \theta)D_y^k[\Lambda(y, \theta)F_1(y, \theta)]\, dy\, d\theta,$$

similar estimates for $D_\xi^k v(\xi, \tau)$ are obtained along the same lines using inequalities (4.30). Formula (4.27) and the estimates from Lemma 4.1 imply (4.32) for $\xi < 0$.

In order to verify these estimates for $\xi \geq 0$, we recast $F_1(\xi, \tau)$ in the form $F_2(\xi, \tau) + \rho(\tau)$, where $F_2(\xi, \tau) = -\int\limits_{\xi}^{\infty} F(y, \tau)\, dy$, and $\rho(\tau) = \int\limits_{-\infty}^{\infty} F(y, \tau)\, dy = O\left(\tau^{-2N+2}\right)$. The integral (4.31) then falls into two integrals. The first satisfies the same estimates as $v(\xi, \tau)$, but for $\xi \geq 0$, while the second integral, as one can easily verify from the heat equation, equals $\Lambda(\xi, \tau) \int\limits_{-\infty}^{\tau} \rho(\theta)\, d\theta$. After substitution into formula (4.27) the corresponding term in $w(\xi, \tau)$ vanishes and affects neither the estimate of $w(\xi, \tau)$, nor that of its derivatives. The validity of (4.32) is hereby proved for both $\xi \leq 0$ and $\xi \geq 0$. The validity of equality (4.28) is evident. ∎

The proof of Theorem 4.2 will be achieved by induction in the index k. Suppose that the existence of $w_{l,i}(\xi, \tau)$ is proved for all $l < k$, $i \leq l - 1$. Then, owing to formula (4.6), the right-hand side of equation (4.5) is already constructed. The asymptotic series $W_{k,j}$ appearing in the right-hand side of equality (4.14) is an f.a.s. of equation (4.5).

Let $X_N(\xi, \tau) = B_N W_{k,j}$ be the partial sum of the series (4.14) containing the subscripts $l \leq N$. Then it satisfies the equation

$$L_3 X_N \equiv \frac{\partial X_N}{\partial \tau} + \frac{\partial}{\partial \xi}(\Gamma X_N) - \frac{\partial^2 X_N}{\partial \xi^2} = E_{k,j}(\xi, \tau) + Z_N(\xi, \tau),$$

where $|Z_N| \leq MR^{-N_1}$ for $\tau \leq -1$, $N_1 \to \infty$ as $N \to \infty$, with similar estimates holding for the derivatives of Z_N. (For brevity, the subscripts k and j in the functions X_N, Z_N and other functions will be omitted until the end of the proof.) Since the partial sum X_N has singularities at the origin and discontinuities on the positive half of the τ-axis, it is convenient to multiply it by a function $\chi(\xi, \tau) \in C^\infty(\mathbf{R}^2)$ vanishing for $\xi^2 + \tau^2 \leq 1$ and for $|\xi| \leq 1$, $\tau \geq 0$, and equal to unity outside the union of the disk $\xi^2 + \tau^2 \leq 4$, and the half-strip $|\xi| \leq 2$, $\tau \geq 2$.

Then $L_3(\chi(\xi, \tau)X_N) = E_{k,j} + \tilde{Z}_N(\xi, \tau)$, where $\tilde{Z}_N \in C^\infty(\mathbf{R}^2)$, and satisfies the same estimates as Z_N.

Now, according to Lemma 4.2, construct a solution of the equation $L_3 W_N = \tilde{Z}_N(\xi, \tau)$ for $\tau \leq -1$, and then extend it for all $\tau \geq -1$ by formulas (4.27), (4.31). The function $w_N = \chi(\xi, \tau)X_N - W_N$ satisfies equation (4.5) and the estimate $|w_N(\xi, \tau) - X_N(\xi, \tau)| \leq MR^{-m}$ for $\tau \leq -1$. It is not difficult to note that the function w_n does not depend on N for N sufficiently large. Indeed, the difference $v_{N,N_1} = w_N - w_{N_1}$ satisfies the homogeneous equation $L_3 v_{N,N_1} = 0$, decays rapidly as $\tau \to -\infty$, and, therefore, equals zero by virtue of the maximum principle. (The change of variables $w = \hat{w}\frac{\Gamma}{\xi}$ results in a parabolic equation for $\hat{w}$ with the zero coefficient at the function $\hat{w}$.) The function w thus constructed satisfies equation (4.5) and the estimates $|w - X_N| \leq M_N R^{-m}$ for $\tau \leq -1$ for any N, where $m \to \infty$ as $N \to \infty$. ∎

5. This is the last subsection of §4, and its theme is the investigation of the asymptotics of the solutions $w_{k,j}$ constructed above as $\tau \to \infty$.

As in the preceding subsection, it is convenient to consider the difference between $w_{k,j}$ and the partial sums of the series $W_{k,j}$. Thus, we have to examine the solution of equation (4.26) for $\tau \geq -1$ with initial function decaying rapidly as $|\xi| \to \infty$ and rapidly decaying right-hand side $F(\xi, \tau)$. However, as compared to §4, two difficulties arise. First, the series (4.14) is discontinuous for $\xi = 0$, $\tau \to \infty$, and the right-hand side F constructed from its partial sums is small only outside the neighborhood $|\xi|\tau^{1/2} < R^\nu$, $\tau \to \infty$. Second, even the smallness of the right-hand side $F(\xi, \tau)$ everywhere for $\tau \geq -1$, $R \to \infty$ does not imply the smallness of the solution in the same domain. The solution has a rather complicated structure for $|\xi|\tau^{1/2} < \text{const}$, $\tau \to \infty$, and it is quite difficult to obtain the exact form of the asymptotics of the functions $w_{k,j}$ as $\tau \to \infty$ from the explicit formulas. We, therefore, restrict out attention to the study of the asymptotics of the functions $w(\xi, \tau)$ for $\xi = 0$, $\tau \to \infty$ in the case where $w(\xi, \tau)$ is a solution of equation (4.26) with the right-hand side decaying rapidly for

$\tau \geq -1$, $R \to \infty$. The form of the asymptotic series for $w_{k,j}(\xi, \tau)$ will be established, as a by-product, in §6 at the end of the analysis.

The main result of the present subsection is the following theorem.

THEOREM 4.3. *Suppose that $w(\xi, \tau)$ is a bounded solution of equation (4.26), and the following estimates are satisfied:*

$$|D^k F(\xi, \tau)| \leq M(\xi^2 + \tau^2 + 1)^{-N},$$
$$|D^k w(\xi, -1)| \leq M(\xi^2 + 1)^{-N} \quad for \ k \leq N,$$

where N is a sufficiently large number (here and in what follows D^k denotes the derivative of order k). Then the function $w(0, \tau)$ has the following asymptotic representation as $\tau \to \infty$:

$$w(0, \tau) = C\tau^{1/2}\left(1 + \sum_{k=1}^{m} \gamma_k \tau^{-2k}\right) + O\left(\tau^{-2m-1}\right), \quad (4.37)$$

where $m \to \infty$ as $N \to \infty$, and γ_k are absolute constants.

We begin the investigation of the solution of equation (4.26) for $\tau \geq -1$ with the proof of some auxiliary lemmas. From the technical point of view it is convenient to consider the intervals of time $-1 \leq \tau \leq T_0$ and $\tau \geq T_0 > 0$ separately.

LEMMA 4.3. *Let $w(\xi, \tau)$ be a bounded solution of equation (4.26) for $-1 \leq \tau \leq T_0$, and suppose that the right-hand side $F(\xi, \tau)$ and the initial function $w(\xi, -1)$ satisfy the estimates*

$$|D^k w(\xi, -1)| + |D^k F(\xi, \tau)| \leq M(1 + \xi^2)^{-N} \quad for \ k \leq N,$$

where N is a sufficiently large number. Then $w(\xi, \tau)$ satisfies the estimates

$$|D^k w(\xi, \tau)| \leq M(1 + \xi^2)^{-m} \quad for \ k \leq m,$$

where $m \to \infty$ as $N \to \infty$.

PROOF. The estimate for the function $w(\xi, \tau)$ can be obtained from the maximum principle using the barrier functions $\exp[\alpha\tau(\xi^2 + A)]$ and $(1 + \xi^2)^{-N}$ for sufficiently large α and A. Here one has to take into account the estimates

$$|\Gamma(\xi, \tau)| \leq M(1 + |\xi|^{1/3}), \qquad \left|\frac{\partial\Gamma}{\partial\xi}(\xi, \tau)\right| < M(1 + |\xi|^{2/3})^{-1}$$

for $-1 \leq \tau \leq \tau_0$ which follow from the asymptotic expansion (4.23).

Differentiating equation (4.26) with respect to ξ, one gets a similar problem for $\partial w/\partial \xi$, and the same barrier functions yield an estimate for $\partial w/\partial \xi$. The estimates for the higher derivatives with respect to ξ are obtained in the same way. The derivatives involving differentiations with respect to τ can be estimated directly from equation (4.26), as well as by differentiating this equation a necessary number of times with respect τ and ξ. ∎

In order to investigate the solution of equation (4.26) for $\tau \geq \tau_0 > 0$, we again use the change of unknown function (4.27) and the explicit formula for the solution of equation (4.28).

LEMMA 4.4. *Suppose that, for $\tau \geq \tau_0 > 0$, the function $w(\xi, \tau)$ is a bounded solution of equation (4.26), where*

$$F(\xi, \tau) = \frac{\partial}{\partial \xi} F_1(\xi, \tau), \qquad w(\xi, \tau_0) = \frac{\partial}{\partial \xi} w_1(\xi), \tag{4.38}$$

and let the following estimates be satisfied:

$$|D^k F_1(\xi, \tau)| \leq M(\xi^2 + \tau^2)^{-N},$$

$$|D^k w_1(\xi)| \leq M(1 + \xi^2)^{-N} \qquad for \ k \leq N,$$

where N is a sufficiently large number. Then $w(\xi, \tau)$ satisfies the estimates

$$|D^k w(\xi, \tau)| \leq M(\xi^2 + \tau^2)^{-m} \qquad for \ k \leq m, \tag{4.39}$$

where $m \to \infty$ as $N \to \infty$.

PROOF. After the transformation (4.27), equation (4.28) yields the following formula for $v(\xi, \tau)$:

$$v(\xi, \tau) = \int_{\tau_0}^{\tau} \int_{-\infty}^{\infty} G(\xi - y, \tau - \theta) \Lambda(y, \theta) F_1(y, \theta) \, dy \, d\theta$$

$$+ \int_{-\infty}^{\infty} G(\xi - y, \tau - \tau_0) \Lambda(y, \tau_0) w_1(y) \, dy = v_1(\xi, \tau) + v_2(\xi, \tau).$$

The integrals v_1 and v_2 are estimated in the same way as in Lemma 4.2. For the function $v_1(\xi, \tau)$ the estimate (4.33) is valid, where R, Q, and S denote the same functions as before.

Consider, to be definite, the case $\xi \geq 0$. Then

$$|v_1(\xi, \tau)| \leq M \int_{\tau_0}^{\tau} \frac{1}{\sqrt{\tau - \theta}} \int_{0}^{\infty} (\theta^2 + y^2)^{-N+1} \exp Q(\xi, y, \tau, \theta) \, dy \, d\theta.$$

For $\xi \geq 0$, $y \geq 0$ the function $Q(\xi, y, \tau, \theta)$ has a single maximum with respect to y, viz., for $y = Y(\xi, \tau, \theta)$. This maximum satisfies relation (4.35), and, consequently, $Q(\xi, Y(\xi, \tau, \theta), \tau, \theta) = S(\xi, \theta)$. One can easily check that $|H(\xi, \tau)| \geq (\tau^{3/2} + \xi)^{1/3}$ whereby it is sufficient to consider just the case where $Y = \xi - (\tau - \theta) H(y, \theta) > 2^{-1}(\tau^{3/2} + \xi)$. Then for $y < \frac{2}{3} Y(\xi, \tau, \theta)$ the function $\exp[Q(\xi, y, \tau, \theta) - S(\xi, \tau)]$ is exponentially small, and for $y \geq \frac{2}{3} Y(\xi, \tau, \theta)$ one has the inequality $Q_{yy} < -[2(\tau - \theta)]^{-1}$, and the integral over this domain is estimated as in Lemma 4.2. The integral $v_2(\xi, \tau)$, and derivatives of the function $v(\xi, \tau)$ are estimated in a similar way. Formula (4.27) implies (4.39). ∎

PROOF OF THEOREM 4.3. First, we assume that $w(\xi, \tau_0) \equiv 0$, and recast the right-hand side in the form $F(\xi, \tau) = F_2(\xi, \tau) + F_3(\xi, \tau)$ where

$$F_3(\xi, \tau) = \rho(\xi) \int_{-\infty}^{\infty} F(y, \tau) \, dy,$$

$\rho(\xi)$ is a compactly supported infinitely differentiable function, $\int_{-\infty}^{\infty} \rho(\xi) \, d\xi = 1$.

Then the function $F_2(\xi, \tau)$ satisfies the conditions of Lemma 4.4. It remains, therefore, to examine the solution of equation (4.26) for which $F(\xi, \tau) = \lambda(\tau) \rho(\xi)$,

$D^k\lambda(\tau) = O\left(\tau^{-N}\right)$, and $w(\xi, \tau_0) = 0$. The solution $w(\xi, \tau)$ is defined by formula (4.27), where

$$v(\xi, \tau) = \int_{\tau_0}^{\tau} \lambda(\theta) \int_{-\infty}^{\infty} G(\xi - y, \tau - \theta)\Lambda(y, \theta)\rho_1(y)\, dy\, d\theta, \qquad (4.40)$$

$\rho_1(y) = \int_{-\infty}^{y} \rho(z)\, dz$. Let $\mathrm{supp}\, \rho \in [-a, a]$. Then

$$v(0, \tau) = \int_{\tau_0}^{\tau} \lambda(\theta) \int_{-a}^{a} G(y, \tau - \theta)\Lambda(y, \theta)\rho_1(y)\, dy\, d\theta$$

$$+ \int_{\tau_0}^{\tau} \lambda(\theta) \int_{a}^{\infty} G(y, \tau - \theta)\Lambda(y, \theta)\, dy\, d\theta = v_1(\tau) + v_2(\tau).$$

The integral v_1 decays rapidly as $\tau \to \infty$. This is proved exactly as in Lemma 4.4.

Substituting the integral (4.18) for $\Lambda(y, \theta)$, and changing the order of integration in $v_2(\tau)$, one obtains

$$v_2(\tau) = \int_{\tau_0}^{\tau} \lambda(\theta) \int_{-\infty}^{\infty} \exp\left(-\frac{z^4}{8} + \frac{z^2\tau}{4}\right) \mathrm{erfc}\left(\frac{a + z(\tau - \theta)}{2\sqrt{\tau - \theta}}\right)\, dz\, d\theta,$$

where $\mathrm{erfc}\, x = \frac{1}{\sqrt{\pi}} \int_{x}^{\infty} \exp(-t^2)\, dt$. The integral over θ should be split into two integrals: one for $\theta > \tau/2$, and the other for $\theta \leq \tau/2$. The first integral is $O\left(\tau^{-N+1} \exp(\tau^2/8)\right)$, while the second equals

$$\int_{\tau_0}^{\tau/2} \lambda(\theta) \int_{-\infty}^{\infty} \exp\left(-\frac{z^4}{8} + \frac{z^2\tau}{4}\right) \mathrm{erfc}\left(\frac{a + z(\tau - \theta))}{2\sqrt{\tau - \theta}}\right)\, dz\, d\theta$$

$$= \int_{\tau_0}^{\infty} \lambda(\theta)\, d\theta \int_{-\infty}^{0} \exp\left(-\frac{z^4}{8} + \frac{z^2\tau}{4}\right) dz + O\left(\tau^{-N-1} \exp\frac{\tau^2}{8}\right)$$

$$= \tau^{1/2} \int_{\tau_0}^{\infty} \lambda(\theta)\, d\theta \exp\left(\frac{\tau^2}{8}\right) \left[\sum_{k=0}^{m} c_k \tau^{-2k} + O\left(\tau^{-2m-2}\right)\right], \qquad c_0 \neq 0.$$

The derivative $(\partial v/\partial \xi)(0, \tau)$ is of the same form as the integral $v(0, \tau)$. The only difference is the factor $-z/2$ appearing in the integrand. Therefore, the asymptotics of $(\partial v/\partial \xi)(0, \tau)$ differs from that of $v(0, \tau)$ by the additional factor $\sqrt{\tau}$.

The explicit form of $\Lambda(\xi, \tau)$ implies that $(\partial\Lambda/\partial\xi)(0, \tau) = 0$, and

$$\Lambda(0, \tau) = \tau^{-1/2} \exp\left(\frac{\tau^2}{8}\right) \sum_{k=0}^{\infty} d_k \tau^{-2k}, \qquad \tau \to \infty, \quad d_0 \neq 0.$$

It follows from this asymptotic expansion, the asymptotics for $\frac{\partial v}{\partial \xi}(0, \tau)$ proved above, and formula (4.27) that, in the case of the zero initial conditions, representation (4.37) holds for the solution w. In formula (4.37) γ_k are the coefficients of the asymptotic expansion of the function $2[\Lambda(0, \tau)]^{-1} \int_0^\infty z \exp\left(-z^4/8 + z^2\tau/4\right) dz$.

It now remains to consider the homogeneous equation (4.26). In this case, as above, one can represent the initial function in the form $w(\xi, \tau_0) = w_2(\xi) + w_3(\xi)$, where $w_3(\xi) = \rho(\xi) \int_{-\infty}^\infty w(z, \tau_0) dz$ thus reducing the problem to the investigation of an integral similar to (4.40). Thus, the asymptotic expansion (4.37) with same coefficients γ_k as those mentioned above is also valid for the solution of the homogeneous equation. ∎

The matching procedure for the series (4.3), series (1.4), and the series to be constructed in the next section causes the right-hand side of an equation of the form (4.26) to decay rapidly as $|\xi| + |\tau| \to \infty$. Theorem 4.3 shows that this is not sufficient for the solution to tend to zero uniformly as $\tau \to \infty$. Therefore, in the sequel we shall additionally match the values of the solutions for $\xi = 0$, $\tau \to \infty$. It turns out that the rapid decay of a solution of equation (4.26) for $\xi = 0$, $\tau \to \infty$, together with the assumptions of Theorem 4.3, ensures the rapid decay of the solution everywhere as $|\xi|+|\tau| \to \infty$. The proof of this fact is the theme of the last theorem of the present section.

THEOREM 4.4. *Suppose that the conditions of Theorem 4.3 are satisfied, and, in addition, $|D^k w(0, \tau)| \le M\tau^{-2N}$, where k and N have the same meaning as in Theorem 4.3. Then the function $w(\xi, \tau)$ satisfies estimates (4.39).*

PROOF. The transformation $w = v \partial \Gamma/\partial \xi$ yields the following equation for $v(\xi, \tau)$:

$$L_4 v = \frac{\partial v}{\partial \tau} + b(\xi, \tau)\frac{\partial v}{\partial \xi} - \frac{\partial^2 v}{\partial \xi^2} = \tilde{F}(\xi, \tau),$$

where $\tilde{F}(\xi, \tau) = F(\xi, \tau)[(\partial \Gamma/\partial \xi)(\xi, \tau)]^{-1}$, and

$$b(\xi, \tau) = \Gamma(\xi, \tau) - 2\Gamma_{\xi\xi}(\xi, \tau)[\Gamma_\xi(\xi, \tau)]^{-1}$$
$$= -2\left\{\Lambda^2\Lambda_{\xi\xi\xi} - 2\Lambda\Lambda_\xi\Lambda_{\xi\xi} + \Lambda_\xi^3\right\}\Lambda^{-1}(\Lambda\Lambda_{\xi\xi} - \Lambda_\xi^2)^{-1}.$$

The function $\tilde{F}(\xi, \tau)$ satisfies the same conditions as $F(\xi, \tau)$. This follows easily from the asymptotics of $\Gamma(\xi, \tau)$ as $|\xi| \to \infty$, $\tau \to \infty$.

The explicit form (4.18) of the function $\Lambda(\xi, \tau)$ makes it possible to find an estimate and the asymptotics of the coefficient $b(\xi, \tau)$:

$$b(\xi, \tau) = \sqrt{\tau}\,\frac{\sinh \eta - (2\tau)^{-2}\sinh 3\eta + \beta_1(\tau, \eta)}{\cosh \eta - (2\tau)^{-2}\cosh 3\eta + \beta_2(\tau, \eta)}$$

for $|\xi| < $ const, where $\eta = 2^{-1}\xi\sqrt{\tau}$, and

$$\beta_i(\tau, \eta) = O(\tau^{-2}\exp|\eta| + \tau^{-4}\exp|3\eta|)(1 + \eta^2),$$

$b(\xi, \tau) < -\mu\sqrt{\tau}$ for $\xi > $ const > 0, where $\mu > 0$.

In what follows we consider only the values $\xi > 0$. The asymptotics of the coefficient $b(\xi, \tau)$ easily implies the following statement.

For any small $\nu > 0$ and sufficiently large τ there exists a constant $\mu > 0$ such that

$$b(\xi, \tau) > -\mu\tau^{-3/2} \quad \text{for } \xi < 2\tau^{-1/2}(\ln 2\tau - \nu),$$
$$b(\xi, \tau) < -\mu\tau^{1/2} \quad \text{for } \xi > 2\tau^{-1/2}(\ln 2\tau + \nu). \tag{4.41}$$

Furthermore,

$$|b(\xi, \tau)| < M_1\tau^{1/2} \tag{4.42}$$

for all ξ if $\tau \geq \tau_0 > 0$.

The estimate of the solution $w(\xi, \tau)$ for $\xi \geq 0$, $\tau \geq \tau_0$ will be obtained with the help of a barrier function $U(\xi, \tau)$. It is sufficient to require that the function $U(\xi, \tau)$ satisfies the following conditions:

$$U(0, \tau) > M_1\tau^{-2N}, \qquad \tau \geq \tau_1 > \tau_0,$$
$$U(\xi, \tau_1) > M_1(1 + \xi^2)^{-N}, \qquad \xi \geq 0, \tag{4.43}$$
$$U(\xi, \tau) < M_2(\xi^2 + \tau^2)^{-m}, \qquad \xi \geq 0, \ \tau \geq \tau_1,$$
$$L_4U > M(\xi^2 + \tau^2)^{-N} \qquad \xi \geq 0, \ \tau \geq \tau_1. \tag{4.44}$$

However, a more convenient way is to construct $U(\xi, \tau)$ not as a smooth, but a piecewise smooth function modifying condition (4.44) accordingly. Denote $\xi_1 = 2\tau^{-1/2}(\ln 2\tau - \nu)$, $\xi_2 = 2\tau^{-1/2}(\ln 2\tau + \nu)$. The function $U(\xi, \tau)$ is continuous everywhere for $\xi \geq 0$, $\tau \geq \tau_0$, smooth everywhere with the exception of the curves $\xi = \xi_1(\tau)$, $\xi = \xi_2(\tau)$ on which the usual conditions for the one-sided derivatives must hold:

$$\frac{\partial U}{\partial \xi}(\xi_i(\tau) - 0, \tau) > \frac{\partial U}{\partial \xi}(\xi_i(\tau) + 0, \tau), \qquad i = 1, 2. \tag{4.45}$$

Inequality (4.44) must hold outside these curves.

Denote $\sigma(\tau) = \left[1 + \xi_1(\tau) - \xi_1^2(\tau)\right]^{-1}(\tau + \xi_2(\tau))^{-N}$, and set

$$U(\xi, \tau) = \begin{cases} \sigma(\tau)(1 + \xi - \xi^2) & \text{for } \xi \leq \xi_1(\tau), \\ (\xi + \tau)^{-N} & \text{for } \xi \geq \xi_2(\tau), \end{cases}$$

$$U(\xi, \tau) = \tau^{-1/2-N}[4\tau^{-1/2}\xi\ln 2\tau - \xi^2 + 4\tau^{-1}(\nu^2 - \ln^2 2\tau)] + (\tau + \xi_2(\tau))^{-N}$$
$$\text{for } \xi_1(\tau) \leq \xi \leq \xi_2(\tau).$$

The function U is continuous on the curves $\xi = \xi_i(\tau)$. To verify this property, it is sufficient to establish that the expression in the square brackets vanishes for $\xi = \xi_i(\tau)$. The validity of inequalities (4.43) is easily verified directly. It remains to check inequalities (4.44), (4.45).

For $\xi < \xi_1(\tau)$, by virtue of (4.41),

$$L_4U = \sigma'(\tau)(1 + \xi - \xi^2) + (1 - 2\xi)\sigma(\tau)b(\xi, \tau) + 2\sigma(\tau)$$
$$> \sigma(\tau)[2 - \mu\tau^{-3/2} - 2|\sigma'(\tau)|[\sigma(\tau)]^{-1}] > \sigma(\tau)$$

for sufficiently large τ.

For $\xi > \xi_2(\tau)$, by virtue of (4.41),

$$L_4U = U\{-N(\xi + \tau)^{-1} - N(\xi + \tau)^{-1}b(\xi, \tau) - N(N + 1)(\xi + \tau)^{-2}\}$$
$$> U(\xi + \tau)^{-1}N(\mu\tau^{1/2} - 1 - (N + 1)(\xi + \tau)^{-1}) > (\xi + \tau)^{-N-1}$$

for sufficiently large τ.

For $\xi_1(\tau) < \xi < \xi_2(\tau)$ one has

$$L_4 U = \tau^{-N-1/2}[(4\tau^{-1/2}\ln 2\tau - 2\xi)b(\xi, \tau) + 2 + O(\tau^{-2}\ln^2 \tau)] + O(\tau^{-N-1}).$$

In the domain under consideration the relation $|\xi - 2\tau^{-1/2}\ln 2\tau| \leq 2\nu\tau^{-1/2}$ holds. It follows hence, and from (4.42) that

$$L_4 U > \tau^{-N-1/2}[2 - 4M_1\nu - M\tau^{-1}] - M\tau^{-N-1} \geq \tau^{-N-1}$$

for τ sufficiently large if one chooses $\nu < (4M_1)^{-1}$.

Thus, (4.44) is satisfied. Let us now check (4.45):

$$\frac{\partial U}{\partial \xi}(\xi_1(\tau) - 0, \tau) - \frac{\partial U}{\partial \xi}(\xi_1(\tau) + 0, \tau)$$

$$= \sigma(\tau)(1 - 2\xi_1(\tau)) - \tau^{-N-1/2}\left(4\tau^{-1/2}\ln 2\tau - 2\xi_1(\tau)\right) > 0$$

for sufficiently large τ,

$$\frac{\partial U}{\partial \xi}(\xi_2(\tau) - 0, \tau) - \frac{\partial U}{\partial \xi}(\xi_2(\tau) + 0, \tau)$$

$$= \tau^{-N-1/2}(4\tau^{-1/2}\ln 2\tau - 2\xi_2(\tau)) + N(\xi_2(\tau) + \tau)^{-N-1}$$

$$= -4\nu\tau^{-N-1} + N(\tau + \xi_2(\tau))^{-N-1} > 0$$

for sufficiently large N and τ. This concludes the process of estimating the function $v(\xi, \tau)$, and hereby the function $w(\xi, \tau)$. The statement of the theorem about the derivatives of the solution $w(\xi, \tau)$ follows from the a priori estimates of derivatives of solutions of a parabolic equation. ∎

§5. Construction of asymptotics in the vicinity of the discontinuity curve

The last domain where the asymptotic expansion of the solution $u(x, t, \varepsilon)$ has to be constructed is a neighborhood of the curve $l = \{x, t: x = s(t), 0 \leq t \leq T\}$. Recall that, as in the problem considered in §2, the series (1.4) does not provide a correct asymptotic expansion near l because its coefficients are discontinuous on l. For a correct description of the asymptotics in the vicinity of l, we introduce, as in §1 (1.19) the new coordinates

$$\zeta = \varepsilon^{-4}(x - s(t)), \quad t. \tag{5.1}$$

It is not entirely evident in what form one should seek the asymptotic expansion of the function $v(\zeta, t, \varepsilon)$. It is clear that, as $\zeta \to \pm\infty$, this series has to be matched to the series (1.4) as $x \to s(t) \pm 0$. This implies that, as in §§1, 2 the terms containing ε^{4k}, $k \geq 0$, must be present in the asymptotic expansion of the function $v(\zeta, t, \varepsilon) \equiv u(x, t, \varepsilon)$. However, the asymptotic expansion must be matched not just to the series (1.4) in a neighborhood of l, but also to the series (4.3) in a neighborhood of the origin (see Figure 36). It is, therefore, natural to use the same gauge sequence for the asymptotics of the function $v(\zeta, t, \varepsilon)$ as in the expression (4.3) for the series W. Accordingly, we set

$$V = \sum_{k=0}^{\infty} \varepsilon^k \sum_{j=0}^{k} v_{k,j}(\zeta, t) \ln^j \varepsilon \tag{5.2}$$

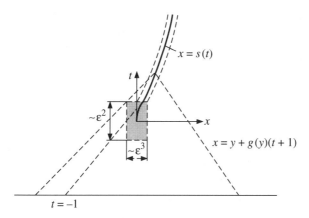

FIGURE 36

and substitute this series into equation (1.11). As a result, one obtains almost the same system as (1.12)–(1.14). To be more precise, the system for $v_{4k,0}$ is of the same form as (1.12)–(1.14) so that in what follows both notations will be used assuming that $v_{4k,0}(\zeta, t) \equiv v_{4k}(\zeta, t)$. For the remaining functions $v_{k,j}(\zeta, t)$ the system is of almost the same form:

$$L_1 v_{k,j} = \frac{\partial}{\partial \zeta} \tilde{G}_{k,j}(v_0, v_{1,0}, v_{1,1}, \ldots , v_{k-1,k-1}) + \frac{\partial v_{k-4,j}}{\partial t} \qquad (5.3)$$

where L_1 is the operator defined in (1.13), and

$$\tilde{G}_{k,j}(v_0, v_{1,0}, v_{1,1}, \ldots , v_{k-1,k-1}) = \sum_{q \geq 2} \frac{\varphi^{(q)}(v_0)}{q!} \sum_{\substack{\sum i_p = k \\ \sum m_p = j}} \prod_{p=1}^{q} v_{i_p, m_p}. \qquad (5.4)$$

The boundary conditions at $\pm\infty$ for the functions $v_{4k}(\zeta, t)$ are given by relations (1.23), (1.22) while for the rest of the functions one has

$$v_{k,j}(\zeta, t) \to 0 \quad \text{as } |\zeta| \to \infty. \qquad (5.5)$$

The existence of solutions for the problems (1.12)–(1.14) for all $t > 0$ was proved in §1 (Theorem 1.1). The existence of solutions for the problems (5.3)–(5.5) for the remaining functions $v_{k,j}(\zeta, t)$ is proved in the same way. For each $t > 0$, the relations $v_{k,j}(\zeta, t) \in \mathfrak{M}^+ \cap \mathfrak{M}^-$ hold if $k \neq 0$ mod 4 or $j > 0$, where the classes $\mathfrak{M}^\pm$ are those defined in §1. As shown in §1, all functions $v_{k,j}(\zeta, t)$ are smooth for $t > 0$ because condition (1.3) is satisfied for these values of t. However, in contrast to §2, in this problem one has $\omega(y, t) \to 0$ as $y \to 0$, $t \to 0$. Hereby the functions $v_{k,j}(\zeta, t)$ have singularities as $t \to 0$. Accordingly, the purpose of the present section is to examine the asymptotics as $t \to 0$ of the solutions $v_{k,j}(\zeta, t)$ virtually constructed in §1.

For the description of the asymptotics of the functions $v_{k,j}(\zeta, t)$ it is convenient to introduce appropriate classes of functions. Everywhere below we denote $\eta = \zeta\sqrt{t}/2$, and begin with the definition of classes of functions in one variable η.

The class $\mathfrak{A}^+$ is the set of functions $v(\eta) \in C^\infty(\mathbf{R}^1)$ such that each of them satisfies the inequalities

$$\left|\frac{d^k v}{d\eta^k}\right| \le M_k e^{-2\eta}(1 + |\eta|^\lambda), \qquad k = 0, 1, 2, \ldots,$$

where the constants M_k and λ depend, in general, on the function v. The class $\mathfrak{A}^-$ is defined in a similar way with $e^{-2\eta}$ replaced by $e^{2\eta}$.

The class $\mathfrak{B}_\alpha$ is the set of functions $v(\zeta, t) \in C^\infty(\mathbf{R}^1 \times (0, T])$ for which the asymptotic expansion

$$v(\zeta, t) = t^\alpha \sum_{j=0}^\infty t^{j/2} \sum_{q=0}^j V_{j,q}(\eta) \ln^q t \tag{5.6}$$

holds as $t \to 0$. Everywhere below the coefficients $V_{j,q}$ will be denoted by $\Pi_{j,q} v$.

It is not difficult to deduce from formulas (1.22), (3.4), (3.14) that $P_m^\pm(\zeta, t) \in \mathfrak{B}_{1/2-2m}$, and $\Pi_{j,q} P_m$ are polynomials of degree no greater than m.

The class $\mathfrak{B}_\alpha^+$ is the subset of functions $v(\zeta, t) \in \mathfrak{B}_\alpha$ such that the following conditions are satisfied:

(a) $\Pi_{j,q} v \in \mathfrak{A}^+$ for all j and q;

(b) for any natural N there exist constants $\lambda > 0$ and $M_{k,l} > 0$ such that

$$\left|\frac{\partial^{k+l}}{\partial\zeta^k \partial t^l}[v(\zeta, t) - B_{N/2+\alpha}v]\right| \le M_{k,l} \exp(-2\eta) t^{\alpha+N/2+k/2-l}(1 + |\eta|^\lambda),$$

$$k, l = 0, 1, 2, \ldots, \tag{5.7}$$

where $B_{N/2+\alpha}v$ is the partial sum of the series (5.6) including all the terms for which $j \le N$.

The class $\mathfrak{B}_\alpha^-$ is defined in the same way with $\mathfrak{A}^+$ replaced by $\mathfrak{A}^-$ and $\exp(-2\eta)$ by $\exp(2\eta)$.

As in §1, we denote by $\tilde{v}(\zeta, t)$ the solution of equation (1.16) given by formula (1.20) for $c_0(t) \equiv 0$ and $a = 0$. The last equality is acceptable because the relations

$$v_0(-\infty, t) = u_0(s(t) - 0, t) > 0,$$
$$v_0(+\infty, t) = u_0(s(t) + 0, t) < 0$$

hold for small $t > 0$. The constant $C(t)$ is defined by equalities (1.17). Thus, the explicit formula for $\tilde{v}(\zeta, t)$ is of the following form:

$$\zeta = \int_0^{\tilde{v}} \left[\varphi(z) - s'(t)z - \varphi\left(P_0^+(t)\right) + s'(t)P_0^+(t)\right]^{-1} dz. \tag{5.8}$$

Theorem 5.1. *The following relations hold:* $\tilde{v}(\zeta, t) \in \mathfrak{B}_{1/2}$,

$$\Pi_{0,0}\tilde{v} = -\frac{1}{2b^2}\tanh\eta, \qquad \Pi_{j,q}\tilde{v} = 0 \quad \text{for } q > 0,$$

$$\tilde{v}(\zeta, t) - P_0^\pm(t) \in \mathfrak{B}_{1/2}^\pm.$$

The proof can be obtained, for example, from the explicit formula (5.8). It follows from Theorem 4.1 and formula (1.18) that $s'(t) = t\Phi_1(t)$, $\varphi\left(P_0^+(t)\right) - s'(t)P_0^+(t) = t/4b^2 + t^2\Phi_2(t)$, where, as before, Φ_i denotes smooth functions. Making the change of variable $z = (2b^2)^{-1}\theta\sqrt{t}$ in the integral (5.8), and denoting $\tilde{v} = (2b^2)^{-1}\sqrt{t}\psi$, $\eta = 2^{-1}s\sqrt{t}$, one obtains the equality

$$\eta = \int_0^\psi \left[\theta^2 - 1 - \sqrt{t}\Phi(\theta, \sqrt{t})\right]^{-1} d\theta.$$

This immediately yields the asymptotic expansion (5.6) for $\alpha = 1/2$ which is uniform with respect to η on any compact set. Here $\Pi_{0,0}\tilde{v}(\eta) = -(1/2b^2)\tanh\eta$, $\Pi_{j,q}\tilde{v}(\eta) = 0$ if $q > 0$.

In order to obtain the asymptotics of the functions $\Pi_{k,0}\tilde{v}(\eta)$ as $\eta \to \pm\infty$ and the estimate (5.7), one has to act somewhat differently. Consider, for example, the values $\zeta > \text{const} \cdot t^{-1/2}$, and make another change of variables in the integral (5.8), viz.: $z = P_0^+(t)\theta$, $\tilde{v} = P_0^+(t)W$, $\eta = 2^{-1}\zeta\sqrt{t}$. The resulting relation is

$$\eta = -(1 + \sqrt{t}\Phi_3(\sqrt{t}))\int_0^W \left[\theta^2 - 1 + \sqrt{t}(1 - \theta)\Phi_4(\sqrt{t}, \theta)\right]^{-1} d\theta,$$

which easily yields estimates (5.7) for $\eta > \text{const}$. The same estimates for $\eta < \text{const}$ are obtained in a similar manner. ∎

One can now obtain the asymptotics of the function $c_0(t)$ which is defined as a solution of equation (1.31). For that purpose we have to examine the asymptotics of the right-hand side of that equation as $t \to 0$. It follows from (1.22), (3.4), and (3.14) that

$$\frac{\partial}{\partial\zeta}P_1^\pm(\zeta, t) = \frac{\partial u_0}{\partial x}(s(t) \pm 0, t) = \frac{\psi'(y_\pm(t))}{\omega(y_\pm(t), t)}$$

$$= \frac{-2^{-1}b^{-2} \pm \sqrt{t}\Phi_1(t) + t\Phi_2(t)}{2t \pm t\sqrt{t}\Phi_3(t) + t^2\Phi_4(t)},$$

$$P_1^\pm(0, t) = u_1(s(t) \pm 0, t)$$

$$= [\omega(y_\pm(t), t)]^{-2}[\omega(y^\pm(t), t)\Phi_5(y_\pm(t)) + y_\pm(t)\Phi_6(y_\pm(t))]$$

$$+ [\omega(y_\pm(t), t)]^{-1}\Phi_7(y_\pm(t))\ln[\omega(y_\pm(t), t)]$$

$$= \pm t^{-3/2}\Phi_8(t) + t^{-1}\Phi_9(t) + (\Phi_{10}(t) \pm \sqrt{t}\Phi_{11}(t))t^{-1}\ln t.$$

Therefore, the sum of the terms in the right-hand side of (1.31) containing no integrals equals $t^{-1/2}[\Phi(t) + \Phi(t)\ln t]$.

The asymptotics of the integrals in the right-hand side of (1.31) is easily computed with the help of Theorem 5.1 and the explicit form of its principal term $\Pi_{0,0}\tilde{v}$. As a result, equation (1.31) takes the form:

$$\frac{d}{dt}[c_0(t)\varkappa(t)] = t^{-1/2}\Phi(t^{1/2}) + t^{-1/2}\Phi(t)\ln t, \tag{5.9}$$

where $\varkappa(t) = P_0^+(t) - P_0^-(t) = \sqrt{t}\Phi(t)$. (It is possible that the first of the functions Φ in equation (5.9) is a smooth function of t, but this fact can be clarified only after a more detailed examination of the form of the functions $\Pi_{k,0}\tilde{v}$ in Theorem 5.1. On the other hand, this fact does not affect the subsequent analysis in any essential way.) Thus, it follows from equation (5.9) that

$$c_0(t) = \Phi_1(\sqrt{t}) + \Phi_2(t)\ln t + c[\varkappa(t)]^{-1}.$$

It is intuitively clear and will be confirmed below that $c = 0$, We will, therefore, assume that the function $c_0(t)$ is defined conclusively:

$$c_0(t) = \Phi_1(\sqrt{t}) + \Phi(t)\ln t. \tag{5.10}$$

This also provides the final definition for the function $v_0(\zeta, t) = \tilde{v}(\zeta + c_0(t), t)$.

THEOREM 5.2. *The following relations hold*: $v_0(\zeta, t) \in \mathcal{B}_{1/2}$,

$$\Pi_{0,0}v_0 = -\frac{\tanh\eta}{2b^2}, \qquad v_0(\zeta, t) - P_0^{\pm}(t) \in \mathcal{B}_{1/2}^{\pm}.$$

The proof consists in applying Theorem 5.1 to the function $\tilde{v}(\zeta + c_0(t)), t)$, and expanding each of the functions $\Pi_{j,0}\tilde{v}$ in a Taylor series at the point $\eta = 2^{-1}\zeta\sqrt{t}$ taking into account the form of (5.10). ■

LEMMA 5.1. *Consider the equation*

$$L_1 v = F(\zeta, t), \tag{5.11}$$

where L_1 *is the operator* (1.12), *and suppose that* $F(\zeta, t) \in \mathcal{B}_\alpha^+ \cap \mathcal{B}_\alpha^-$. *Then a solution of equation* (5.11) *such that*

$$v(\zeta, t) \xrightarrow[|\zeta|\to\infty]{} 0 \tag{5.12}$$

exists for $t > 0$ *if and only if*

$$\int_{-\infty}^{\infty} F(\zeta, t)\,d\zeta = 0. \tag{5.13}$$

Provided this condition is satisfied, there exists a solution $v^*(\zeta, t) \in \mathcal{B}_{\alpha-1}^+ \cap \mathcal{B}_{\alpha-1}$. *Any solution of the problem* (5.11), (5.12) *equals* $v^*(\zeta, t) + c(t)Z(\zeta, t)$, *where* $Z(\zeta, t) = (\partial\tilde{v}/\partial\zeta)(\zeta, t)$ *and* $\tilde{v}(\zeta, t)$ *is defined by formula* (5.8).

The proof of the necessity is obtained by integrating the equation from $-\infty$ to $+\infty$, while the sufficiency follows from the formula for the solution:

$$v(\zeta, t) = Z(\zeta, t)\left\{\int_0^\zeta F_1(\zeta_1, t)[Z(\zeta_1, t)]^{-1}\,d\zeta_1 + c(t)\right\} = v^*(\zeta, t) + c(t)Z(\zeta, t).$$

Here $F_1(\zeta, t) = \int\limits_{-\infty}^{\infty} F(\zeta_1, t)\,d\zeta_1 \in \mathfrak{B}_{\alpha-1/2}$ by virtue of (5.13), $Z(\zeta, t) \in \mathfrak{B}_1^+ \cap \mathfrak{B}_1^-$
by Theorem 5.1. The same theorem implies that

$$[Z(\zeta, t)]^{-1} = -\frac{2b^2 \cosh^2 \eta}{t} \sum_{j=0}^{\infty} t^{j/2} X_j(\eta),$$

where the functions $X_j(\eta)$ are of slow growth, $X_j(\eta) \in C^\infty(\mathbf{R}^1)$, and the asymptotic expansion is uniform with respect to η, and can be differentiated term-by-term. The explicit form of $v^*(\zeta, t)$ now implies that $v^*(\zeta, t) \in \mathfrak{B}_{\alpha-1}^+ \cap \mathfrak{B}_{\alpha-1}^-$. ∎

LEMMA 5.2. *Suppose that the functions* $P^+(\zeta, t)$ *and* $P^-(\zeta, t)$ *belong to the class* $\mathfrak{B}_\alpha$, $F(\zeta, t) - L_1 P^\pm(\zeta, t) \in \mathfrak{B}_\alpha^\pm$, *where* L_1 *is the operator* (1.12). *Then a solution of the problem*

$$L_1 v = F(\zeta, t), \qquad v - P^\pm \in \mathfrak{B}_{\alpha-1}^\pm \tag{5.14}$$

exists if and only if condition (1.33) *is satisfied.*

PROOF. The fact that condition (1.33) is necessary and sufficient for the solvability of the equation $L_1 v = F(\zeta, t)$ was proved in Lemma 1.1. It remains to check that $v - P^\pm \in \mathfrak{B}_{\alpha-1}^\pm$ which is an immediate consequence of the explicit formula (1.34). ∎

THEOREM 5.3. *For the solutions of the system* (1.12), (1.13), (5.3) *subject to conditions* (1.23), (5.5) *the following statements hold*:

(1) $v_{k,i}(\zeta, t) \in \mathfrak{B}_{n_k}$, *where* $n_k = 1/2 - 2[k/4]$;

(2) *if* $i > 0$ *or* $k \not\equiv 0 \bmod 4$, *then*

$$v_{k,i}(\zeta, t) \in \mathfrak{B}_{n_k}^+ \cap \mathfrak{B}_{n_k}^-;$$

(3) *if* $k = 4m$, *where* $m \geq 0$ *is an integer, then*

$$v_{4m}(\zeta, t) - P_m^\pm(\zeta, t) \in \mathfrak{B}_{1/2-2m}^\pm;$$

(4) *for* $k > 0$, *if all functions* $v_{k_1,j}$ *for* $k_1 < k$ *are already found, then each of the functions* $v_{k,j}$ *is determined uniquely up to the summand* $c_{k,i}[\varkappa(t)]^{-1} Z(\zeta, t)$, *where* $\varkappa(t) = P_0^+(t) - P_0^-(t)$.

PROOF. For $k = 0$ the assertion of the theorem follows from Theorem 5.2. For $k = 1$, $i = 0$ equation (5.3) is a homogeneous one. Hence $v_{1,0}(\zeta, t) = c_{1,0}(t) Z(\zeta, t)$, where $Z(\zeta, t) = (d\tilde{v}/d\zeta)(\zeta, t) \in \mathfrak{B}_1^+ \cap \mathfrak{B}_1^-$ by virtue of Theorem 5.1. The coefficient $c_{1,0}(t)$ is defined from equation (5.3) for $k = 5$. Indeed, Lemma 5.1 implies the relation $\int\limits_{-\infty}^{\infty} (\partial v_{1,0}/\partial\zeta)(\zeta, t) = 0$. Hence $c_{1,0}(t) = c_{1,0}[\varkappa(t)]^{-1}$, where $c_{1,0} = \text{const}$. The form of $\varkappa(t)$ (see formula (1.22), Theorem 3.1, and relation $u_0(x, t) = \psi(y)$) implies that $v_{1,0}(\zeta, t) \in \mathfrak{B}_{1/2}^+ \cap \mathfrak{B}_{1/2}^-$.

Now, in exactly the same way, the statement of the theorem is proved successively for $v_{1,j}$, $v_{2,j}$, and $v_{3,j}$. The functions $v_{1,j}$, $v_{2,j}$, and $v_{3,j}$ are constructed in such a way that the problems (5.3), (5.5) for $k = 5$, $k = 6$, and $k = 7$ are solvable. For the function $v_{4k,0}(\zeta, t) \equiv v_{4k}(\zeta, t)$ the problem (1.13), (1.23) also has a solution in view of the choice of $c_0(t)$.

The proof now proceeds by induction as in Theorem 1.1. A nonessential difference is that new functions $v_{k,j}(\zeta, t)$ arise for $j > 0$, and for $j = 0$, $k \not\equiv 0 \bmod 4$, while the essential difference is that one has to examine the asymptotics of the solutions obtained as $t \to 0$, i.e., to verify conditions (2), (3).

Thus, let us assume that the assertions of the theorem hold for $k < 4n$, where n is a natural number. Suppose also that for $4n \leq k < 4(n + 1)$ the problems (1.13), (1.23), (5.3), (5.5) are solvable, and that some solutions $v_{k,i}^*(\zeta, t)$ of these problems satisfy conditions (1)–(3). We now show that for $4n \leq k < 4(n + 1)$ the functions $v_{k,i}(\zeta, t)$ are defined up to the summand $c_{k,i}[\varkappa(t)]^{-1} Z(\zeta, t)$ in such a way that conditions (1)–(4) of the theorem are satisfied, and the problems (1.13), (1.23), (5.3), (5.5) for $4(n + 1) \leq k < 4(n + 2)$ can be solved.

Apply Lemma 5.3 to the function $v_{4n+4}(\zeta, t)$ taking for $P^{\pm}(\zeta, t)$ the polynomials $P_{n+1}^{\pm}(\zeta, t)$ from formulas (1.22). The functions $P_k^{\pm}(\zeta, t)$ satisfy the recurrence system (1.24). Therefore,

$$L_1 P_{n+1}^{\pm} = \frac{\partial}{\partial \zeta} \left\{ \left[\varphi'(P_0^{\pm}) - \varphi'(v_0) \right] P_{n+1}^{\pm} \right\} + \frac{\partial P_n^{\pm}}{\partial t}$$
$$+ \frac{\partial}{\partial \zeta} \sum_{q \geq 2} \frac{\varphi^{(q)}(P_0^{\pm})}{q!} \sum_{\sum i_p = n+1} \prod_{p=1}^{q} P_{i_p}^{\pm}. \tag{5.15}$$

From the inductive hypothesis, formulas (3.14), (1.22) and the form of equations (1.13) we now conclude that the conditions of Lemma 5.2 are satisfied for $\alpha = -\frac{1}{2} - 2n$. According to this lemma the problems (1.13), (1.23) are solvable if and only if condition (1.33) holds, where $P^{\pm} = P_{n+1}^{\pm}(\zeta, t)$,

$$F(\zeta, t) = \frac{\partial}{\partial \zeta} G_{n+1}(v_0, v_4, \ldots, v_{4n}) + \frac{\partial v_{4n}}{\partial t},$$

and G_{n+1} is defined in formula (1.14). Since the solution of the problem (1.13), (1.23) for $k = 4n$ can be written in the form $v_{4n}^*(\zeta, t) + c_{4n,0}(t) Z(\zeta, t)$, the inductive hypothesis and the form of (5.15) for $L_1 P_{n+1}^{\pm}$ imply that condition (1.33) can be written as follows:

$$\frac{d}{dt} \int_{-\infty}^{\infty} c_{4n,0}(t) Z(\zeta, t) \, d\zeta = \frac{d}{dt} [c_{4n,0}(t) \varkappa(t)] = f_n(t),$$

where the function $f_n(t)$ admits the asymptotic expansion

$$f_n(t) = t^{-2n-1} \sum_{j=0}^{\infty} t^{j/2} \sum_{q=0}^{j} f_{n,j,q} \ln^q t, \qquad t \to 0.$$

Hereby the function $c_{4n,0}(t)$ is defined up to the summand $\text{const} \cdot [\varkappa(t)]^{-1}$, the function $v_{4n}(\zeta, t)$ constructed above satisfies condition (3) of the present theorem, the problem (1.13), (1.23) for $k = 4n + 4$, $i = 0$ is solvable, and its solution $v_{4n+4}^*(\zeta, t)$ satisfies conditions (1), (3).

In a similar manner, one constructs the remaining functions $v_{4n,i}(\zeta, t)$ for $i > 0$, and the functions $v_{k,i}(\zeta, t)$ for $4n < k < 4(n + 1)$, and ensures the solvability of the corresponding problems (5.3), (5.5) for $v_{4n+4,i}^*(\zeta, t)$ for $i > 0$ and $v_{k,i}^*(\zeta, t)$ for $4(n + 1) < k < 4(n + 2)$. For this it is sufficient to use the induction hypothesis and Lemma 5.1. ∎

§6. Construction of the uniform asymptotic expansion

In fact, a uniform asymptotic expansion of the solution $u(x, t, \varepsilon)$ has almost been constructed in the preceding sections. Outside the curve l this is the outer expansion (1.4), in a neighborhood of the origin it is the series (4.3), and, finally, in a neighborhood of the curve for $t > 0$ it is the series (5.2) (see Figure 36). By construction, the series (1.4) is matched both to the series (4.3) and the series (5.2). Two parts of the job have to be completed. First, the coefficients of the series (5.2) are not yet definitively determined (the degree of arbitrariness is given by Theorem 5.3). Second, the matching of the series (4.3) to (5.2) is not yet established. Both tasks are closely related and will be solved simultaneously. It is the matching condition for the series (5.2) and (4.3) that makes it possible to determine all $v_{k,j}(\zeta, t)$ uniquely.

We begin by replacing each term of the series (5.2) with its asymptotics as $t \to 0$ and pass to the inner variables $\xi = \varepsilon^{-3}x$, $\tau = \varepsilon^{-2}t$. We have to take into account that

$$\eta = \frac{\zeta\sqrt{t}}{2} = \frac{x - s(t)}{\varepsilon^4}\sqrt{t} = \frac{\xi\sqrt{\tau}}{2} - \sum_{k=2}^{\infty} a_k \varepsilon^{2k-3}\tau^{k+1/2}.$$

For each function $v_{k,j}(\zeta, t)$ for $4m \le k < 4(m+1)$ one obtains the asymptotic expansion

$$v_{k,j}(\zeta, t) = \varepsilon^{1-4m}\tau^{1/2-2m} \sum_{\substack{i \ge l \ge 0 \\ j \ge 5i}} \varepsilon^i P_{j,m,l}(\tau^{1/2})[\ln(\varepsilon^2\tau)]^l V_{m,i,l}(z), \qquad (6.1)$$

where $z = \xi\sqrt{\tau}/2$. Here $P_{j,m,l}$ are polynomials of degree not greater than j. One has to isolate the summand in the expression for $v_{k,j}(\zeta, t)$ which is not yet determined uniquely. According to Theorems 5.1 and 5.3 this summand equals to

$$c_{k,j}[\varkappa(t)]^{-1}Z(\zeta, t) = c_{k,j}\varepsilon\tau^{1/2}\left(-\frac{1}{8b^2\cosh^2 z} + \sum_{\substack{i \ge 1, \\ l \le 5i}} \varepsilon^i \overline{P}_l(\sqrt{\tau})\overline{V}_i(z)\right),$$
$$(6.2)$$

where the polynomials $\overline{P}_l$, and the functions $\overline{V}_i$ are determined by the asymptotics of $\varkappa$ and Z uniquely. Thus, it is convenient to represent the asymptotics of the function $v_{k,j}$ as the sum of the series (6.1) and (6.2), where all P_j and $V_{m,i,l}$ are determined uniquely, provided all $v_{q,s}$ for $q < k$ are fixed. Substituting these asymptotic expansions into the series

(5.2), and regrouping its terms, one obtains the relation

$$v(\zeta, t, \varepsilon) = \sum_{k=1}^{\infty} \varepsilon^k \sum_{j=0}^{k-1} U_{k,j}(\xi, \tau) \ln^j \varepsilon$$

$$+ \sum_{k=0}^{\infty} \varepsilon^{k+1} \sum_{j=0}^{k} c_{k,j} \tau^{1/2} \left(-\frac{1}{8b^2 \cosh^2 z} \right.$$

$$\left. + \sum_{\substack{i \geq 1, \\ l \leq 5i}} \varepsilon^i \overline{P}_l(\sqrt{\tau}) \overline{V}_i(z) \right) \ln^j \varepsilon, \qquad (6.3)$$

where $U_{k,j}$ are asymptotic series as $\tau \to \infty$. The coefficients of these series can easily be found from (6.1). Combining both sums in (6.3), we arrive at the relation

$$v(\zeta, t, \varepsilon) = \sum_{k=1}^{\infty} \varepsilon^k \sum_{j=0}^{k-1} \tilde{U}_{k,j}(\xi, \tau) \ln^j \varepsilon. \qquad (6.4)$$

The asymptotic series $\tilde{U}_{k,j}$ depend on the constants $c_{k,j}$. Each term of the series is of the form $\tau^{r/2} \ln^n \tau V(z)$, where r and n are integers, $n \geq 0$, and the function $V(z)$ is asymptotically polynomial at $\pm\infty$.

It remains to show that the constants $c_{k,j}$ may be chosen in such a way that $\tilde{U}_{k,j}(\xi, \tau)$ are asymptotic expansions as $\tau \to \infty$ of the functions $w_{k,j}(\xi, \tau)$ constructed in §4. First, we note that $v(\zeta, t, \varepsilon)$ is, by construction, an asymptotic solution of equation (0.1). Therefore, for any choice of $c_{k,j}$, the series $\tilde{U}_{k,j}(\xi, \tau)$ are f.a.s. of the same recurrence system as the functions $w_{k,j}(\xi, \tau)$, i.e., the system (4.4)–(4.6). Furthermore, both series (4.3) and (5.2) are matched to the series (1.4) as $|\xi| \to \infty$ and $|\zeta| \to \infty$, respectively. Hence, $w_{k,j}(\xi, \tau)$ and $\tilde{U}_{k,j}(\xi, \tau)$ are close to each other as $|\xi| \to \infty$. To be more precise, for $\tau \geq 1$, $|\xi|\sqrt{\tau} > \tau^{\alpha}$, where $\alpha > 0$, the difference between $w_{k,j}(\xi, \tau)$ and the N th partial sum of the series $\tilde{U}_{k,j}(\xi, \tau)$ is of the order of magnitude $O((|\xi| + |\tau|)^{-N_1})$, where $N_1 \to \infty$ as $N \to \infty$.

The functions $w_{1,0}(\xi, \tau)$ and $\tilde{U}_{1,0}(\xi, \tau)$ satisfy the nonlinear equation (4.4) and have to be considered separately.

LEMMA 6.1. *The series $\tilde{U}_{1,0}(\xi, \tau)$ constructed above is an asymptotic expansion of the function $w_{1,0}(\xi, \tau)$, as $\tau \to \infty$, $|\xi|\tau^{1/2} < \tau^{\alpha}$.*

PROOF. The series $\tilde{U}_{1,0}(\xi, \tau)$ does not depend on the coefficients $c_{k,j}$. Theorems 5.2 and 5.3 imply that

$$\tilde{U}_{1,0}(\xi, \tau) = \sqrt{\tau} \left(-\frac{\tanh z}{2b^2} + \sum_{k=1}^{\infty} \tau^{-2k} q_k(z) \right), \qquad (6.5)$$

where $z = \xi\sqrt{\tau}/2$, and $|q_k(z)| \leq M_k(1 + |z|^k)$. The explicit form of $w_{1,0}(\xi, \tau) = (1/2b^2)\Gamma(\xi, \tau)$ (see (4.17), (4.18), (4.23)) implies that the asymptotic expansion of this function is of the same form. This justifies the choice of $c_0(t)$ made before Theorem 5.2 in formula (5.10) which ensures that the principal term of the asymptotics for $w_{1,0}$ coincides with the first term of the series $\tilde{U}_{1,0}$. It remains to show that other terms $q_k(z)$ are also determined uniquely. Substituting the series (6.5) into equation (4.4), one obtains the system of differential equations

$$\frac{1}{2}(q_k \tanh z)' + \frac{1}{4}q_k'' = \frac{z}{2}q_{k-2}' - \frac{2k-1}{2}q_{k-1} + b^2 \left(\sum_{j=1}^{k-1} q_j q_{k-j} \right)', \qquad (6.6)$$

where $q_0(z) = - \tanh z / 2b^2$. At infinity, the functions $q_k(z)$ exponentially approach the given polynomials. If all the preceding q_j are known, the solution of equation (6.6) is determined up to the summand $c_k(\cosh z)^{-2}$. The constant c_k is determined from the next equation for q_{k+1}, and the condition that $q_{k+1}(z)$ must have given asymptotics at $\pm\infty$. ∎

THEOREM 6.1. *There exist constants $c_{k,j}$ such that the series $\tilde{U}_{k,j}(\xi, \tau)$ defined in formula (6.4) are asymptotic expansions of the functions $w_{k,j}(\xi, \tau)$ as $\tau \to \infty$, $|\xi|\tau^{1/2} < \tau^\alpha$, $\alpha < 2$.*

PROOF. For $k = 1$, $j = 0$ the validity of the asymptotics is established in the preceding lemma. The proof now proceeds by induction in k. All the other $w_{k,j}$ and $\tilde{U}_{k,j}$ satisfy the linear equations (4.5). Furthermore, as mentioned above, for $|\xi|\tau^{1/2} > \tau^\alpha$, $\alpha > 0$, the functions $w_{k,j}$ and the series $\tilde{U}_{k,j}$ asymptotically coincide with the series (4.14). The functions $w_{k,j}(\xi, \tau)$ are defined for all ξ and τ, while the series $\tilde{U}_{k,j}(\xi, \tau)$ are asymptotic only for $\tau \to \infty$, $|\xi|\tau^{1/2} < \tau^\alpha$, $\alpha < 2$. We, therefore, consider for $\tau \geq \tau_0$ the functions $U^*_{k,j,N}$, i.e., the partial sums of the composite expansions constructed in the usual way from the series $\tilde{U}_{k,j}$ and (4.14). The difference $U^*_{k,j,N}(\xi, \tau) - w_{k,j}(\xi, \tau)$ satisfies equation (4.26). By the inductive hypothesis and the construction procedure for $U^*_{k,j,N}$, the right-hand side of the equation and the initial function satisfy the condition of Theorem 4.3. Consequently, for $\xi = 0$ this difference has the asymptotic expansion (4.39). Choosing the constant $c_{k,j}$ in such a way that the coefficient of $\tau^{1/2}$ vanishes, we conclude from Theorem 4.3 that other coefficients of the asymptotic expansion for $\xi = 0$ vanish up to a sufficiently large number depending on N. It now follows from Theorem 4.4 that for $\tau \geq \tau_0 > 1$ the absolute value of the difference under consideration together with that of the derivatives of sufficiently high order does not exceed $M(\xi^2 + \tau^2)^{-N_1}$. By consecutively choosing the constants $c_{k,j}$ in this way we arrive at the conclusion of the theorem. ∎

The construction of the asymptotic expansion of the solution $u(x, t, \varepsilon)$ is completed. Note that, as a by-product, we have clarified the question about the asymptotics of the solutions $w_{k,j}(\xi, \tau)$ as $\tau \to \infty$ which remained open in §4. It follows from Theorem 6.1 that for $\tau > 0$, $|\xi|\tau^{1/2} > \tau^\beta$, $\beta > 0$, $R(\xi, \tau) \to \infty$ these functions have asymptotic expansions (4.14). For $|\xi|\tau^{1/2} < \tau^\alpha$, $\alpha < 2$, $\tau \to \infty$, the functions $w_{k,j}(\xi, \tau)$ admit asymptotic expansions $\tilde{U}_{k,j}(\xi, \tau)$ with the constants $c_{k,j}$ chosen as in the proof of Theorem 6.1.

THEOREM 6.2. *For the solution $u(x, t, \varepsilon)$ of the problem (0.1), (0.2) the following asymptotic expansions are valid: the series (4.3) in the domain $\{x, t: |x|^2 + |t|^3 < \varepsilon^{\nu_1}\}$, the series (5.2) in the domain $\{x, t: \varepsilon^{\nu_3} < t \leq T, |x - s(t)| < \varepsilon^{\nu_2}\}$, the series*

(1.4) *in the domain*

$$\{x, t: -1 \le t \le T, \, |t|^3 + x^2 > \varepsilon^{3-\nu_3}\} \cap \{x, t: 0 \le t \le T, \, |x - s(t)| > \varepsilon^{4-\nu_4}\}.$$

Here ν_i are arbitrary positive numbers, and the coefficients $u_k(x, t)$, $w_{k,j}(\xi, \tau)$, and $v_{k,j}(\zeta, t)$ are constructed in Theorems 3.2, 4.2, 5.3, and 6.1. The asymptotics (4.3), (5.2), and (1.4) mean that the absolute value of the difference between $u(x, t, \varepsilon)$ and a partial sum of the corresponding asymptotic expansion does not exceed $M_N \varepsilon^{\alpha N}$, where N is the number of the partial sum, and $\alpha > 0$ depends on the corresponding ν_i.

PROOF. Taking partial sums of the series listed in the theorem, one can easily form the composite expansion $T_N(x, t, \varepsilon)$. The matching condition for the series (1.4), (4.3), and (5.2) implies that the function T_N satisfies equation (0.1) everywhere in Ω to within $O(\varepsilon^{N_1})$. By construction, T_N satisfies condition (0.2). Therefore, $|T_N(x, t, \varepsilon) - u(x, t, \varepsilon)| < M\varepsilon^{N_1}$. The explicit form of the series under consideration and the asymptotics of their coefficients yield $N_1 > \alpha N$. ∎

REMARK 1. All the assertions and proofs are extended, with minor modifications, to the equation

$$\frac{\partial u}{\partial t} + \varphi_0(x, t, u)\frac{\partial u}{\partial x} + \varphi_1(x, t, u) = \varepsilon\frac{\partial^2 u}{\partial x^2}, \qquad \text{where} \quad \frac{\partial \varphi_0}{\partial u} > 0. \qquad (6.7)$$

REMARK 2. Since equation (0.1) is of divergence type, the expansion (5.2) for it is of a simpler form, namely: $v_{k,j}(\zeta, t) \equiv 0$ for $j > 0$, and also for $k \not\equiv 0 \bmod 4$.

This fact can be proved by integrating both sides of the equation over a domain whose boundary encircles the origin. One should substitute the series (1.4) and (5.2) into the resulting integral along the boundary thus making it clear that the series (5.2), as well as (1.4), includes only integer powers of ε. The asymptotic expansion (5.2) for equation (6.7) is apparently of a general, more complicated form.

Write out the first uniform approximation of the solution $u(x, t, \varepsilon)$. Let $\chi(z) = 1$ for $z \ge 0$, $\chi(z) = 0$ for $z < 0$, and

$$\hat{u}(x, t, \varepsilon) = u_0(x, t) + v_0(\zeta, t) + \varepsilon w_{1,0}(\xi, \tau) - \chi(\zeta)u_0(s(t) + 0, t)$$
$$- (1 - \chi(\zeta))u_0(s(t) - 0, t) - \varepsilon H(\xi, \tau) - 2\sqrt{t}\chi(\zeta) + \sqrt{t}.$$

Here ξ, τ, and ζ are the inner variables introduced above, the functions $u_0(x, t)$, $v_0(\zeta, t)$, $w_{1,0}(\xi, \tau)$, and $H(\xi, \tau)$ are those defined in §§1, 3, 4. The difference $\hat{u}(x, t, \varepsilon) - u(x, t, \varepsilon)$ is of the order of magnitude $O(\varepsilon^4)$ for $|x| + |t| > \text{const} > 0$, and $O(\varepsilon^2|\ln\varepsilon|)$ for $|\xi| + |\tau| < \text{const}$. The approximation error in the intermediate domains has the order of magnitude between $\varepsilon^2 \ln \varepsilon$ and ε^4.

§7. Asymptotics of the flame wave

This section differs from preceding ones mainly by the fact that it only gives (a brief) construction of the asymptotics and does not provide its justification. The place of this section in Chapter VI is partly warranted by the fact that the physical problem under consideration in its original formulation is described by a system of quasilinear parabolic equations. However, under far-reaching simplifying conditions, the problem reduces to the investigation of an ordinary differential equation, and it might have been more appropriate to consider it in Chapter II. We now proceed with the description of the problem.

The simplest model of gas flame in a one-dimensional medium reduces to the system of equations

$$c\frac{\partial T}{\partial t} = \lambda\frac{\partial^2 T}{\partial x^2} + Qw, \qquad \frac{\partial a}{\partial t} = D\rho\frac{\partial^2 a}{\partial x^2} - w, \qquad (7.1)$$

where $T(x, t)$ stands for the temperature, and $a(x, t)$ for the concentration of the burning gas mixture. Here the specific heat capacity c, the thermal conductivity λ, and the heat of reaction Q are constant. The reaction rate w is related to the concentration by the equality $w = a^\beta k_0 \exp(-E/RT)$, where $k_0 = \text{const}$, R is the gas constant. The reaction order β and the activation energy E are also assumed to be constant. Another assumption is that the condensed matter diffusion coefficient D is zero, and one looks for a traveling wave type solution with a constant velocity $\varkappa$:

$$T(x, t) = T(x - \varkappa t), \qquad a(x, t) = a(x - \varkappa t). \qquad (7.2)$$

For $x \to \infty$ (where the reaction has not yet started) one has $T \to T_0$, $a \to 1$, while for $x \to -\infty$ the limit values of these functions equal T_1 and zero. We now have to substitute the functions (7.2) into equations (7.1) (for $D = 0$), and integrate the first equation once. The resulting equations are

$$\lambda T' = c\varkappa(T_1 - T) - Q\varkappa a, \qquad \varkappa a' = w, \qquad (7.3)$$

yielding, in particular, $T_1 - T_0 = Qc^{-1}$. Denoting $z = (T_1 - T)(T_1 - T_0)^{-1}$, and taking z for the independent variable, one arrives at the single differential equation

$$\frac{\varkappa^2 c}{\lambda}\frac{da}{dz} = \frac{a^\beta k_0}{a - z}\exp\left(-\frac{E}{R(T_1 - zQc^{-1})}\right) \qquad (7.4)$$

and the boundary conditions: $a(0) = 0$, $a(1) = 1$.

The small parameter in the problem is the value $\varepsilon = RcT_1^2(EQ)^{-1}$ describing the relative width of the chemical reaction zone (more details on the physical interpretation of the problem and references thereto can be found in [58]).

Now make the change of variable $z = \varepsilon\xi$, and introduce the following notation for the two dimensionless parameters

$$b = Q(cT_1)^{-1} < 1, \qquad \nu = c\varkappa^2(\lambda k_0\varepsilon)^{-1}\exp(E/RT_1)$$

together with a new notation for the unknown function showing its dependence on ε: $a(z) \equiv u(\xi, \varepsilon)$. Equation (7.4) and the boundary conditions take the form

$$\nu\frac{du}{d\xi} = \frac{u^\beta}{u - \varepsilon\xi}\exp\left(-\frac{\xi}{1 - b\varepsilon\xi}\right), \qquad (7.5)$$

$$u(0, \varepsilon) = 0, \qquad u(\varepsilon^{-1}, \varepsilon) = 1.$$

Here, along with the unknown function $u(\xi, \varepsilon)$, one also has to find the constant $\nu(\varepsilon)$ related to the unknown velocity $\varkappa$ of the wave of form (7.2).

The last subtlety consists in the fact that the problem (7.5) has no solution, as well, of course, as the problem (7.3) with the above conditions at infinity. In fact, for temperatures close to T_0 ($z \approx 1$) the solution is not of the form (7.2). Nevertheless, equations (7.3) and (7.5) do correctly describe the physical process in the main flame area, and one of the possible models consists in varying the rate of reaction w for $T \sim T_0$—the right-hand side of equation (7.5) is assumed to vanish for $z \geq z_0$ (i.e., for $\xi \geq z_0 \varepsilon^{-1}$), where $0 < z_0 < 1$.

Thus, the final mathematical statement of the problem is as follows:

$$\nu \frac{du}{d\xi} = \frac{u^\beta}{u - \varepsilon\xi} \left[\exp\left(-\frac{\xi}{1 - b\varepsilon\xi}\right) \right] \chi\left(\xi - \frac{z_0}{\varepsilon}\right), \qquad \varepsilon > 0, \qquad (7.6)$$

$$u(0, \varepsilon) = 0, \qquad u(\infty, \varepsilon) = 1, \qquad (7.7)$$

where $\chi(t) \in C^\infty(\mathbf{R}^1)$, $0 < z_0 < 1$, $\chi(t) \equiv 1$ for $t \leq -1$, $\chi(t) \equiv 0$ for $t \geq 0$. The function $u(\xi, \varepsilon)$ and parameter $\nu(\varepsilon)$ are unknown, and it is required to find their asymptotics as $\varepsilon \to 0$. It turns out that this asymptotics is independent of both z_0 and the form of the function $\chi(t)$.

We consider only the case $1 < \beta < 2$, and write out the asymptotic series for $\nu(\varepsilon)$ and $u(\xi, \varepsilon)$:

$$\nu(\varepsilon) = \sum_{k=0}^\infty \nu_k \varepsilon^k, \qquad \varepsilon \to 0, \qquad (7.8)$$

$$u(\xi, \varepsilon) = \sum_{k=0}^\infty \varepsilon^k u_k(\xi), \qquad \varepsilon \to 0. \qquad (7.9)$$

On inserting these series into (7.6) one obtains the recurrence system of equations

$$\nu_0 u_0' = u_0^{\beta-1} e^{-\xi}, \qquad (7.10)$$

$$lu_1 \equiv \nu_0 u_1' - (\beta - 1)u_0^{\beta-2} e^{-\xi} u_1 = -\nu_1 u_0' + u_0^{\beta-2} e^{-\xi}(\xi - bu_0\xi^2), \qquad (7.11)$$

$$lu_k = -\nu_k u_0' - \sum_{i=1}^{k-1} \nu_i u_{k-i}' + u_0^{\beta-1} f_k(\xi, u_0, \ldots, u_{k-1}), \qquad k \geq 2. \qquad (7.12)$$

Equation (7.10) is solved explicitly. The explicit form of $u_0(\xi)$ implies that conditions (7.7) are satisfied if and only if $u_0(\xi) = (1 - e^{-\xi})^{1/\nu_0}$, $\nu_0 = 2 - \beta$. This value of ν_0 and the form of the function $u_0(\xi)$ will remain fixed until the end of the section. Then the rest of the functions $u_k(\xi)$ for $k \geq 1$ must satisfy the equalities

$$u_k(0) = 0, \qquad (7.13)$$

$$u_k(\infty) = 0. \qquad (7.14)$$

The only solution of equation (7.11) satisfying condition (7.14) is given by the function

$$u_1(\xi) = \nu_1 \nu_0^{-2} e^{-\xi} u_0^{\beta-1}(\xi) - \nu_0^{-1} u_0^{\beta-1}(\xi) \int_\xi^\infty [\theta(u_0(\theta))^{-1} - b\theta^2] e^{-\theta} \, d\theta. \quad (7.15)$$

A similar formula is obtained for any $k \geq 2$:

$$u_k(\xi) = \nu_k \nu_0^{-2} e^{-\xi} u_0^{\beta-1}(\xi) - \nu_0^{-1} u_0^{\beta-1}(\xi)$$
$$\times \int_\xi^\infty \left[f_k(\theta) - u_0^{1-\beta}(\theta) \sum_{i=1}^{k-1} \nu_i u'_{k-i}(\theta) \right] d\theta. \quad (7.16)$$

The solution $u_1(\xi)$ satisfies condition (7.13) for any ν_1, and, therefore, it is not possible to determine ν_1 from (7.15). The second approximation brings even more trouble. If $\beta > 3/2$ then for any ν_1 and ν_2 the function $u_2(\xi)$ has singularity at zero of the type $\xi^{(3-2\beta)/(2-\beta)}$. The ensuing functions $u_k(\xi)$ also have increasing singularities at zero. For $4/3 < \beta < 3/2$ the functions $u_k(\xi)$ exhibit singularities starting with $u_3(\xi)$, for $5/4 < \beta < 4/3$ singularities start with $u_4(\xi)$, etc.

Thus, the problem (7.6), (7.7) is a bisingular one. Let us look for another scale, and another f.a.s. near the point $\xi = 0$. It is natural to choose the scales in such a way that both terms in the denominator of (7.6), i.e., u and $\varepsilon\xi$ are of the same order. Therefore,

$$\xi = \varepsilon^\alpha \eta, \qquad u(\xi, \varepsilon) \equiv \varepsilon^{1+\alpha} v(\eta, \varepsilon). \quad (7.17)$$

Equating the orders of both sides of equation (7.6), one obtains $1 = (\beta - 1)(1 + \alpha)$, i.e.,

$$\alpha = (2 - \beta)(\beta - 1)^{-1}. \quad (7.18)$$

After the change of variables (7.17), (7.18), equation (7.6) assumes the form (taking into account that in the domain under consideration $\chi \equiv 1$)

$$\nu \frac{dv}{d\eta} = \frac{v^\beta}{v - \eta} \exp\left(-\varepsilon^\alpha \frac{\eta}{1 - b\varepsilon^{1+\alpha}\eta} \right). \quad (7.19)$$

The first condition in (7.7) is, naturally, preserved:

$$v(0, \varepsilon) = 0, \quad (7.20)$$

and as $\eta \to \infty$ the series for v must be matched to the series (7.9). The form of (7.19) makes it clear that the f.a.s. v must include not just integer powers of ε, but also the terms with coefficients of the form $\varepsilon^{k+\alpha i}$. Similar terms must also be present in the asymptotics of $\nu(\varepsilon)$, and, consequently, in the f.a.s. u.

Thus, the series (7.8), (7.9) have to be replaced with an asymptotics of a more complicated form involving terms like $\varepsilon^{k+\alpha i} u_{k,i}(\xi)$, $\varepsilon^{k+\alpha i} v_{k,i}(\eta)$.

Moreover, the asymptotics of the functions $u_{k,i}(\xi)$ for $\xi \to 0$ may, in general, include logarithmic terms. Owing to this fact, the factors $\ln^j \varepsilon$ may appear, as we have repeatedly seen in the preceding chapters, in the f.a.s. u and v as well. We, therefore, write down the final form of the series:

$$\nu = \sum_{\substack{k \geq 0, i \geq 0 \\ 0 \leq j \leq i+k-1}} \nu_{k,i,j}\, \varepsilon^{k+\alpha i} \ln^j \varepsilon, \tag{7.21}$$

$$U = \sum_{\substack{k \geq 0, i \geq 0 \\ 0 \leq j \leq i+k-1}} \varepsilon^{k+\alpha i} \ln^j \varepsilon\, u_{k,i,j}(\xi), \tag{7.22}$$

$$V = \sum_{\substack{k \geq 0, i \geq 0, \\ 0 \leq j \leq i+k-1}} \varepsilon^{k+\alpha i} \ln^j \varepsilon\, v_{k,i,j}(\eta). \tag{7.23}$$

For the convenience of notation we denote $\nu_{k,i,0}$ by $\nu_{k,i}$ using the same notation for the coefficients of the remaining series. Furthermore, $\nu_{k,0} = \nu_k$, $u_{k,0}(\xi) \equiv u_k(\xi)$, where $u_k(\xi)$ are solutions of equations (7.11), (7.12) defined by formulas (7.15), (7.16). Equation (7.6) easily yields the equations for the other $u_{k,i,j}(\xi)$. We restrict our attention to writing out the equations for $u_{k,i}(\xi)$:

$$lu_{k,i} = u_0^{\beta-1} f_{k,i}(\xi, u_0, u_1, \ldots) - \sum_{\substack{0 \leq l \leq k, 0 \leq p \leq i, \\ l+p>0}} \nu_{l,p} u'_{k-l,i-p}(\xi), \qquad i > 0. \tag{7.24}$$

(The calculation shows that for $i+k \leq 2$ the coefficients of the logarithmic terms in the asymptotics of $u_{k,i}(\xi)$ and $v_{k,i}(\eta)$ vanish. Hence for $i+k \leq 2$ one has $j = 0$. So far no one was able to either prove or disprove the presence of logarithmic terms in the subsequent terms of the series (7.21)–(7.23).)

The equations and boundary condition for $v_{k,i,j}(\eta)$ follow from (7.23) and (7.19), (7.20). We again limit ourselves to the functions $v_{k,j}(\eta)$:

$$\nu_0 v_0' = v_0^\beta (v_0 - \eta)^{-1}, \qquad v_0 \equiv v_{0,0}, \tag{7.25}$$

$$\begin{aligned} l_1 v_{k,i} &\equiv \nu_0 v'_{k,i} - \left[\beta v_0^{\beta-1}(v_0-\eta)^{-1} - v_0^\beta(v_0-\eta)^{-2}\right] v_{k,i} \\ &= \varphi_{k,i}(\eta, v_0, \ldots) - \sum_{\substack{0 \leq l \leq k, 0 \leq p \leq i, \\ l+p>0}} \nu_{l,p} v'_{k-l,i-p}(\eta), \qquad i+k > 0, \end{aligned} \tag{7.26}$$

$$v_{k,i,j}(0) = 0. \tag{7.27}$$

Now, as usual, we have to find the asymptotics of the functions $u_{k,i,j}(\xi)$ as $\xi \to 0$, the asymptotics of the functions $v_{k,i,j}(\eta)$ as $\eta \to \infty$, and examine the degree of arbitrariness these functions may be chosen with. The difference from the preceding problems lies in the fact that in matching the series (7.22), (7.23) one has also to determine the coefficients of the numerical series (7.21). Moreover, the coefficients $\nu_{k,i,j}$ are the only arbitrary numbers that have to be determined. Indeed, the functions $u_{k,i,j}(\xi)$ are uniquely determined from their respective equations and from the condition that they should decay at infinity by formulas essentially coinciding with (7.15), provided all $\nu_{l,p,s}$ for $l \leq k$, $p \leq i$, $s \leq j$ are known. As we shall see below, the functions $v_{k,i,j}(\eta)$ are also determined from their equations and condition (7.27) uniquely. Denote

$$\gamma = \alpha^{-1} = (\beta - 1)(2 - \beta)^{-1}$$

and write out some useful relations between the exponents:

$$\nu_0^{-1} = (1 - \beta)^{-1} = 1 + \gamma, \qquad \beta(2 - \beta)^{-1} = 1 + 2\gamma,$$
$$(\beta - 1)(1 + \alpha) = 1, \qquad (\beta - 1)(1 + \gamma) = \gamma.$$

The explicit form of $u_0(\xi)$ implies that

$$u_0(\xi) = \xi^{1+\gamma} \left(1 - \frac{\xi}{2} + \frac{\xi^2}{6} - \cdots \right)^{\gamma+1} = \xi^{1+\gamma} \left(1 + \sum_{s=1}^{\infty} d_s \xi^s \right), \qquad \xi \to 0.$$

We now begin with the examination of the asymptotics of $v_0(\eta)$. Equation (7.25) can be solved explicitly in quadratures. Taking into account condition (7.27), one has

$$\eta = \alpha \exp \left(\alpha v_0^{1-\beta} \right) \int_{v_0^{1-\beta}}^{\infty} \zeta^{(1-\beta)^{-1}} \exp(-\alpha\zeta) \, d\zeta.$$

This easily yields the asymptotics as $v_0 \to \infty$. Denoting $x = \alpha v_0^{1-\beta}$, one has for noninteger α:

$$\eta = \alpha^\alpha x^{-\alpha} \{ 1 - x(\alpha - 1)^{-1} - x^\alpha \Gamma(1 - \alpha)$$
$$- x^{1+\alpha} \Gamma(1 - \alpha) + x^2 (\alpha - 1)^{-1} (\alpha - 2)^{-1} + \ldots \}, \qquad x \to 0, \tag{7.28}$$

where $\Gamma(z)$ is the Euler function. Reversing this equality results in the asymptotics

$$v_0(\eta) = \eta^{1+\gamma} \left(1 + \sum_{i+j>0} c_{i,j} \eta^{-i-\gamma j} \right), \qquad \eta \to \infty, \tag{7.29}$$

where $c_{1,0} = \alpha^{\alpha-1}(1 + \alpha)\Gamma(1 - \alpha)$, $c_{0,1} = (\alpha + 1)(\alpha - 1)^{-1}$, etc.

There is no difficulty in writing out explicit formulas for the remaining coefficients $c_{i,j}$.

Thus, we see that in their principal terms the series (7.22) and (7.33) are matched automatically:

$$A_{1+\alpha,\eta} A_{0,\xi} U = A_{1+\alpha,\eta} u_0(\varepsilon^\alpha \xi) = \varepsilon^{1+\alpha} \eta^{1+\gamma},$$
$$A_{0,\xi} A_{1+\alpha,\eta} (\varepsilon^{1+\alpha} V) = A_{0,\xi} (\varepsilon^{1+\alpha} v_0(\eta)) = \xi^{1+\alpha} = \varepsilon^{1+\alpha} \eta^{1+\gamma}.$$

For the subsequent matching of the series (7.22), (7.23) it is convenient to use Table 10. Its structure repeats that of the preceding tables and needs no additional explanation. (Everywhere in this section we assume that β, α, and γ are irrational numbers so that the coefficients of the asymptotics exhibit no singularities (which was the case in (7.28) for natural α). A remark about rational β is to be found at the end of this section.)

So far only the first column and the first row of the table ($v_0(\eta)$ and $u_0(\xi)$) are determined whereby the principal term of the asymptotics of the function $u_1(\xi)$ must equal $c_{0,1}\xi$, and the principal term of the asymptotics of the function $u_{1,0}(\xi)$ equals $c_{1,0}\xi^\gamma$. It follows from equation (7.24), where $f_{0,1} \equiv 0$ and the right-hand side equals $-\nu_{0,1}u_0'$, that

$$u_{0,1}(\xi) = \nu_{0,1}\nu_0^{-2}u_0^{\beta-1}(\xi)e^{-\xi}.$$

Hence $\nu_{0,1} = \nu_0^2 c_{1,0}$.

Formula (7.16) for $k = 1$, $f_1(\xi) = e^{-\xi}(\xi u_0^{-1}(\xi) - b\xi^2)$ makes it possible to find the asymptotics of $u_1(\xi)$ as $\xi \to 0$. The second term in this formula equals

$$(\alpha+1)(\alpha-1)^{-1}\xi + Cu_0^{\beta-1}(\xi) + O\left(\xi^2 + \xi^{1+\gamma}\right), \qquad \xi \to 0.$$

Since $u_0^{\beta-1}(\xi) \sim \xi^\gamma$ as $\xi \to 0$, the constant ν_1 must be chosen in such a way that the expansion of the function $u_1(\xi)$ contains no term ξ^γ. Hereby ν_1, $\nu_{0,1}$ are found, and $u_1(\xi)$, $u_{0,1}(\xi)$ completely determined. Their asymptotics are written out in the second and third rows of the table.

We now go back to the functions $v_{k,i}(\eta)$. For $i + k > 0$ they satisfy the linear equations (7.26). The solutions of these equations satisfying conditions (7.27) are of the form

$$v_{k,i}(\eta) = \nu_0^{-1} \exp Z(\eta) \int_0^\eta \tilde{\varphi}_{k,i}(\tau) \exp(-Z(\tau))\,d\tau, \qquad (7.30)$$

where $\tilde{\varphi}_{k,i}$ is the right-hand side of equation (7.26), and

$$Z(\eta) = \frac{1}{\nu_0} \int_{\bar{\eta}}^\eta [v_0(\theta)]^{\beta-1}(v_0(\theta) - \theta)^{-2}[\beta(v_0(\theta) - \theta) - v_0(\theta)]\,d\theta. \qquad (7.31)$$

The explicit form of $v_0(\theta)$ easily implies that $Z(\eta) \to \infty$ as $\eta \to 0$. An examination by induction of the asymptotics of $\tilde{\varphi}_{k,i}(\eta)$ as $\eta \to 0$ easily shows that the integral in (7.30) converges, and, therefore, the solution of the problem (7.26), (7.27) is unique and given by this formula. The same holds for the functions $v_{k,i,j}(\eta)$ for $j > 0$. We will no longer dwell on this question, and proceed to the examination of the asymptotics of $v_{k,i,j}(\eta)$ as $\eta \to \infty$.

It follows from (7.31) that, for an appropriate choice of $\overline{\eta}$,

$$\exp Z(\eta) = \eta^\gamma \left(1 + \sum_{i+k>0} b_{k,i}\eta^{-k-\gamma i}\right), \qquad \eta \to \infty,$$

where $b_{0,1} = \gamma^{-1}(1-\gamma)^{-1}$, $b_{1,0} = (\beta-1)c_{1,0}$, and $c_{1,0}$ is the same as in formula (7.29).

Since ν_1 and $\nu_{1,0}$ are already determined, the functions $v_1(\eta)$ and $v_{1,0}(\eta)$ are determined uniquely. Their asymptotics as $\eta \to \infty$ are given in the second and third columns of Table 10 (the subscripts of the constants c are omitted).

TABLE 10

$\varepsilon^{1+\alpha}V$ / U	$\varepsilon^{1+\alpha}v_0$	$\varepsilon^{2+\alpha}v_1$	$\varepsilon^{1+2\alpha}v_{0,1}$	$\varepsilon^{3+\alpha}v_2$	$\varepsilon^{2+2\alpha}v_{1,1}$	$\varepsilon^{1+3\alpha}v_{0,2}$	...
u_0	$\varepsilon^{1+\alpha}\eta^{1+\gamma}$ / $\xi^{1+\gamma}$	—	$\varepsilon^{1+2\alpha}c\eta^{2+\gamma}$ / $c\xi^{2+\gamma}$	—	—	$\varepsilon^{1+3\alpha}c\eta^{3+\gamma}$ / $c\xi^{3+\gamma}$	...
εu_1	$\varepsilon^{1+\alpha}c_{0,1}\eta$ / $\varepsilon c_{0,1}\xi$	$\varepsilon^{2+\alpha}c\eta^{1+\gamma}$ / $\varepsilon c\xi^{1+\gamma}$	$\varepsilon^{1+2\alpha}c\eta^2$ / $\varepsilon c\xi^2$	—	$\varepsilon^{2+2\alpha}c\eta^{2+\gamma}$ / $\varepsilon c\xi^{2+\gamma}$	$\varepsilon^{1+3\alpha}c\eta^3$ / $\varepsilon c\xi^3$	...
$\varepsilon^\alpha u_{0,1}$	$\varepsilon^{1+\alpha}c_{1,0}\eta^\gamma$ / $\varepsilon^\alpha c_{1,0}\xi^\gamma$	—	$\varepsilon^{1+2\alpha}c\eta^{1+\gamma}$ / $\varepsilon^\alpha c\xi^{1+\gamma}$	—	—	$\varepsilon^{1+3\alpha}c\eta^{2+\gamma}$ / $\varepsilon^\alpha c\xi^{2+\gamma}$	...
$\varepsilon^2 u_2$	$\varepsilon^{1+\alpha}c_{0,2}\eta^{1-\gamma}$ / $\varepsilon^2 c_{0,2}\xi^{1-\gamma}$	$\varepsilon^{2+\alpha}c\eta$ / $\varepsilon^2 c\xi$	$\varepsilon^{1+2\alpha}c\eta^{2-\gamma}$ / $\varepsilon^2 c\xi^{2-\gamma}$	$\varepsilon^{3+\alpha}c\eta^{1+\gamma}$ / $\varepsilon^2 c\xi^{1+\gamma}$	$\varepsilon^{2+2\alpha}c\eta^2$ / $\varepsilon^2 c\xi^2$	$\varepsilon^{1+3\alpha}c\eta^{3-\gamma}$ / $\varepsilon^2 c\xi^{3-\gamma}$	...
$\varepsilon^{1+\alpha}u_{1,1}$	$\varepsilon^{1+\alpha}c_{1,1}$ / $\varepsilon^{1+\alpha}c_{1,1}$	$\varepsilon^{2+\alpha}c\eta^\gamma$ / $\varepsilon^{1+\alpha}c\xi^\gamma$	$\varepsilon^{1+2\alpha}c\eta$ / $\varepsilon^{1+\alpha}c\xi$	$\varepsilon^{3+\alpha}c\eta^{2\gamma}$ / $\varepsilon^{1+\alpha}c\xi^{2\gamma}$	$\varepsilon^{2+2\alpha}c\eta^{1+\gamma}$ / $\varepsilon^{1+\alpha}c\xi^{1+\gamma}$	$\varepsilon^{1+3\alpha}c\eta^2$ / $\varepsilon^{1+\alpha}c\xi^2$	...
$\varepsilon^{2\alpha}u_{0,2}$	$\varepsilon^{1+\alpha}c_{2,0}\eta^{\gamma-1}$ / $\varepsilon^{2\alpha}c_{2,0}\xi^{\gamma-1}$	—	$\varepsilon^{1+2\alpha}c\eta^\gamma$ / $\varepsilon^{2\alpha}c\xi^\gamma$	—	—	$\varepsilon^{1+3\alpha}c\eta^{\gamma+1}$ / $\varepsilon^{2\alpha}c\xi^{\gamma+1}$	...
...	...	...	...	...	...	...	...

Now we can proceed to the determination of $u_2(\xi)$, $u_{1,1}(\xi)$, $u_{0,2}(\xi)$. The corresponding constants ν_2, $\nu_{1,1}$, $\nu_{0,2}$ are found from the terms of the form ξ^γ given in the table (such is the principal term of the asymptotics of the first summand in formula (7.16) with the yet unknown coefficient ν_2, and similar terms for $u_{k,i}(\xi)$). The function $u_2(\xi)$ and all the subsequent $u_k(\xi)$ can be found directly from the fact that the coefficient of ξ^γ must vanish regardless of the construction procedure for $v_{k,i,j}(\eta)$. Hence the asymptotics of the function $u_k(\xi)$ as $\xi \to 0$ includes no logarithmic terms. The asymptotics of the other functions $u_{k,i}$ for $i > 0$ may, in general, involve terms of the form $\xi^\mu \ln \xi$. (As we have already mentioned, there are no such terms in Table 10.)

On the next step one defines the functions $v_2(\eta)$, $v_{1,1}(\eta)$, and $v_{0,2}(\eta)$. The process then continues by induction. If the asymptotics of the function $u_{k,i}(\xi)$ for $\xi \to 0$ turns out to include a term of the form $\xi^\gamma \ln \xi$ (formula (7.16) shows that this is the principal logarithmic term), then the change of variable (7.17) turns it into the term $\varepsilon\eta^\gamma(\ln \eta + \alpha \ln \varepsilon)$. The second summand is corrected by introducing the term $\varepsilon^{k+i\alpha} \ln \varepsilon \, u_{k,i,l}(\xi)$. Further, the exponents in the powers of $\ln \varepsilon$ may grow as in the problem of Chapter II, §2, (see Table 2), and the series U and V take the form (7.22), (7.23). We give no further details of the construction and no proof of the fact that the terms of the asymptotics appearing in all squares of Table 10, with the exception of those containing $c\xi^\gamma$, automatically coincide. The proof can be carried out along the same lines as in §§1 and 2 of Chapter II. Thus, the asymptotic expansion of the solution of the problem (7.6), (7.7) is constructed to any power of ε, and it is uniform everywhere for $\xi \geq 0$. The main point is the construction of the asymptotics of $\nu(\varepsilon)$, and, consequently, the wave velocity $\varkappa(\varepsilon)$.

In conclusion a few comments about rational values of the parameter β. This asymptotics for $\beta = \beta_0$ can be easily obtained by passing to the limit as $\beta \to \beta_0$, taking care that the partial sum contains a necessary number of terms. For example, for $\beta_0 = 3/2$ ($\alpha_0 = 1$), in order to obtain the asymptotics to within $o(\varepsilon^2)$, one has to consider the terms of order 1, ε, ε^α. The resulting equality is

$$\nu(\varepsilon, \beta) = (2 - \beta)\{1 + \varepsilon[\alpha(\psi(1) - \psi(1 - \alpha^{-1})) - 2b] + \varepsilon^\alpha \alpha^\alpha \Gamma(1 - \alpha)\}$$
$$+ O\left(\varepsilon^2 + \varepsilon^{2\alpha}\right), \qquad \varepsilon \to 0,$$

where $\psi(z) = \Gamma'(z)/\Gamma(z)$. Passing to the limit as $\beta \to 3/2$ ($\alpha \to 1$), one has

$$\nu\left(\varepsilon, \frac{3}{2}\right) = \frac{1}{2}\left[1 + \varepsilon \ln \frac{1}{\varepsilon} + \varepsilon(\Gamma'(1) - 2b)\right] + O\left(\varepsilon^2 \ln \frac{1}{\varepsilon}\right).$$

Notes and Comments on Bibliography

Chapter I

What we now call singular perturbation problems for differential equations have been considered for a long time going back at least to the last century. References to such works can be found, for example, in [85]. Nevertheless, a sufficiently general theory appeared relatively recently. The Cauchy problem for systems of differential equations was considered in [116], [117] (more details on the development of this line can be found in [121]–[123], [127]).

Rigorous mathematical analysis of singular perturbation boundary value problems for partial differential equations appeared only in the 1950s (see [69], [94], [95], [55]). A detailed investigation of boundary value problems of the type considered in Examples 1–3 has been made in [124]–[126]. Problems of this type are often called boundary layer type problems. The same name is sometimes also applied to the problems considered in this book and called bisingular. In [124] the situation described in Chapter I is called that of regular degeneracy. There is no accepted terminology. It is apparently convenient to call the problems described in Chapter I problems with exponential boundary layer (or a similar term taking into account the fact that coefficients of the outer expansion are smooth functions, while the boundary layer functions decay exponentially at infinity). The method suggested in [124]–[126] is often called the Vishik-Lyusternik method. Its subsequent development is described in [119]. Problems with corner boundary layers of exponential type similar to Example 4 are considered in [11]–[14]. Taken together, problems with exponential boundary layer and bisingular problems can naturally be called problems of boundary layer type thus singling them out among a large number of other questions in the theory of differential equations with a small parameter which are not treated in this book.

Among other branches of the asymptotic approach to differential equations with small parameter we note the investigation of equations with rapidly oscillating coefficients and boundary value problems in perforated domains (see, for example, [7], [129]). The Krylov-Bogolyubov-Mitropolskii method of averaging [9] is not considered at all.

There are closer problems, also not considered here, for differential equations whose coefficients are smooth and of slow variation, while the solutions are nevertheless rapidly oscillating. The method used for the investigation of these problems is usually called the WKB method, and, roughly speaking, reduces to the study of the "fast" phase and "slow" amplitude of oscillations. The method is described in the books [33], [28], [71]–[73]. The resulting equation for the amplitude also often turns out to be bisingular, and can be analyzed using the method of matched asymptotic expansions given above.

Chapter II

The method of matched asymptotic expansions arose, under different names given in the Introduction, in mechanics, and made it possible to construct the first terms of the asymptotics, to solve the arising paradoxes, etc. We only note the articles [56], [57], [63], [103]. The history of the question can be found in the monographs [121], [15], [85], [62]. For ordinary differential equations, the method has been used in different situations in [19], [101], [83]. For partial differential equations a rigorous justification of the asymptotics has been obtained relatively recently (see [1], [40], [42], [65], [77], [113], [27] et al.) For the problems of short-wave diffraction mentioned above a rigorous mathematical investigation was carried out in [2]–[5] et al.

§1.   Example 5 is of an educational nature and is provided for the explanation of the technique developed in the last 10 to 12 years (see [35], [36], [40], [42], [47], [49], [65], [66], [90]–[92], [109]–[111]). This method is quite close to the one described in [121], [29], but in our view is much more convenient and consistent. Another, but essentially close approach is developed in [77]–[81].

For elliptic partial differential equations the problem mentioned in the Remark to §1 is considered in [45].

§2.   Problems similar to Example 6 are considered in [22].

§3.   This section describes the contents of the research thesis of F. M. Sattarova made at the Bashkir State University in 1979.

Chapter III

§1.   The problem is of an auxiliary, illustrative nature. In [36] the construction of the asymptotics of solutions having singularities at a point is set forth in the form convenient for the purposes of the problem under consideration. Much earlier, the asymptotics of Green's function has been constructed in a somewhat different form in [32]. In the case where the singularity of the solution is located near the boundary, the asymptotics of Green's function for an elliptic second order equation was studied in [47].

§2.   The section presents, in a somewhat simplified form, the paper [35]. This paper considers the problem in the case of a general elliptic second order equation. The two-dimensional problem of the flow past a thin solid body was treated in detail in [121]. Exterior boundary value problems for thin solid bodies have been considered in [78], [80], [81], [106], [25]–[27].

§**3.** The exposition follows that of [36] where the same problem is considered for the equation with variable coefficients. The method can also be applied to elliptic problems of higher order, as well as to other problems with singular perturbations in boundary conditions. Problems of these kind have been studied in [31], [77], [79]. In the so called critical cases, as in the problem of §3, rational functions of $\ln \varepsilon$ appear as the gauge functions. Note that such gauge functions are typical for a rather general situation (see [120, Chapter 9, §3]).

§**4.** The section presents, in a simplified form, the results of [38].

Chapter IV

§**1.** The results of this section have been obtained in [42]. However, the approach used in [42] is far from perfect. The treatment in this book is based on the methods developed in [66] which is more natural. Such an approach makes it possible to obtain the asymptotics for a wide range of problems. The analysis of boundary value problems, for which the characteristic of the limit first order equation coincides with a part of the boundary, was conducted in [66] for the three-dimensional case, and in [48], [49] for a system of elliptic equations. Note that the asymptotics of the boundary value problems (1.1), (1.2) for constant coefficients was earlier investigated in [16] by directly analyzing the explicit formulas for a solution.

§**2.** The exposition mainly follows the article [67]. In the examination of the inner expansion the methods of [43] are used. For domains with the non-smooth boundary, the asymptotics of the solution of an elliptic equation with a small parameter was studied in [115] (where there is no "inner" boundary layer), in [68], [77], [86], [87].

§**3.** The problem considered in this section was discussed back in [124], some estimates were given in [20], but a complete solution was obtained only in [65]. A similar analysis for an elliptic equation of higher order was conducted in [110].

§**4.** The asymptotics of the solution of the problem considered in this section was first obtained in [40], where both the exposition and techniques are far from perfect. In this section we use the technique developed in the subsequent papers ([66], [109], etc.). For an elliptic equation of higher order the asymptotics was studied, in a similar situation, in [111]. The explicit formulas mentioned in the remark to §4 were obtained in [37].

§**5.** The estimates of subsection 2 of this section were obtained in [110].
We also note that the method of matched asymptotic expansions is used not only in the case of elliptic equations but for a wide range of other boundary value problems. The asymptotics of solutions of pseudodifferential equations was studied in [93], in particular, for singular integral equations it was

considered in [91], [92], for hyperbolic equations in [89]. Two other examples are considered in the last chapters of the present book.

Chapter V

In the case where the initial data decay at infinity faster than any polynomial, a problem which is slightly more general than (0.1), (0.2) was studied in [50]. In this paper the method of averaging [9] was used, which yields only the first terms of the asymptotics. Earlier, the method of averaging was applied to the analysis of the asymptotics of solutions for similar problems with periodic initial data (see [52], [54], [105], [112]).

For solutions exhibiting asymptotic stability at infinity, the method of matched asymptotic expansion in the form demonstrated in the present chapter proved more convenient. A detailed presentation of these results for general hyperbolic systems is given in [51], for other problems in [53]. These papers include the assumption that initial functions tend to their limits faster than any polynomial. The analysis of the asymptotics in Chapter V in the case where initial functions tend to their limits at infinity as powers is due to L. A. Kalyakin and is published here for the first time.

Chapter VI

The problem formulated in this chapter attracted mathematicians for a long time (see [34]). The most comprehensive study of the Cauchy problem for the limit equation in the most general case was made in the works by O. A. Oleĭnik ([96], [98]). Here the limit transition for the solutions of perturbed equations for $\varepsilon \to 0$ was considered and justified. However, the construction of the complete asymptotic expansion of the solution requires stronger restrictions on the initial function. This problem has been considered from different standpoints in the papers [6], [46], [104], [114], [8] etc.). The asymptotics in the vicinity of the discontinuity line of the limit equation, as well as the so-called soliton-like solutions, was recently studied for a wide range of problems ([74]–[76], [18], et al.). These problems are more difficult than the one considered in Chapter VI, still more so because one usually considers the effects of small dispersion instead of small dissipation. However, the analysis of the behavior of the solution in a neighborhood of the "gradient catastrophe" point is much more complicated, and has not been done yet.

§1.   The results of this section are well known and are of an auxiliary nature.

§2.   The problem presented in this section was published in [46], [88].

§3–6.   These sections present the results of [39].

§7.   The results of these section were obtained jointly with S. I. Khudyayev [41].

Bibliography

1. K. I. Babenko, *Theory of perturbations of stationary flows of viscous incompressible fluids at small Reynolds numbers*, Preprint IPM, Moscow, 1975; English transl., Selecta Math. Soviet. **3** (1983/84), 2, 111–149.

2. V. M. Babich, *The strict justification of the short-wave approximation in the three-dimensional case*, Zap. Nauchn. Sem. Leningrad. Otdel. Mat. Inst. Steklov. (LOMI) **34** (1973), 23–51. (Russian)

3. V. M. Babich and V. S. Buldyrev, *Asymptotics methods in short wave diffraction problems*, "Nauka", Moscow, 1972. (Russian)

4. V. M. Babich and S. A. Egorov, *Solution of the caustic problem using the method of local expansions*, Topics in Dynamic Theory of Propagation of Seismic Waves, vol. vol. 12, "Nauka", Leningrad, 1973, pp. 4–14. (Russian)

5. V. M. Babich and N. Ya. Kirpichnikova, *The boundary layer method in diffraction problems*, Izdat. Leningrad. Univ., Leningrad, 1974; English transl., Springer-Verlag, Berlin and New York, 1979.

6. N. S. Bakhvalov, *Asymptotic behavior for small ε of the solution of the equation $u_t + (\varphi(u))_x = \varepsilon u_{xx}$ corresponding to a rarefaction wave*, Žh. Vyčisl. Mat. i Mat. Fiz. (3) **6** (1966), 521–526.

7. N. S. Bakhvalov and G. P. Panasenko, *Homogenization averaging of processes in periodic media. Mathematical problems of the mechanics of composite materials*, "Nauka", Moscow, 1984; English transl., Kluwer, Dordrecht, 1989.

8. V. N. Bogaevskiĭ and A. Ya. Povzner, *Algebraic methods in nonlinear perturbation theory*, "Nauka", Moscow, 1987. (Russian)

9. N. N. Bogolyubov and Yu. A. Mitropol'skiĭ, *Asymptotic methods in the theory of nonlinear oscillations*, "Nauka", Moscow, 1974; English transl. of 2nd ed., Internat. Monographs on Advanced Math. and Physics, Hindustan, Delhi, 1961; Gordon and Breach, New York, 1962.

10. N. G. de Bruijn, *Asymptotic methods in analysis*, Bibliotheca Mathematica, 4, North Holland, Amsterdam, Noordhoff, Groningen, and Interscience, New York, 1958.

11. V. F. Butuzov, *On the asymptotics of the solution of the equation $\mu^2 \Delta u - k^2(x, y)u = f(x, y)$ in a rectangular domain*, Differentsial'nye Uravneniya (9) **9** (1973), 1654–1660; English transl. in Differential Equations **9** (1973).

12. _____, *On the asymptotics of the solution of singularly perturbed equations of elliptic type in a rectangular domain*, Differentsial'nye Uravneniya (6) **11** (1975), 1030–1041; English transl. in Differential Equations **11** (1975).

13. _____, *On the construction of boundary layer functions in some singular perturbation problems of elliptic type*, Differentsial'nye Uravneniya (10) **13** (1977), 1829–1835. (Russian)

14. _____, *Corner boundary layer in mixed singular perturbation problems for hyperbolic equations*, Mat. Sb. (3) **104** (1977), 460–485; English transl., Math. USSR-Sb. **33** (1977) 3, 403–426.

15. J. Cole, *Perturbation methods in applied mathematics*, Ginn-Blaisdell, Waltham, Mass., 1968.

16. L. P. Cook, G. S. S. Ludford, *The behavior as* $\varepsilon \to +0$ *of solution to* $\varepsilon \nabla^2 w = \partial w / \partial y$ *on the rectangle* $0 \le x \le |y| \le 1$, SIAM J. Math. Anal. (1) **4** (1973), 161–184.

17. R. Courant, *Methods of mathematical physics*. Vol. II: *Partial differential equations*, Interscience, New York, 1962.

18. S. Yu. Dobrokhotov and V. P. Maslov, *Multiphase asymptotics of nonlinear partial differential equations with a small parameter*, Sov. Sci. Rev. **3** (1982), 221–311.

19. A. A. Dorodnicyn, *Asymptotics of the solution of the van der Pol equation*, Prikl. Mat. Mekh. (3) **11** (1947), 313–328; English transl., Amer. Math. Soc. Transl. (1) **4** (1962), 1–23.

20. W. Eckhaus, *Boundary layers in linear elliptic singular perturbation problems*, SIAM Rev. (2) **14** (1972), 226–231.

21. K. V. Emel'yanov, *On the asymptotics of the solution of the first boundary value problem for the equation* $\varepsilon u'' + xa(x)u' - b(x)u = f(x)$, Application of the Method of Matched Asymptotic Expansions to Boundary Value Problems for Differential Equations, Ural. Nauchn. Centr AN SSSR, Sverdlovsk, 1979, pp. 5–14. (Russian)

22. _____, *On the solution of the first boundary value problem for an ordinary differential equation with a small parameter*, Differential Equations with a Small Parameter, Ural. Nauchn. Centr AN SSSR, Sverdlovsk, 1980, pp. 3–7. (Russian)

23. A. Erdélyi, *Asymptotic expansions*, Dover, New York, 1956.

24. M. A. Evgrafov, *Asymptotic estimates and entire functions*, "Nauka", Moscow, 1979; English transl. of 2nd ed., Gordon and Breach, New York, 1961.

25. M. V. Fedoryuk, *Asymptotics of the solution of the Dirichlet problem for the Laplace and Helmholts equations outside a thin cylinder*, Izv. Akad. Nauk SSSR Ser. Mat. (1) **45** (1981), 167–186; English transl., Math. USSR-Izv. **18** (1982), 1, 145–162.

26. _____, *Asymptotic behavior of the solution of the problem of scattering by a cylinder with large perturbation*, Trudy Moscov. Mat. Obšč. **48** (1985), 150–162; English transl., Trans. Moscow Math. Soc. **48** (1986), 167–174.

27. _____, *The Dirichlet problem for the Laplace operator in the exterior of a thin solid revolution*, Trudy Sem. S. L. Soboleva **1** (1980), 113–131; English transl., Amer. Math. Soc. Transl. (2) **126** (1985), 61–76.

28. _____, *Asymptotics methods for linear ordinary differential equations*, "Nauka", Moscow, 1983. (Russian)

29. L. E. Fraenkel, *On the method of matched asymptotic expansions*. Parts 1–3, Proc. Cambridge Philos. Soc. **65** (1969), 209–231, 233–251, 263–284.

30. A. Friedman, *Partial differential equations of parabolic type*, Holt, Rinehart and Winston, 1964.

31. R. R. Gadyl'shin, *Asymptotics of the eigenvalue of a singularly perturbed selfadjoint elliptic problem with a small parameter in the boundary conditions*, Differentsial'nye Uravneniya (4) **22** (1986), 640–652; English transl., Differential Equations **22** (1986), 4, 474–483.

32. J. Hadamard, *Recherches sur les solutions fondamentales et l'integration des équation linéaires aux dérivées partielles*, Ann. Sci. École Norm. Super. **21** (1904), 531–566.

33. J. Heading, *Phase-integral methods*, Methuen's Physical Monographs, London, 1961.

34. E. Hopf, *The partial differential equation* $u_t + u u_x = \mu u_{xx}$, Comm. Pure Appl. Math. (3) **3** (1950), 201–230.

35. A. M. Il'in, *Boundary value problem for elliptic second order equation in a domain with a narrow slit*. I. *Two-dimensional case*, Mat. Sb. (4) **99** (1976), 514–537; English transl., Math. USSR-Sb. **28** (1976), 4, 459–480.

36. _____, *Boundary value problem for elliptic second order equation in a domain with a narrow slit*. II. *Domain with a small hole*, Mat. Sb. (2) **103** (1977), 265–284; English transl., Math. USSR-Sb. **32** (1977), 2, 227–244.

37. _____, *Asymptotic behavior of a boundary value problem on the half-line for a parabolic equation*, Application of the Method of Matched Asymptotic Expansions to Boundary Value Problems for Differential Equations, Ural. Nauchn. Centr AN SSSR, Sverdlovsk, 1979, pp. 81–92. (Russian)

38. _____, *Study of the asymptotic behavior of the solution of an elliptic boundary value problem in a domain with a small hole*, Trudy Sem. Petrovsk. **6** (1981), 57–82. (Russian)

39. _____, *Cauchy problem for a single quasi-parabolic equation with a small parameter*, Dokl. Akad. Nauk SSSR (3) **283** (1985), 530–534; English transl., Soviet Math. Dokl. **32** (1985), 1, 133–136.

40. A. M. Il'in, Yu. P. Gor'kov, and E. F. Lelikova, *Asymptotics of solutions of elliptic equations with a small parameter at higher derivatives in a neighborhood of a singular characteristic of the limit equation*, Trudy Sem. Petrovsk. **1** (1975), 75–133. (Russian)

41. A. M. Il'in and S. I. Khudyayev, *On the asymptotics of a stationary ignition wave in a condensed medium*, Chim. Fiz. (4) **8** (1989).

42. A. M. Il'in and E.F Lelikova, *The method of matched asymptotic expansions for the equations* $\varepsilon\Delta u - a(x, y)u_y = f(x, y)$, Mat. Sb. (4) **96** (1975), 568–583; English transl., Math. USSR-Sb. **25** (1975), 4, 533–548.

43. _____, *Asymptotics of solutions of some elliptic equations in unbounded domains*, Mat. Sb. (3) **119** (1982), 307–324; English transl., Math. USSR-Sb. **47** (1983), 2, 295–313.

44. A. M. Il'in and O. A. Oleĭnik, *Asymptotic behavior of solutions of the Cauchy problem for some quasilinear equations for large time values*, Mat. Sb. (2) **51** (1960), 191–216. (Russian)

45. A. M. Il'in and K. Kh. Nasirov, *A method of matching asymptotic expansions for an elliptic boundary problem with a small parameter*, Differential Equations with a Small Parameter, Ural. Nauchn. Centr AN SSSR, Sverdlovsk, 1980, pp. 8–15. (Russian)

46. A. M. Il'in and T. N. Nesterova, *Asymptotics of the solution of the Cauchy problem for a single quasilinear equation with a small parameter*, Dokl. Akad. Nauk SSSR (1) **240** (1978), 11–13; English transl., Soviet Math. Dokl. **19** (1978), 3, 529–532.

47. A. M. Il'in and B. I. Suleĭmanov, *Asymptotic behavior of the Green's function for a second-order elliptic equation near the boundary of a domain*, Izv. Akad. Nauk SSSR Ser. Mat. (6) **47** (1983), 149–165; English transl., Math. USSR-Izv. **23** (1984), 3, 579–596.

48. L. A. Kalyakin, *Asymptotics of the solution of two linear MHD equations with singular perturbation. II. Rectilinear flow in a channel with a rectangular protrusion*, Differentsial'nye Uravneniya (10) **15** (1979), 1873–1887; English transl. in Differential Equations **15** (1979).

49. _____, *Asymptotics of the solution of two linear MHD equations with singular perturbation. I. Standard problem in elliptic layer*, Differentsial'nye Uravneniya (10) **18** (1982), 1724–1738; English transl. in Differential Equations **18** (1982).

50. _____, *Asymptotic decay into simple waves of the solution of a perturbed hyperbolic system of equations*, Differential Equations with a Small Parameter, Ural. Nauchn. Centr AN SSSR, Sverdlovsk, 1980, pp. 36–49. (Russian)

51. _____, *Long-wave asymptotics of the solution of a hyperbolic system of equations*, Mat. Sb. (5) **124** (1984), 96–120; English transl. Math. USSR-Sb. **52** (1985), 1, 91–114.

52. _____, *Long-wave asymptotics of the solution of the Cauchy problem for a system of equations with nonlinear perturbation*, Dokl. Akad. Nauk SSSR (1) **283** (1985); English transl., Soviet Math. Dokl. **32** (1985), 1, 9–13.

53. _____, *Long-wave asymptotics of solutions of nonlinear systems of equations with dispersion*, Dokl. Akad. Nauk SSSR (4) **288** (1986), 809–813; English transl., Soviet Math. Dokl. **33** (1986), 3, 769–774.

54. _____, *Asymptotic integration of a perturbed hyperbolic system of equations in the class of conditionally periodic functions*, Trudy Moskov. Mat. Obshch. **49** (1986), 56–70; English transl., Trans. Moscow Math. Soc. **49** (1987), 57–72.

55. S. L. Kamenomostskaya, *The first boundary value problem for an equation of elliptic type with a small parameter at higher derivatives*, Izv. Akad. Nauk SSSR Ser. Mat. (5) **19** (1955), 345–360. (Russian)

56. S. Kaplun, *The role of coordinated systems in boundary-layer theory*, Z. Angew. Math. Phys. **5** (1954), 111–135.

57. S. Kaplun and P. Lagerstrom, *Asymptotic expansions of Navier-Stokes solutions for small Reynolds numbers*, J. Math. Mech. **6** (1957), 585–593.

58. S. I. Khudyaev, *On the asymptotic theory of a stationary ignition wave*, Chim. Fiz. (5) **6** (1987), 681–691.

59. V. A. Kondrat'ev, *Boundary value problems for elliptic equations in the domains with conic or corner points*, Trudy Moskov. Mat. Obshch. **16** (1967), 209–292; English transl., Trans. Moscow Math. Soc. **16** (1967), 227–314.

60. O. A. Ladyzhenskaya, V. A. Solonnikov, and N. N. Ural'ceva, *Linear and quasi-linear equations of parabolic type*, "Nauka", Moscow, 1967; English transl., Transl. Math. Monographs, 23, Amer. Math. Soc., Providence, RI, 1968.

61. O. A. Ladyzhenskaya and N. N. Ural'ceva, *Linear and quasi-linear elliptic equations*, "Nauka", Moscow, 1973; English transl. of 1st ed., Academic Press, New York, 1968.

62. P. A. Lagerstrom, *Matched asymptotic expansions: ideas and techniques*, Springer, New York, 1988.

63. P. A. Lagerstrom and J. D. Cole, *Examples illustrating expansion procedures for the Navier-Stokes equations*, J. Rat. Mech. Anal. **4** (1955), 817–882.

64. N. N. Lebedev, *Special functions and their applications*, "Fizmatgiz", Moscow–Leningrad, 1963; English transl., Prentice-Hall,, Englewood Cliffs, NJ, 1965, 1965.

65. E. F. Lelikova, *On the asymptotics of the solution of an elliptic second order equation with a small parameter at higher derivatives*, Differentsial'nye Uravneniya (10) **12** (1976), 1852–1865; English transl. in Differential Equations **12** (1976).

66. _____, *The method of matched asymptotic expansions for the equation $\varepsilon\delta u - au_z = f$ in a parallelepipedon*, Differentsial'nye Uravneniya (9) **14** (1978), 1638–1648; English transl. in Differential Equations **14** (1978).

67. _____, *Asymptotic behavior of the solution of an elliptic equation with a small parameter in a domain with the piece-wise smooth boundary*, Application of the Method of Matched Asymptotic Expansions to Boundary Value Problems for Differential Equations, Ural. Nauchn. Centr AN SSSR, Sverdlovsk, 1979, pp. 40–57. (Russian)

68. _____, *Asymptotic behavior of the solution of an elliptic equation with small parameter in a region with a conic point*, Differentsial'nye Uravneniya (2) **19** (1983), 305–318; English transl., Differential Equations **19** (1983), 2, 231–243.

69. N. Levinson, *The first boundary value problem for $\varepsilon\delta u + A(x, y)u_x + B(x, y)u_y + C(x, y)u = D(x, y)$ for small ε*, Ann. of Math. (2) **51** (1950), 428–445.

70. S. A. Lomov, *Introduction into the general theory of singular perturbations*, "Nauka", Moscow, 1981; English transl. to be published, Amer. Math. Soc., Providence, RI.

71. V. P. Maslov, *Perturbation theory and asymptotic methods*, Izdat. MGU, Moscow, 1965. (Russian)

72. _____, *Complex W.K.B. method in non-linear equations*, "Nauka", Moscow, 1977. (Russian)

73. V. P. Maslov and M. V. Fedoryuk, *Semi-classical approximation in quantum mechanics*, "Nauka", Moscow, 1976; English transl., D. Reidel, Dordrecht-Boston, 1981.

74. V. P. Maslov and G. A. Omel'yanov, *Asymptotic soliton form solutions of equations with small dispersion*, Uspekhi Mat. Nauk (3) **36** (1981), 63–126; English transl., Russian Math. Surveys **36** (1981), 3, 73–150.

75. V. P. Maslov, G. A. Omel'yanov, and V. A. Tsupin, *Asymptotic behavior of certain differential, pseudo-differential and dynamical systems with small dispersion*, Mat. Sb. (2) **122** (1983), 197–219; English transl., Math. USSR-Sb. **50** (1985), 1, 191–213.

76. V. P. Maslov and G. A. Omel'yanov, *Conditions Hugoniot type for infinitely narrow solutions of the simple wave equation*, Sibirsk. Mat. Zh. (5) **24** (1983), 172–182; English transl., Siberian Math. J. **24** (1983), 5, 787–795.

77. V. G. Maz'ya, S. A. Nazarov, and B. A. Plamenevskiĭ, *Asymptotic behavior of solutions of elliptic boundary value problems under singular perturbations of the domain*, Izdat. Tbilis. Univ., Tbilisi, 1981. (Russian)

78. _____, *Dirichlet problem in domains with thin cross connections*, Sibirsk. Mat. Zh. (2) **25** (1984), 161–179; English transl., Siberian Math. J. **25** (1984), 2, 297–313.

79. _____, *Asymptotic expansions of eigenvalues of boundary value problems for the Laplace operator in domains with small openings*, Izv. Akad. Nauk SSSR Ser. Mat. (2) **48** (1984), 347–371; English transl., Math. USSR-Izv. **24** (1985), 2, 321–346.

80. _____, *Boundary value problems in a domain with narrow bridges*, Funktsional. Anal. i Priložhen. (2) **17** (1982), 20–29. (Russian)

81. _____, *The asymptotic behavior of solutions of the Dirichlet problem in a domain with a cut out thin tube*, Mat. Sb. (2) **116** (1981), 187–217; English transl., Math. USSR-Sb. **44** (1983), 2, 167–194.

82. C. Miranda, *Partial differential equations of elliptic type*, Springer, New York, 1970.

83. E. F. Miščenko and N. Kh. Rozov, *Differential equations with small parameters and relaxation oscillations*, "Nauka", Moscow, 1975; English transl., Plenum Press, New York-London, 1980.

84. A. H. Nayfeh, *Introduction to perturbation techniques*, Wiley-Interscience, New York, 1981.

85. _____, *Perturbation methods*, Wiley-Interscience, New York, 1973.

86. S. A. Nazarov, *The Vishik-Lyusternik method for elliptic boundary value problems in regions with conic points. I. Problem in a cone*, Sibirsk. Mat. Zh. (4) **22** (1981), 142–163; English transl., Siberian Math. J. **22** (1981), no. 4, 594–611 (1982).

87. _____, *The Vishik-Lyusternik method for elliptic boundary value problems in regions with conic points. II. Problem in a bounded domain*, Sibirsk. Mat. Zh. (5) **22** (1981), no. 5, 132–152; English transl., Siberian Math. J. **22** (1981), 5, 753–769 (1982).

88. T. N. Nesterova, *Asymptotic behavior of solution of the Burgers equation in the neighborhood of the merging of two lines of a discontinuity*, Differential Equations with a Small Parameter, Ural. Nauchn. Centr AN SSSR, Sverdlovsk, 1980, pp. 66–86. (Russian)

89. _____, *The method of matching asymptotic expansions for the solution of a hyperbolic equation with a small parameter*, Mat. Sb. (4) **120** (1983), 546–555; English transl., Math. USSR-Sb. **48** (1984), 2, 541–550.

90. V. Yu. Novokshenov, *Asymptotics of the solution of a single elliptic equation with discontinuous boundary conditions*, Differentsial'nye Uravneniya (10) **12** (1976), 1625–1637; English transl. in Differential Equations **12** (1976).

91. _____, *Asymptotics of the solution of a singular integral equation with a small parameter*, Mat. Sb. (3) **100** (1976), 455–475; English transl., Math. USSR-Sb. **29** (1976), 3, 411–429.

92. _____, *Singular integral equation with a small parameter on a finite closed interval*, Mat. Sb. (4) **105** (1976); English transl., Math. USSR-Sb. **34** (1978), 4, 475–502.

93. _____, *Asymptotics with respect to a small parameter of the solution of an elliptic pseudo-differential equation in half-space*, Differential Equations with a Small Parameter, Ural. Nauchn. Centr AN SSSR, Sverdlovsk, 1980, pp. 87–110. (Russian)

94. O. A. Oleĭnik, *On the second boundary value problem for an elliptic equation with a small parameter at higher derivatives*, Dokl. Akad. Nauk SSSR (5) **79** (1951), 735–737. (Russian)

95. _____, *On equations of elliptic type with a small parameter at higher derivatives*, Mat. Sb. (1) **31 (73)** (1952), 104–117. (Russian)

96. O. A. Oleĭnik, *On the Cauchy problem for non-linear equations in the class of discontinuous functions*, Dokl. Akad. Nauk SSSR (3) **95** (1954), 451–454. (Russian)

97. F. Olver, *Introduction to asymptotics and special functions*, Academic Press, New York and London, 1974.

98. O. A. Oleĭnik, *Discontinuous solutions of non-linear differential equations*, Uspekhi Mat. Nauk (3) **12** (1953), 3–73. (Russian)

99. O. A. Oleĭnik and T. D. Ventsel, *The first boundary value problem and the Cauchy problem for quasi-linear equations of parabolic type*, Mat. Sb. (1) **41** (1957), 105–128. (Russian)

100. I. G. Petrovskiĭ, *Lectures on partial differential equations*, "Fizmatgiz", Moscow, 1961. (Russian)

101. L. S. Pontryagin, *Asymptotic behavior of solutions of systems of differential equations with a small parameter at higher derivatives*, Izv. Akad. Nauk SSSR Ser. Mat. (5) **21** (1957), 605–626. (Russian)

102. L. Prandtl, *Über Flussigkeitsbewegung bei sehr kleiner Reibung*, Proc. Internat. Congr. Math. (Heidelberg, 1904), Leipzig, 1905, 484–491.

103. I. Proudman and J. R. A. Pearson, *Expansions at small Reynolds numbers for the flow past a sphere and a circular cylinder*, J. Fluid Mech. (3) **2** (1957), 237–262.

104. V. I. Pr'yažinskiĭ and V. G. Sushko, *Asymptotics with respect to a small parameter of some solutions of the Cauchy problem for a single quasi-linear parabolic equation*, Dokl. Akad. Nauk SSSR (2) **247** (1979), 283–285; English transl., Soviet Math. Dokl. **20** (1979), 4, 698–700.

105. M. I. Rabinovich and A. A. Rosenblum, *On the asymptotic methods of solving non-linear partial differential equations*, Prikl. Mat. Mekh. **36** (1972), 330–343. (Russian)

106. M. D. Ramazanov, *Problem of the flow of an inviscid incompressible fluid around a thin wing with the sharp rear edge*, Mathematical Analysis and Related Topics in Mathematics, "Nauka", Novosibirsk, 1978, pp. 224–236. (Russian)

107. E. Ya. Riekstyn'sh, *Asymptotic expansion of integrals*. Vol. I, "Zinatne", Riga, 1974. (Russian)

108. B. L. Roždestvenskiĭ and N. N. Yanenko, *Systems of quasi-linear equations and their applications to gas dynamics*, "Nauka", Moscow, 1978; English transl., Amer. Math. Soc., Providence, RI, 1983.

109. Yu. Z. Shaygardanov, *On the asymptotics of the solution of the boundary value problem for a single parabolic equation of the fourth order*, Differentsial'nye Uravneniya (4) **15** (1979), 668–680; English transl. in Differential Equations **15** (1979).

110. _____, *Asymptotic behavior be with respect to the small parameter of the solution of the Dirichlet problem for an elliptic problem of order $2m$ degenerating into the equation of the first order*, VINITI, Dep. No 5011–81, Moscow, 1981. (Russian)

111. _____, *Asymptotic behavior with respect to the parameter of the solution of a high-order elliptic equation in a neighborhood of the line of the degeneracy of the limit equation*, Differentsial'nye Uravneniya (4) **21** (1985), 706–715; English transl., Differential Equations **21** (1985), 4, 482–490.

112. A. L. Shtaras, *Asymptotics integration of weakly linear partial differential equations*, Dokl. Akad. Nauk SSSR (3) **237** (1977), 525–528; English transl., Soviet Math. Dokl. **18** (1977), 6, 1462–1466.

113. Ya. S. Soĭbelman, *Asymptotic behavior of the capacitance of a condensor with plates of arbitrary form*, Sibirsk. Mat. Zh. (6) **25** (1984), 167–181; English transl., Siberian Math. J. **25** (1984), 6, 966–978.

114. V. G. Sushko and E. A. Lapshin, *Asymptotic expansions of solutions of some problems related to non-linear acoustics*, Interaction of One-Dimensional of Waves in Media without Dispersion, Izdat. MGU, Moscow, 1983, pp. 224–236. (Russian)

115. N. M. Temme, *Analytical methods for a singular perturbation problem in a sector*, SIAM J. Math. Anal. (6) **5** (1974), 876–887.

116. A. N. Tikhonov, *On the dependence of solutions of differential equations on a small parameter*, Mat. Sb. (2) **22** (1948), 193–204. (Russian)

117. _____, *On systems of differential equations containing a parameter*, Mat. Sb. (1) **27** (1950), 147–156. (Russian)

118. A. N. Tikhonov and A. A. Samarskiĭ, *Equations of mathematical physics*, "Nauka", Moscow, 1966; English transl., Pergamon, Oxford, and Macmillan, New York, 1963; 2nd ed., Vols. 1, 2, Holden-Day, San Francisco, Ca., 1964, 1967.

119. V. A. Trenogin, *Development and application of the asymptotic Lusternik-Vishik method*, Uspekhi Mat. Nauk (4) **25** (1970), 123–156; English transl., Russian Math. Surveys **25** (1970), 4, 119–156.

120. B. R. Vainberg, *Asymptotic methods in equations of mathematical physics*, Izdat. MGU, Moscow, 1982; English transl., Gordon & Breach, New York, 1989.

121. M. D. Van Dyke, *Perturbation methods in fluid mechanics*, Academic Press, New York, 1964.

122. A. B. Vasil'eva, *The development of the theory of ordinary differential equations with a small parameter at higher derivatives between 1966 and 1976*, Uspekhi Mat. Nauk (6) **31** (1976), 102-122; English transl., Russian Math. Surveys **31** (1976), 6, 109–131.

123. A. B. Vasil'eva and V. F. Butuzov, *Asymptotic expansions of solutions of singularly perturbed equations*, "Nauka", Moscow, 1973. (Russian)

124. M. I. Vishik and L. A. Lyusternik, *Regular degeneration and boundary layer for linear differential equations containing a small parameter*, Uspekhi Mat. Nauk (5) **12** (1957), 3–122. (Russian)

125. _____, *Solution of some perturbation problems in the case of matrices and self-adjoint and non-self-adjoint differential equations*, Uspekhi Mat. Nauk (3) **15** (1960), 3–78. (Russian)

126. _____, *Asymptotic behavior of solutions of linear differential equations with large or rapidly varying coefficients and boundary conditions*, Uspekhi Mat. Nauk (4) **15** (1960), 3–95. (Russian)

127. A. B. Vasil′eva and V. M. Volosov, *On the works of A. N. Tikhonov and his successors on ordinary differential equations containing a small parameter*, Uspekhi Mat. Nauk (2) **22** (1967), 149–167. (Russian)

128. J. Whitham, *Linear and non-linear waves*, Wiley, New York London, 1974.

129. V. V. Zhikov, S. M. Kozlov, O.A.Oleinik, and Ha Tyen Ngoan, *Averaging and G-convergence of differential operators*, Uspekhi Mat. Nauk (5) **34** (1979), 65–133; English transl., Russian Math. Surveys **34** (1979), 5, 69–147.

Subject Index

Asymptotic series 2
Asymptotic expansion of a function 3
Bisingular problem 6, 27, 28, 32, 63, 86, 124
Boundary layer 4, 18, 21, 26, 120
Boundary layer functions 5, 20, 21, 26
Composite asymptotic expansion 8, 45, 140, 227
Corner boundary layer 24, 141, 190
Formal asymptotic solution (f.a.s.) 1, 14, 15, 16, 20, 22, 32
Formal series 10, 14, 41
Gauge sequence 2, 18, 41, 53, 96
Gradient catastrophe 216
Green's function 74, 128, 142
Hugoniot condition 222
Inner asymptotic expansion (inner expansion) 5, 18, 20, 163, 197
Inner variables 5, 17, 163
Intermediate boundary layer 47, 52
Interior layer 4
Method of matched asymptotic expansions 6, 32
Outer asymptotic expansion (outer expansion) 4, 18, 20, 218
Pearcey function 239
Residual in the boundary conditions 14, 16, 19, 20, 24, 141
Singular characteristic 120, 202, 205
Singular perturbation of the boundary 63
Singular perturbation problem 4
Uniform asymptotic expansion 3, 7, 22, 43, 64, 94, 159
Whitney fold 237

Recent Titles in This Series

(Continued from the front of this publication)

(See the AMS catalogue for earlier titles)